Nanomaterials for Smart Manufacturing

This textbook discusses the realm of nanomaterials science, offering a detailed and comprehensive exploration of materials at the nanoscale level. The book breaks traditional boundaries to provide a dynamic investigation of the rapidly evolving field of nanomaterials science. It offers a thorough examination of the subject, encompassing essential aspects such as the properties of nanomaterials, various synthesis methods, characterization techniques, and their diverse applications across multiple disciplines.

Nanomaterials for Smart Manufacturing: Fundamental Principles, Theory, Case Studies, and Advanced Applications explores the practical aspects of nanomaterial fabrication, moving beyond theoretical concepts. This textbook is divided into four parts, covering fundamental concepts to advanced applications and case studies, and discusses a wide array of techniques including traditional bottom-up methods like chemical vapor deposition to cutting-edge top-down approaches such as lithography, offering students valuable insights into the precise fabrication of nanomaterials. It covers topics ranging from strength and toughness at the nanoscale to thermal conductivity and nanoscale heat transfer, providing a comprehensive understanding of how these materials perform under various conditions. Additionally, the textbook showcases the tangible impact of nanomaterials across industries such as electronics, medicine, energy, and catalysis. Real-world case studies, examples, and reader comprehension questions illustrate the revolutionary role of nanomaterials in shaping our world.

This definitive guide is tailored for students, researchers, and industry professionals seeking to unravel the mysteries of the nanoworld. Serving as a bridge between the macro and micro worlds, this textbook offers a holistic understanding of nanomaterials.

Figure slides are available to instructors for qualified textbook adoption.

Innovations in Smart Manufacturing for Long-Term Development and Growth
Series Editors: Atul Babbar, Gursel Alici, Yu Dong, Ankit Sharma

Advances in Pre- and Post-Additive Manufacturing Processes
Innovations and Applications
Naveen Mani Tripathi and Ankit Sharma

Handbook of 3D Printing in Pharmaceutics
Innovations and Applications
Prakash Katakam, Ranvijay Kumar, Nishant Ranjan and Atul Babbar

Intelligent Machinery Fault Diagnostics and Prognostics
The Future of Smart Manufacturing
Deepam Goyal, Ankit Sharma and Mohamad Abou Houran

Sensor Technologies and Wireless Communications in Industry 5.0
A Solution for Smart Manufacturing
Edited by Ramkumar K.R., Bhupendra Singh, Lei Pan and Ankit Sharma

Handbook of Smart Manufacturing Technologies
Industrial Engineering in the Digital Age
Edited by Mahendra U. Gaikwad, Ankit Sharma, Muthuramalingam Thangaraj and Sachin Salunkhe

Handbook of Bio-Based Materials for Smart Manufacturing
Edited by Devendra Singh, Naveen Mani Tripathi, Dharminder Singh and Ankit Sharma

Smart Manufacturing with AIoT
Signal Processing, Computer Vision, and Data Analytics
Edited by Keyurkumar Patel, Ankit D. Oza, Ankit Sharma, Archana Mantri, and Hreetabh Kishore

Handbook of Zirconia-Based Materials
Manufacturing, Electrochemical, and Biomedical Engineering
Edited by Jashanpreet Singh, Hitesh Vasudev, Ankit Sharma and Sunaina

Nanomaterials for Smart Manufacturing
Fundamental Principles, Theory, Case Studies, and Advanced Applications
Edited by Dinesh Deshwal, Anil Kumar Narwal, Vivek Srivastava, Atul Babbar and Jonathan M. Weaver

For more information about this series, please visit: www.routledge.com/Innovations-in-Smart-Manufacturing-for-Long-Term-Development-and-Growth/book-series/CRCISMLDG

Nanomaterials for Smart Manufacturing

Fundamental Principles, Theory, Case Studies, and Advanced Applications

Edited by
Dinesh Deshwal
Anil Kumar Narwal
Vivek Srivastava
Atul Babbar
Jonathan M. Weaver

CRC Press
Taylor & Francis Group
Boca Raton London New York

CRC Press is an imprint of the
Taylor & Francis Group, an **informa** business

Designed cover image: Shutterstock

First edition published 2026
by CRC Press
2385 NW Executive Center Drive, Suite 320, Boca Raton FL 33431

and by CRC Press
4 Park Square, Milton Park, Abingdon, Oxon, OX14 4RN

CRC Press is an imprint of Taylor & Francis Group, LLC

ISBN: 978-1-032-90866-3 (hbk)
ISBN: 978-1-032-90871-7 (pbk)
ISBN: 978-1-003-56023-4 (ebk)

DOI: 10.1201/9781003560234

Typeset in Times
by SPi Technologies India Pvt Ltd (Straive)

Access the Instructor and Student Resources: www.routledge.com/9781032908663

Contents

PART 4 Specialized Applications and Case Studies

Preface

Hopefully, this book provides the theoretical basis as well as real-life examples, serving as a bridge to the unique applications and offering a fresh, integrated view for study, research, or business in this area. This book introduces the basic principles of nanomaterials, and then elaborates on their distinct mechanical, thermal, and chemical properties. Readers travel into the detailed territory of nanoscience, gaining a complete understanding of nanomaterial formation. In practice, this book introduces and confronts the advanced characterization techniques necessary for the study of nanomaterials. Best practices and state-of-the-art applications follow in case studies, explaining how nanomaterials are pioneering smart manufacturing worldwide. Every chapter sets forth the potential of nanotechnology and its ability to vastly change the way manufacturing will be done in the future. Particularly, the analytical and messy nature of the topic might require attention to sustainability and scalability, both in terms of a solution that can integrate nanomaterials within industrial ecosystems sustainably while remaining environmentally and economically viable.

More than a scientific base, *Nanomaterials for Smart Manufacturing: Fundamental Principles, Theory, Case Studies, and Advanced Applications* is an open door toward the infinite potential of nanotechnology. For students who want to understand the fundamentals, researchers who want novel ideas, and industry professionals who want practical implementation, this book provides a broad and future-oriented view on the crossroads of nanotechnology and manufacturing. May this book be a curiosity builder, an innovation driver, and a small contribution to the evolution of smart manufacturing where nanoworld collides with the upcoming industry.

About the Editors

Dinesh Deshwal is Head of the Mechanical Engineering Department at Shree Guru Gobind Singh Tricentenary University, Gurugram and is pursuing a Ph.D. from Deenbandhu Chhotu Ram University of Science and Technology in the Department of Mechanical Engineering. He earned his master's degree from the same institute in 2014, laying the foundation for a career dedicated to research and education. With a passion for advancing knowledge, Mr. Deshwal has contributed significantly to the field of nanomaterials. His publications reflect a commitment to exploring diverse aspects of mechanical engineering and nanotechnology. Beyond publications, Mr. Deshwal holds more than ten Indian patents, showcasing a remarkable dedication to innovation and technological advancement. His academic journey is complemented by a teaching role at Shree Guru Gobind Singh Tricentenary University, Gurgaon, where he imparts knowledge and contributes to the academic community. Driven by a commitment to excellence, Mr. Deshwal continues to shape the landscape of mechanical engineering and nanotechnology through his research, publications, and innovative contributions to the field.

Anil Kumar Narwal is an accomplished Assistant Professor in the Mechanical Engineering Department at Deenbandhu Chhotu Ram University of Science and Technology, in Murthal, Sonipat, Haryana, India. Dr. Narwal has made significant contributions to the field of mechanical engineering, particularly in the areas of dynamics, soft contact rolling, and nanotechnology-enabled sensing. With a strong academic background and extensive research experience, Dr. Narwal has authored several research papers published in reputable journals and presented at international conferences. His research focuses on the dynamics of soft contact rolling, utilizing innovative approaches such as the multiband graph approach to model and analyze complex interactions between rigid bodies and soft materials. Furthermore, he has made significant contributions to the emerging field of bio-mass sensing using boron nitride nanotubes (BNNTs). His papers on the analysis of surface deviation impact on bio-mass sensing application and resonant frequency analysis of wavy BNNT-based bio-mass sensors highlight his innovative research endeavors in this area. His work has earned him recognition within the academic community, positioning him as a leading researcher in the field.

Vivek Srivastava is a visionary academic professional with a proven track record of excellence in higher education, research, and fostering industry-academia collaborations. He is a Fellow of the Institution of Engineers India (IEI) and has been instrumental in securing significant grants, including a ₹2 crore grant from Forvia Hella. He also played a key role as the architect behind the MoU between Hella Automotive and SGT University. Currently serving as the Dean and Professor at the Faculty of Engineering and Technology, Shree Guru Gobind Singh Tricentenary University, Gurugram, Dr. Srivastava brings over 20 years of expertise in academic administration, policy-making, and research. His leadership has been pivotal in promoting advanced learning strategies and bridging the gap between academia and industry. Dr. Srivastava's extensive experience includes the implementation of policies such as assessment and accreditation, online learning management systems, and the National Education Policy (NEP). His deep understanding of accreditation processes, including NAAC's Quality Indicator Framework (QIF), positions him as a resourceful leader in institutional development. A strong advocate for innovative pedagogy, Dr. Srivastava has developed Problem-Based Learning (PBL) modules for MATLAB and engineering topics to bridge the gap between theoretical concepts and practical applications. Dr. Srivastava's multifaceted contributions to academia, research, and innovation exemplify his dedication to empowering individuals and communities through education, meaningful collaborations, and impactful research.

Atul Babbar serves as the Deputy Dean (Research and Development) at SGT University, Gurugram. He completed his Doctor of Philosophy (Ph.D.) at the Thapar Institute of Engineering and Technology, Patiala, India. Dr. Babbar has been actively involved in teaching and guiding students in academics and research. His research interests include composites, manufacturing, abrasive machining technologies, hybrid machining processes, biomedical and additive manufacturing, among others. He has authored over 50 research articles published in Web of Science (SCI/SCIE) and Scopus-indexed journals. Dr. Babbar has filed 17 Indian patents and 4 international patents, of which 5 Indian patents have been accepted, and 4 international patents have been granted. He has also reviewed research articles for several peer-reviewed Web of Science and Scopus-indexed journals.

Dr. Babbar has published numerous international books and book chapters and has chaired various international conferences under reputed publishing partners such as Elsevier, Springer, and Sage. He is an Associate Member of the Institution of Engineers (AMIE), India, further highlighting his professional accomplishments and dedication to the field.

Jonathan M. Weaver is currently working as a Professor in the Department of Mechanical Engineering at the University of Detroit Mercy, United States. He has published many articles in various high-quality peer-reviewed journals. He has reviewed articles for many international and national peer-reviewed journals. He has supervised many post-graduates and doctoral students. He is a certified professional member of ICAgile. He teaches statics, dynamics, vehicle dynamics, robotics, machine design, DOE, mechanical measurements laboratory, CAE, innovation, product planning and development, systems engineering, and systems architecture. His research interests relate to robotics, vehicle dynamics, design of experiments, robust design, innovation and creativity, and the product development process.

Contributors

Ankit Agarwal
Department of Mechanical Engineering, Dronacharya College of Engineering, Gurugram,Haryana, India

Manish Baboo Agarwal
Sharda School of Basic Sciences and Research, Sharda University Agra, Agra, Uttar Pradesh, India

Prem Narayan Ahirwar
School of Mechanical Engineering, MPU, Bhopal, India

Amandeep
Twente Pathway College, University of Twente, Enschede, Netherlands

Abinet Gosaye Ayanie
Department of Mechanical Engineering, Adama Science and Technology University, Adama, Ethiopia

Sandeep Bansal
Department of Mechanical Engineering, Shree Guru Gobind Singh Tricentenary University, Gurugram, Haryana, India

Mayank Choubey
Department of Mechanical Engineering, Shree Guru Gobind Singh Tricentenary University, Gurugram, Haryana, India

Leliso Hobicho Dabelo
Department of Mechanical Engineering, College of Engineering and Technology, Mattu University, Metu, Ethiopia

Chayan Das
Department of Mechanical Engineering, Shree Guru Gobind Singh Tricentenary University, Gurugram, Haryana, India

Dinesh Deshwal
Department of Mechanical Engineering, Shree Guru Gobind Singh Tricentenary University, Gurugram, Haryana, India

Suman Dhingra
Department of Chemical Engineering, University of Twente, Enschede, Netherlands

Corey Glickman
MIT Technology Review Board, Cambridge, United States

Navneet Goyal
Department of Mechanical Engineering, Sant Longowal Institute of Engineering and Technology, Sangrur, Punjab, India

Belhi Guerira
Laboratory of Mechanical Engineering (LGM), University of Biskra, Biskra, Algeria

Shaman Gupta
Department of Mechanical Engineering, M.M. University Mullana,
Ambala, Haryana, India

Asad Habeeb
Department of Mechanical Engineering, Shree Guru Gobind Singh Tricentenary University,
Gurugram, Haryana, India

Boussehel Hamida
Laboratory of Mechanical Engineering (LGM), Department of Mater Sciences, Faculty of Exact Sciences and Sciences of Nature and Life,
University of Biskra, Biskra, Algeria

George Jacob
General Motors, Warren Technical Center,
Warren, MI, United States

Gurpreet Kaur
Department of Physics, Lovely Professional University,
Phagwara, Punjab, India

Pushpinder Kumar
Department of Mechanical Engineering, Shree Guru Gobind Singh Tricentenary University,
Gurugram, Haryana, India

Rajni
Department of Chemistry, Central Synthetic Chemistry Lab
SGT University
Gurugram, Haryana, India

Sneh Lata
Central Synthetic Chemistry Lab, Department of Chemistry, Faculty of Applied and Basic Sciences, SGT University,
Gurugram, Haryana, India

Manu Mehrotra
Sharda School of Basic Sciences and Research, Sharda University Agra,
Agra, Uttar Pradesh, India

Ankit Mehta
Central Synthetic Chemistry Lab, Department of Chemistry, Faculty of Applied and Basic Sciences, SGT University,
Gurugram, Haryana, India

Shweta Mishra
SGT College of Pharmacy, SGT University,
Gurugram, Haryana, India

Sunny Misra
Department of Mechanical Engineering, Shree Guru Gobind Singh Tricentenary University,
Gurugram, Haryana, India

Pankaj Mittal
Department of Applied Sciences, Anand Engineering College,
Agra, Uttar Pradesh, India

Monika
Department of Mechanical Engineering, Shree Guru Gobind Singh Tricentenary University,
Gurugram, Haryana, India

Anil Kumar Narwal
Department of Mechanical Engineering, Deenbandhu Chhotu Ram University of Science and Technology,
Murthal, Sonipat, Haryana, India

Manish Kumar Panday
Department of Applied Sciences, Anand Engineering College,
Agra, Uttar Pradesh, India

Bharti Raina
Department of Prosthodontics, Faculty of Dental Sciences, Shree Guru Gobind Singh Tricentenary University,
Gurugram, Haryana, India

Atul Raj
Skill Development and Industrial Training, ITI Butana,
Sonipat, Haryana, India

Lakshay Sain
Department of Mechanical Engineering, Deenbandhu Chhotu Ram University of Science and Technology,
Murthal, Sonipat, Haryana, India

Pardeep Sharma
Department of Mechanical Engineering, Deenbandhu Chhotu Ram University of Science and Technology,
Murthal, Sonipat, Haryana, India

Vijay Sharma
Department of Pharmaceutics, SGT College of Pharmacy, SGT University,
Budhera, Gurugram, Haryana, India

Maitree Singh
Department of Physics, Lovely Professional University,
Phagwara, Punjab, India

Manpreet Singh
Department of Mechanical Engineering, Chitkara University Institute of Engineering and Technology,
Rajpura, Punjab, India

Deepika Singh Singraur
Department of Mechanical Engineering, Shree Guru Gobind Singh Tricentenary University,
Gurugram, Haryana, India

Vivek Srivastava
Department of Mechanical Engineering, Shree Guru Gobind Singh Tricentenary University,
Gurugram, Haryana, India

Gaurav Tripathi
Bharat Electronics Limited,
Ghaziabad, Uttar Pradesh, India

Ekta Verma
MET, Faculty of Pharmacy,
Moradabad, India

Yudhveer Kumar Verma
Department of Mechanical Engineering, Dronacharya College of Engineering,Gurugram, Haryana, India

Part 1

Fundamentals of Nanomaterials

1 Properties and Behavior of Nanomaterials

Mechanical, Thermal, and Chemical Aspects

Dinesh Deshwal, Monika, Anil Kumar Narwal, Vivek Srivastava and Deepika Singh Singraur

1.1 INTRODUCTION

Nanomaterials, by definition, are materials with nanoscale dimensions that range from 1 to 100 nm, constituting a paradigm of non-interference with the science of materials by virtue of their distinguished, exceptional properties differing from those traditionally encountered by bulk substances. The very minute size of the particles, along with the high surface area-to-volume ratio, and specific quantum effects combine to provide exceptional mechanical, thermal, and chemical properties, thus enabling various applications. Therefore, such properties are the cornerstone features necessary to yield effective and efficient performance in smart manufacturing (Balaya 2008).

In considering the properties of nanomaterials, it is important to appreciate the differences between structures, among which are nanoparticles, nanotubes, nanowires, and nanofilms, each embodying a different set of physical and chemical properties (Buzea and Pacheco 2017). For example, the strength and thermal conductivity of the carbon nanotubes (CNTs) are elevated while the metal nanoparticles, like gold and silver, impart higher catalytic activity, optical properties, and antibacterial activity. The possibility is in the tunability: they can be varied by changing the size, shape, and composition toward desired properties or signatures for certain industrial applications. Since many industries like electronics, pharmaceuticals, automotive, and aerospace are pushing forward, technology has appeared as a blast of really fast advancement over the last few decades. Nanomaterials in smart manufacturing support precision sensing, real-time monitoring, environmental protection, and energy efficiency. Sensors based on nanomaterials might, for example, be capable of detecting tiny changes in the environment or in material characteristics to contribute to quality control and efficiency. Moreover, they serve in the realm of composites to lighten weight with strength everywhere; the only option is to cut the weight and not compromise with the strength of the vehicle or aircraft (Ayanda et al. 2024).

DOI: 10.1201/9781003560234-2

At the nanoscale, the behavior of materials diverges from that of bulk materials due mainly to the growth in surface area and quantum confinement effects, resulting in phenomena and properties that are unique and indispensable in a variety of smart manufacturing applications.

a. ***Surface area-to-volume ratio***
   Nanomaterials demonstrate an extraordinary surface area-to-volume ratio, far exceeding that of bulk materials, due to their structural characteristics. The increase in material surface area appears to boost reactivity, catalytic efficiency, and adsorption capability. Nanomaterials with their elevated surface area generate greater reaction site density per volume unit, which dramatically advances catalytic process speeds. Catalytic reactions used in environmental applications and fuel cell functionality substantially benefit from the high surface area-to-volume ratio of nanomaterials (Sharma et al. 2015).
b. ***Quantum effects***
   Quantum effects gain dominance in materials when their dimensions achieve the electron's de Broglie wavelength, leading to the formation of discrete energy levels that affect electronic properties. Nanomaterials exhibit numerous electronic, magnetic, and optical properties because of quantum confinement effects. The dimensionally restricted quality of semiconductor nanoparticles creates band gaps that modify their capture abilities and absorptive properties, making possible optoelectronic and photonic applications (Vysikaylo 2012).
c. ***Mechanical properties***
   The mechanical attributes of nanomaterials surpass conventional expectations by delivering improved tensile strength and flexibility in addition to greater material hardness. The remarkable strength and equal ductility of CNTs and graphene enable their pairing with composite materials. Stronger nanocomposite structures become possible while remaining light because of this exceptional characteristic, making them ideal for aerospace and automotive frameworks that require durability combined with weight reduction advantages (Kumar, Sharma, and Dixit 2021).
d. ***Thermal conductivity and stability***
   The thermal properties of nanomaterials vary according to structure and composition. The excellent thermal conductivity of CNTs and graphene is exemplified by applications such as electronics and high-performance computing. On the other hand, materials such as silicon nanowires and certain ceramic nanocomposites exhibit lower thermal conductivity, which is used in thermal insulation applications. Furthermore, particular nanomaterials, like boron nitride nanotubes (BNNTs), can withstand extremely high temperatures and are thermally stable, meaning that they can survive under extreme temperatures without degrading, and are well suited for use in extreme manufacturing conditions (Maselugbo, Harrison, and Alston 2022).

The role of nanomaterials in smart manufacturing is transformative, offering unprecedented opportunities to enhance efficiency, precision, and sustainability, thermal and chemical properties arising from their nanoscale dimensions; nanomaterials enable advancements in sensor technology, catalysis, surface engineering, composite materials, and energy storage. As research continues, the understanding and development of nanomaterials will further integrate them into the fabric of smart manufacturing, supporting the evolution of industrial practices and paving the way for future technological advancements.

1.2 MECHANICAL PROPERTIES OF NANOMATERIALS

Nanomaterials display mechanical properties fundamentally different from bulk materials due to their nanoscale dimensions, unique atomic configurations, and substantial surface area-to-volume ratio. These properties, including tensile strength, elasticity, hardness, flexibility, and surface-related effects, are critical for applications requiring high strength, flexibility, and durability, such as aerospace, automotive, electronics, and biomedical devices. This provides an in-depth analysis of these mechanical properties, focusing on materials such as CNTs, BNNTs, silicon carbide (SiC) nanocomposites, graphene, and transition metal dichalcogenides (TMDs).

1.2.1 Tensile Strength, Elasticity, and Hardness

Tensile strength is one of the most important properties for nanomaterials, particularly for those applications necessitating the strongest and lightest of materials. Nevertheless, the mechanical behavior of CNTs and BNNTs at the nanoscale is predominantly controlled by strong covalent bonding and the negligible level of structural defects, which leads to tensile strengths significantly beyond those of conventional materials (Gao et al. 2017) (Table 1.1).

***CNTs*:** Figure 1.1 shows CNTs consisting of carbon atoms arranged in a tubular structure that can be zigzag, chiral, or armchair. CNTs exhibit tensile strengths ranging from 50 to 100 GPa and a Young's modulus of the order of 1 TPa, about 100 times greater than steel by weight. This high tensile strength makes CNTs highly valuable for reinforcing composite materials where lightweight construction and high durability are required, such as in the aerospace, automotive, and sporting goods industries (Yıldırım, Ataberk, and Ekrem 2021).

TABLE 1.1
Mechanical Properties of Selected Nanomaterials

Nanomaterial	Tensile Strength (GPa)	Young's Modulus (TPa)	Density (g/cm³)
CNTs	50–100	~1.0	~1.3
BNNTs	33–60	~0.7	~2.1
Graphene	130	~1.0	~2.3
SiC nanoparticles	~40	~0.4	~3.2

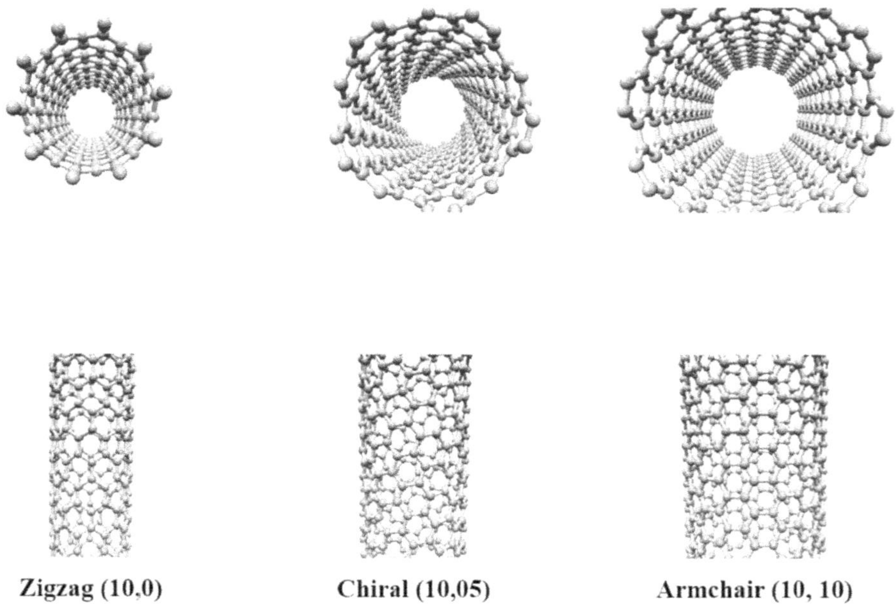

FIGURE 1.1 Different chiral structures of CNTs (e.g., zigzag, chiral, and armchair).

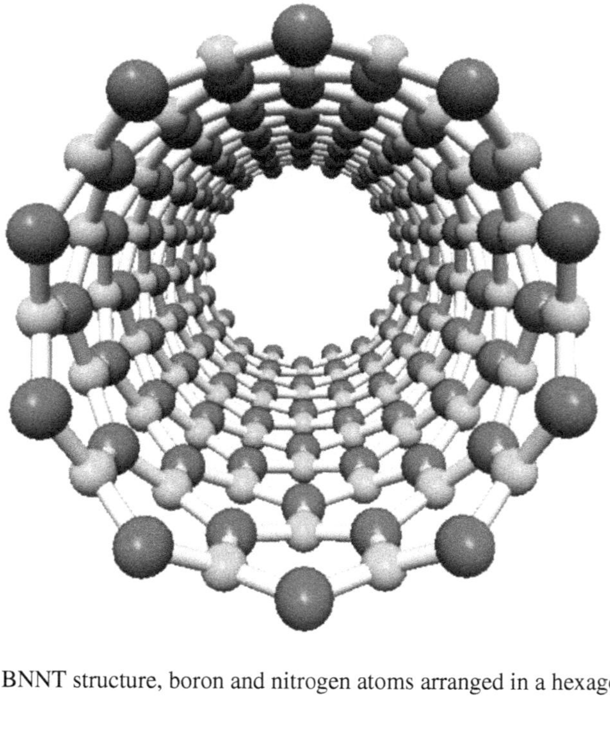

FIGURE 1.2 BNNT structure, boron and nitrogen atoms arranged in a hexagonal lattice.

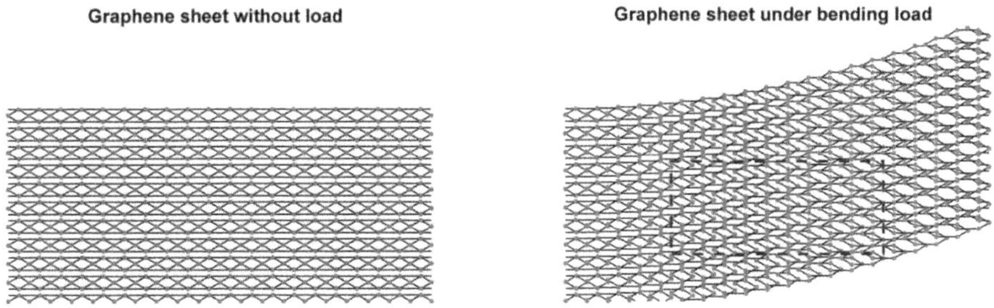

FIGURE 1.3 Graphene under mechanical strain, highlighting its ability to stretch without breaking.

BNNTs**:** BNNTs are structurally similar to CNTs, with boron and nitrogen atoms arranged in a hexagonal lattice as shown in Figure 1.2. While their tensile strength is slightly lower than that of CNTs (33–60 GPa), BNNTs exhibit unique advantages due to their thermal stability and electrical insulation properties. They are highly resistant to oxidation, even at temperatures up to 900°C, making them suitable for extreme environments. These properties have led to the use of BNNTs in applications requiring both high strength and thermal stability, such as aerospace components and high-temperature sensors (Mohammad 2012).

SiC nanocomposites**:** Known for hardness and thermal stability, SiC nanocomposites significantly enhance hardness, toughness, and resistance to thermal shock when incorporated into ceramic matrices. SiC nanoparticles improve Vickers hardness in ceramic composites, making them ideal for cutting tools, wear-resistant coatings, and other high-stress applications (Zhao 2014).

The second important property of nanomaterials is elasticity, enabling them to survive mechanical deformations without sustaining permanent damage. Strong covalent bonds enable materials such as CNTs and BNNTs to undergo significant elastic deformation, making them capable of enduring bending, stretching, or any other kinds of strain. The flexibility is essential for applications in flexible electronics, high-performance sports equipment, or any other application where the material needs to be resilient.

Graphene**:** Graphene's Young's modulus of about 1 TPa allows it to stretch up to 20% of its original length without breaking, as shown in Figure 1.3. The flexibility is also good for such applications where conventional materials would fail under bending or stretching, such as in flexible electronics, wearable devices, etc. Additionally, graphene's electronic properties remain stable under deformation, making it an ideal material for flexible electronics (Ogunbiyi and Yang 2024).

1.2.2 Flexibility and Deformation

Applications that need products to resist repeated mechanical stress, bending, or stretching require materials which possess both flexibility and deformable properties. The atomic dimensions and in-plane bond strength of nanomaterials like graphene

TABLE 1.2
Flexibility and Deformation Characteristics of 2D Nanomaterials

2D Nanomaterial	Young's Modulus (TPa)	Maximum Strain (%)	Thickness (nm)
Graphene	~1.0	~20	0.34
Molybdenum disulfide (MoS_2)	~0.33	~10	~0.65
Tungsten disulfide (WS_2)	~0.27	~9	~0.7

and TMDs create a material platform with excellent flexibility properties (Xu et al. 2013). Table 1.2 showcases the various 2D nanomaterials' Young's modulus, maximum strain, and thickness.

***TMDs*:** TMDs like MoS_2 and WS_2, due to their layered structure held together by van der Waals forces, are highly flexible and resilient under mechanical stress. Their structural configuration allows for easy interlayer sliding, resulting in high flexibility, which is valuable in applications requiring electrical conductivity alongside mechanical compliance, such as in nanoelectrical mechanical systems (NEMS) (Ali et al. 2023).

1.2.3 Surface Effects and Size Dependency

At the nanoscale, surface effects significantly influence mechanical properties like adhesion, friction, and elasticity. Nanomaterials display a distinct surface area-to-volume ratio structure because many atoms remain either at or close to the material surface, according to Table 1.3. The unique surface characteristics of nanomaterials account for behavioral phenomena that bulk materials do not exhibit.

Adhesion and friction: The special characteristics of nanomaterial surfaces lead to distinctive adhesive and friction behaviors. Nanoscale materials deviate from traditional bulk material friction patterns because the contact area together with surface roughness determines their friction behavior. Research conducted with atomic force microscopy (AFM) demonstrates that particles of smaller dimensions exhibit improved adhesive forces by virtue of surface energy properties. Nanomaterials demonstrate an attractive property suitable for applications such as NEMS where exact friction control is necessary.

TABLE 1.3
Nanomaterials Exhibit Different Combinations of Surface Region and Friction Classifications (Zoller et al. 2021)

Material	Surface Area (m^2/g)	Frictional Force (nN)
CNTs	~150–200	Varies with orientation
Nanoparticles	~50–2000	Higher in smaller particles

1.2.4 Theoretical Models for Surface and Size Effects

Theoretical models help explain how nanomaterials interact with one another as well as explain their mechanical properties. The understanding of nanoscale behavioral patterns requires these models to design materials with specific bonding characteristics and mechanical performance properties.

a. ***Derjaguin–Landau–Verwey–Overbeek (DLVO) Theory*:**
   The interaction and stability forces between nanoparticles are explained through DLVO theory by the combination of van der Waals attraction forces with electrostatic repulsive forces. The model functions as the primary explanation for manifesting surface energy alongside adhesion properties in nanomaterials. Nanocomposite applications benefit from understanding these interfacial bonding forces in practice (W. Wu, Giese, and van Oss 1999).
b. ***Johnson–Kendall–Roberts (JKR) Theory*:**
   The JKR theory describes the adhesion and deformation of soft elastic bodies under applied loads. Surface forces take precedence over bulk properties at nanoscale contact mechanics, making this theory essential for such analysis. Researchers use this theory to study how nanomaterials bond to surfaces because surface bonding effects are essential for coating performance (Zhang and Ru 2019).
c. ***Derjaguin–Muller–Toporov (DMT) Theory*:**
   Building upon the JKR model, DMT theory adds contact pressure analysis for rigid elastic bodies. The model serves to explain surface interactions at minimal force levels which underscore applications such as AFM and surface-sensitive technologies. Model theories help explain nanomaterial behavior across different surface conditions and interaction scenarios which provides valuable design information for systems utilizing these materials (Biderman and Bui 2021).

1.3 THERMAL PROPERTIES OF NANOMATERIALS

The behavior of nanomaterials must be understood through their thermal properties when used in electronics alongside energy storage systems and thermal management components. The distinctive heat behavior of nanomaterials results from their microscopic dimensions, atomic arrangements, and enlarged surface area which modify their heat-conducting abilities, structural integrity, thermoelectric characteristics, and material transition temperatures. High-performance materials like graphene and CNTs along with BNNTs and the thermoelectric material Bi_2Te_3 receive detailed analysis for their thermal properties throughout this section.

1.3.1 Thermal Conductivity and Stability

The measurement of heat transmission capabilities in materials depends on their thermal conductivity quality. Nanomaterials, particularly graphene and CNTs, conduct heat exceptionally well thanks to their specific atomic facial arrangement that allows efficient phonon transport. The outstanding thermal stability attributes of BNNTs

TABLE 1.4
Thermal Conductivity

Nanomaterial	Thermal Conductivity (W/m·K)	Maximum Operating Temperature (°C)
Graphene	~5000	>600
CNTs	~3000	500–700
BNNTs	200–300	900–1000

allow the materials to remain intact under high temperatures. Table 1.4 shows the thermal conductivity and maximum operating temperature of various nanomaterials.

***Graphene*:** With thermal conductivity values reaching up to 5000 W/m·K, graphene is one of the most thermally conductive materials. This high conductivity arises from its two-dimensional lattice of carbon atoms bonded in a hexagonal structure, which allows phonons (quasiparticles that carry heat) to travel with minimal scattering as shown in Figure 1.4. Graphene's thermal conductivity is advantageous in applications such as electronic cooling, where rapid heat dissipation is crucial. For example, incorporating graphene as a heat-spreading layer in smartphones and computers helps maintain device performance and longevity by preventing overheating. Additionally, graphene remains thermally stable at temperatures exceeding 600°C, making it suitable for high-temperature applications (Balandin 2011).

***CNTs*:** CNTs have thermal conductivities up to 3000 W/m·K along the tube's axis, making them another excellent material for heat dissipation.

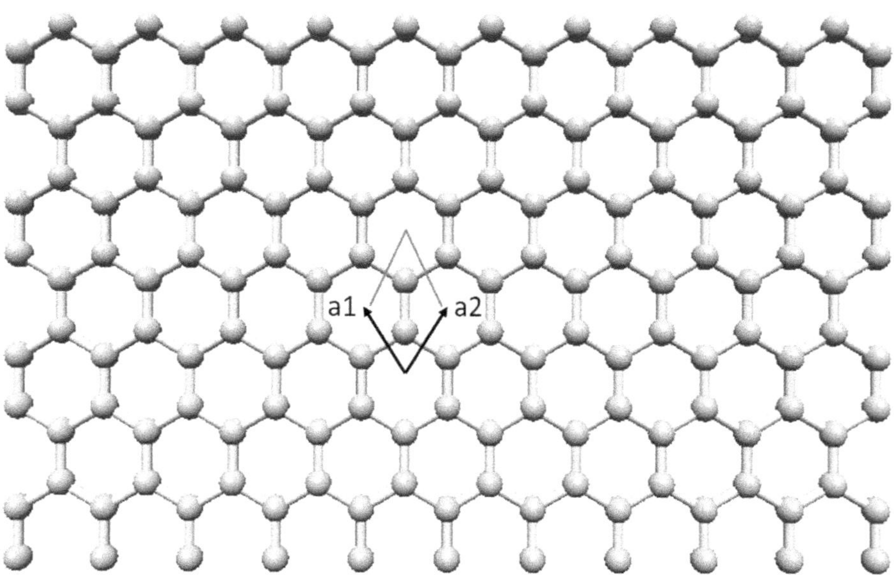

FIGURE 1.4 Graphene structure phonon movement.

CNTs' high thermal conductivity is due to the carbon-carbon covalent bonds in their tubular structure, facilitating efficient heat transfer along the nanotube. CNTs are often used in thermal interface materials (TIMs), where they provide a conductive pathway between heat-generating components and heat sinks. For instance, electronic devices commonly use CNT-based TIMs to improve thermal management. However, it is important to note that CNTs' thermal conductivity is direction-dependent (anisotropic) and is lower in directions perpendicular to the tube axis (Han and Fina 2011).

***BNNTs*:** Although BNNTs have lower thermal conductivity (200–300 W/m·K) than CNTs, they offer superior thermal stability. BNNTs can withstand temperatures of up to 1000°C without undergoing oxidation, which is a significant advantage over other nanomaterials that degrade or lose structural integrity at high temperatures. BNNTs' thermal stability is due to their hexagonal boron-nitrogen lattice structure, which is resistant to thermal expansion and oxidation. The capacity of BNNTs to operate at increased temperatures makes them suitable for aerospace and defense applications in thermal coatings and structural framework use (Zandieh et al. 2024).

1.3.2 Thermoelectric Effects

The property of thermoelectric materials transforms thermal energy difference into electrical energy. Through the Seebeck effect, thermoelectric materials can utilize excess heat to create electricity and demonstrate their utility for heat energy restoration and thermal management operations. Scientists employ Bi_2Te_3 nanomaterials as standard components in thermoelectric devices because they demonstrate exceptional energy conversion capability at nanometer dimensions.

***Bismuth telluride*:** Within the extensive range of thermoelectric materials, Bi_2Te_3 stands out because it demonstrates exceptional properties comprising a superior Seebeck coefficient and a customized thermoelectric power factor (P) that equals $S^2\sigma$. Here S represents the Seebeck coefficient while σ indicates electrical conductivity. Thermoelectric effectiveness of Bi_2Te_3 improves at nanoscale levels due to intensified phonon scattering that produces decreased thermal conductivity yet preserves high electrical conductivity. Thermoelectric optimization relies on maintaining proper equilibrium between material electrical properties and thermal properties. The thermoelectric material Bi_2Te_3 finds extensive applications in thermal management systems to develop portable cooling devices alongside waste heat recovery systems which use its ability to transform excess heat into usable electrical energy. Builders use thermoelectric material Bi_2Te_3 in automotive exhaust systems to change unused heat from the engine to electricity which enhances the efficiency of automobiles. Similarly, Bi_2Te_3 thermoelectric modules are used in spacecraft to generate power from the thermal gradient between sunlight and the shadowed side of the spacecraft (Amin Bhuiyan, Mamur, and Dilmaç 2021).

1.3.3 Thermal Expansion and Melting Points

Nanomaterials exhibit unique behaviors in terms of thermal expansion and melting points, primarily due to their small size and increased surface area as mentioned in Table 1.5. The Gibbs–Thomson effect describes how the melting point of a material decreases as particle size reduces, which has significant implications in nanotechnology and catalysis.

***Thermal expansion*:** Thermal expansion refers to the increase in a material's dimensions as temperature rises. At the nanoscale, materials like aluminum oxide (Al_2O_3) and metals such as gold and silver exhibit higher thermal expansion coefficients than their bulk counterparts. This is because atoms at the surface of nanoparticles are less constrained and can move more freely under thermal stress. The high thermal expansion properties of nanomaterials matter most in applications which need tight dimensional control such as microelectronic systems. Device engineering relies on thermal expansion management techniques to avoid both device warping and operational failure brought on by heat fluctuations (Ajayan, Schadler, and Braun 2003).

***Melting point reduction*:** Nanoscale materials undergo a documented Gibbs–Thomson change which reduces their melting points. The surface-to-volume ratio increases proportionally with decreasing particle size, thus elevating surface energy levels. The elevated surface energy simplifies the atomic escape process from surfaces and causes the melting point to decrease. Attributes of gold nanoparticles measuring 10 nm in diameter result in 300°C below normal melting points for bulk gold. Modern applications in sintering and catalysis heavily depend on the Gibbs–Thomson effect to form materials with ideal functional properties (Schlexer et al. 2019).

The lower melting point characteristics of nanoparticles benefit catalytic applications because reduced activation energy leads to enhanced reaction rates. Lower melting platinum nanoparticles function as fuel cell catalysts because they boost efficient reaction rates within controlled processes.

TABLE 1.5
Melting Points and Thermal Expansion of Selected Nanomaterials

Material	Bulk Melting Point (°C)	Melting Point at 10 nm (°C)	Thermal Expansion Coefficient (10^{-6}/°C)
Gold (G. Wu et al. 2009)	1064	~750	14.2
Silver (Teo and Calabrese 1976)	962	~835	18.9
Aluminum oxide (Al_2O_3) (Yi and Cui 2023)	2072	~1600	8.1

1.3.4 Theoretical Models in Thermal Properties

Several theoretical models provide insights into the unique thermal behavior of nanomaterials:

a. ***Phonon scattering model*:** Nanoscale thermal conductivity experiences decreased effectiveness because phonons collide with material surfaces, thus reducing the material's thermal conductivity capacity. The model demonstrates particular relevance to thermoelectric materials since lower thermal conductivity leads to enhanced thermoelectric efficiency. Structural defects together with interfaces in Bi_2Te_3 nanoparticles enhance phonon scattering phenomena which preserves an optimal electrical to thermal conductivity ratio (Pernot et al. 2010).
b. **Gibbs–Thomson Effect**: The dependence of nanoparticles' melting points on their dimensions finds its explanation through surface melting. Nanoscale particles demonstrate lower melting points thanks to both elevated surface energy and increased atomic movements as described in the Gibbs–Thomson equation. The model brings important advantages to applications in nanoscale melting because it allows better control over particle melting that enhances reaction efficiencies like catalysis (N. Wu et al. 2021).
c. ***Thermoelectric power factor model*:** The product $S^2 \sigma$ exists as the power factor P describes the connection between the Seebeck coefficient and electrical conductivity while determining total thermoelectric performance. Enhanced phonon scattering at the nanoscale enhances thermoelectric behaviors that improve the energy conversion performance of materials including Bi_2Te_3. Designing optimized thermoelectric nanostructured materials depends on the power factor model for system optimization (Yang et al. 2024).

Nanomaterials demonstrate suitable thermal attributes that make them ideal candidates for applications demanding efficient heat dissipation alongside energy conversion and temperature stabilization. These nanomaterials lead innovation in electronics cooling, waste heat recovery, and catalysis applications due to their thermal characteristics, size-dependent melting points, and thermoelectric properties.

1.4 CHEMICAL PROPERTIES AND REACTIVITY OF NANOMATERIALS

Nanomaterials obtain their applications for catalysis sensing environmental remediation and energy systems through their chemical properties. The unique chemical behaviors of nanomaterials emerge because of their big surface area and high surface energy as well as their many reactive sites existing at the nanoscale. This section investigates nanomaterials' chemical attributes through explorations of their enhanced reactivity together with inertness, corrosion resistance features, and details about functionalization opportunities.

The chemical properties of both metal oxides and carbon-based nanomaterials nearly always exhibit increased reactivity beyond their bulk material versions.

Nanomaterials exhibit enhanced chemical behavior due to factors such as increased surface area ratio and elevated surface energy combined with greater surface atomic presence, creating unbound bonds that form reactive areas (Shaikh et al. 2018).

1.4.1 High Surface Area and Surface Energy

The small size of nanoparticles using metal oxides and noble metals, including gold and platinum, enables high surface area exposure relative to their volume dimensions. The material's reactivity benefits from its large active surface which gives many atoms at the surface available for chemical interactions. The chemical inertness of bulk gold changes when the material becomes nanoparticles because such small-scale gold features demonstrate enhanced catalytic potentials. The high reactivity of these nanoparticles makes them suitable for catalytic purposes because they facilitate carbon monoxide (CO) transformation to carbon dioxide (CO_2) during the operation of catalytic converters. The conversion technology works to minimize pollution and create better air conditions (Wittstock and Bäumer 2014).

1.4.2 Increased Catalytic Activity in Metal Oxides

The catalytic activity of metal oxides, including iron oxide (Fe_2O_3), titanium dioxide (TiO_2), and manganese oxide (MnO_2), reaches significant levels at the nanoscale for environmental applications. Industrial applications which rely on iron oxide nanoparticles depend on these particles for toxic gas oxidation functions focused on NOx and CO. Carbon dioxide violates metal oxide TiO_2 functions as a reactive component in photocatalysis through ultraviolet (UV) light-based organic pollutant elimination. Upon incorporating TiO_2 nanomaterials into environmental remediation efforts, they demonstrate excellent photocatalytic potential for wastewater treatment and air purification applications (Long et al. 2015).

1.4.3 Enhanced Reactivity in Carbon-Based Nanomaterials

Electron transfer reactions benefit from carbon-based nanomaterials including CNTs and graphene because their conjugated structure and large surface areas drive catalytic activity. Table 1.6 demonstrates how CNTs serve in fuel cells to provide a catalyst supportive framework which enhances key redox reactions supporting efficient fuel cell operations. Scientists are researching graphene-based catalysts as suitable materials for reducing nitrogen oxide pollutants because they show promise for use

TABLE 1.6
Catalytic Properties of Selected Nanomaterials

Nanomaterial	Application	Catalytic Activity
Gold nanoparticles	CO Oxidation	High reactivity for CO oxidation on supports
Iron oxide nanoparticles	Environmental remediation	Oxidation of CO and NOx
CNTs	Fuel cells	Redox reactions, high electron transfer

in environmentally friendly energy technology systems and air quality purification equipment (Liu, Zhang, and Chen 2014).

1.4.4 Size-Dependent Reactivity of Nanoparticles

The size of nanoparticles significantly impacts their reactivity. As particle size decreases, the surface-to-volume ratio increases, leading to more active sites on the particle surface. For example, gold nanoparticles smaller than 5 nm are highly effective in catalyzing CO oxidation, whereas larger particles do not exhibit the same level of reactivity. This size-dependent reactivity is due to quantum effects, which alter the electronic structure of small particles, making them more susceptible to interactions with other molecules (Min and Friend 2007).

1.4.5 Inertness and Corrosion Resistance

While many nanomaterials are engineered for high reactivity, others are valued for their stability, chemical inertness, and corrosion resistance. Nanomaterials like BNNTs and SiC offer excellent stability, making them suitable for applications in harsh environments.

a. ***Chemical stability of BNNTs*:** BNNTs, structurally similar to CNTs but composed of boron and nitrogen atoms in a hexagonal lattice, exhibit remarkable chemical stability. This stability is due to the strong B–N bonds, which are resistant to chemical reactions and oxidation. The tubular structure of BNNTs demonstrates stability at temperatures exceeding 900°C which contrasts with the degradation of numerous nanomaterials during this range. BNNTs demonstrate outstanding suitability for high-temperature utilization through their stable structure at temperatures surpassing 900 °C. Therefore, such materials find critical use in harsh temperature environments like aerospace parts. Their inert nature makes BNNTs valuable in medical applications because biological fluids and tissue show no chemical reactions with them (Golberg et al. 2010).
b. ***Corrosion resistance in SiC and AL_2O_3*:** The corrosion resistance properties of SiC and Al_2O_3 nanoparticles are well recognized in the technical community. The high resilience of SiC against oxidation allows its use for protective coatings along with components that function properly in challenging environmental conditions below 1200°C. The chemical attack resistance demands SiC as a material choice for protective coatings that shield gas turbines and equipment exposed to high-stress conditions. Applications that need tough, durable coatings exhibiting strong corrosion resistance employ Al_2O_3 for seafront usage and electronic implementations. Al_2O_3 develops an oxide layer which creates a protective barrier that extends the material's service lifetime together with dependable performance (Chen et al. 2024). A comparison chart of various nanomaterials with respect to their minimum and maximum corrosion rates is shown in Figure 1.5.

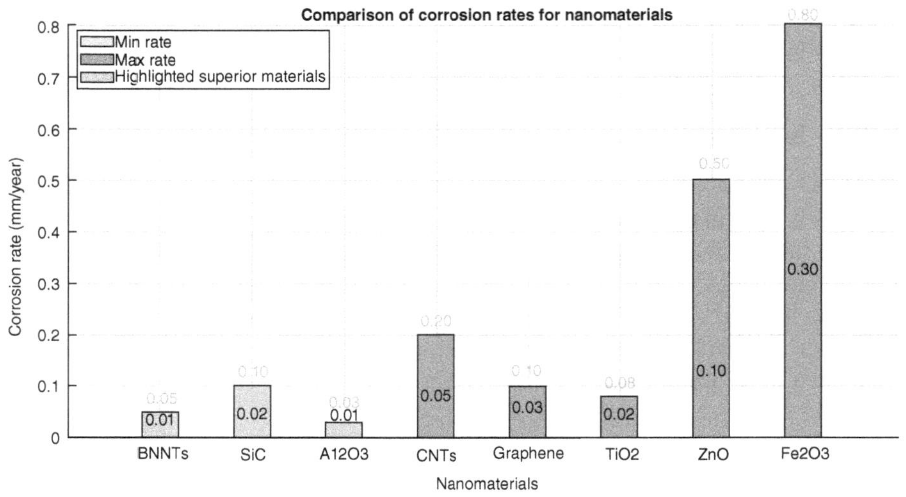

FIGURE 1.5 A comparison chart showing corrosion rates of different nanomaterials, BNNTs, SiC, and Al_2O_3, illustrating their superior corrosion resistance.

1.5 FUNCTIONALIZATION POTENTIAL

The surface chemistry modification known as functionalization creates targeted interactions which boost nanomaterial effectiveness across sensing technologies, catalytic transformations, and drug delivery processes. The addition of functional groups or surface coating with specific molecules helps nanomaterials become more responsive during targeted interactions.

1.5.1 Functionalization Techniques

Three predominant techniques exist for nanomaterial functionalization: chemical vapor deposition (CVD) and plasma treatment along with ligand attachment methods.

a. ***CVD*:** Reactive gases in CVD provide the formation of thin functional layers on nanomaterial surfaces. Environmental group attachment to CNTs using CVD methods results in carboxyl (-COOH) and hydroxyl (-OH) modifications that increase dispersal in solvents while improving biological system compatibility (Dubey et al. 2021).
b. ***Plasma treatment*:** Plasma treatment enables surface modification of nanomaterials, including graphene and CNTs, through functional group addition. A plasma is produced during this procedure, enabling reactive plasma species to interact with nanomaterial surfaces for additional functionalization opportunities (Hosseini and Ghaffarzadeh 2022).
c. ***Ligand attachment*:** The attachment of ligands represents a common processing method for metal nanoparticles, such as gold particles, and Table 1.7 illustrates the binding groups that attach to metal surfaces. The method enables high precision surface interaction management that serves biomedical needs for targeted protein and cell delivery systems (Guan, Harris, and Sun 2023).

TABLE 1.7
Functionalization Methods for Select Nanomaterials

Nanomaterial	Functionalization Method	Applications
CNTs	CVD, plasma treatment	Sensors, biomedical applications
Gold nanoparticles	Ligand attachment	Drug delivery, imaging
Silica nanoparticles	Surface coating	Catalysis, biosensing

1.5.2 Applications of Functionalized Nanomaterials

Sensing and detection: Functionalized nanomaterials achieve superior performance in sensing applications through their detection methods which combine sensitive response and precise analyte recognition. Gold nanoparticles coated with antibodies function as a rapid and accurate diagnostic tool because they enable precise spatial interaction with microscopic proteins or pathogens. Gas sensors based on modified CNTs detect toxic gases through electrical conductivity modifications during gas adsorption to detect ammonia and carbon monoxide (Zhou et al. 2015).

Catalysis: The catalytic properties of nanomaterials experience substantial improvement when active sites are made available through functionalization methods. Platinum nanoparticles loaded with ligands enable oxygen reduction reaction catalysis at fuel cells. The process results in enhanced catalytic performance which simultaneously lowers platinum consumption to cut the total catalyst expense (Ren et al. 2020).

Biomedical applications: Drug delivery systems utilizing functionalized nanoparticles achieve targeted cell or tissue delivery which increases therapeutic efficiency together with reduced systemic effects. Nanoparticles consisting of magnetic cores receive functionalization with biocompatible polymers as well as targeting molecules that direct them toward selecting cancer cells for binding. The targeted method yields drug release control which boosts treatment performance alongside minimizing damage to healthy tissue structures (Pan et al. 2012).

1.5.3 Challenges in Functionalization

The functionalization process both expands nanomaterial usage possibilities and creates difficulties during implementation. Uniform functionalization across curved CNT surfaces remains a difficult goal, and limiting the functionalization degree requirements is essential for maintaining critical properties. Functionalization treatments sometimes modify the electronic and mechanical characteristics of base nanomaterials which can reduce their functional efficiency across various applications. Through combination methods and sequential step procedures, researchers strive to overcome existing difficulties with functionalization while attaining the best possible results (Balasubramanian and Burghard 2005).

1.6 BEHAVIOR OF NANOMATERIALS UNDER DIFFERENT CONDITIONS

Nanomaterials present distinct response patterns together with enhanced performance characteristics under environmental changes, including temperature variations, pressure changes, and chemical encounters. The unique behaviors of nanomaterials contrast strongly with bulk materials because of their diminished scale, elevated surface area properties, and quantum-based scale effects. The quantum properties of nanomaterials create distinctive magnetic, optical, and catalytic features together with their small-scale presence. This section provides a detailed explanation of the behavior of nanomaterials, focusing on their responses to temperature changes, pressure variations, and interactions with different chemicals. It also presents a comprehensive analysis of the quantum confinement effects that influence their operational characteristics.

1.6.1 Temperature Effects

Nanomaterial characteristics respond notably to different temperature levels because of both stability reductions and functional changes. Nanomaterials possess high surface area-to-volume ratios that create a lower melting point than bulk materials. Gold and silver nanomaterials show notable melting point changes because of their tiny size due to the Gibbs–Thomson effect. Two and a half nanometer gold nanoparticles exhibit decreased melting temperatures below standard bulk gold behavior, enabling reactive catalytic conditions that might trigger metal nanoparticles to merge. The surface area reduction of nanoparticles in automotive exhaust catalytic applications diminishes their catalytic performance due to sintering effects. Surface modifications and alloying of stabilizing elements prevent such effects on high surface area through extreme heat conditions. The behavior of nanoparticles shows unique responses to thermal expansion conditions. Each nanoparticle shows its own unique behavior during thermal expansion (Sankar et al. 2020). The thermal expansion coefficients of Al_2O_3 and SiC nanoparticles surpass those found in bulk Al_2O_3 and SiC. The thermal resistance properties of these nanostructures make them suitable for aerospace applications and microelectronics devices (Nandihalli, Mori, and Kleinke 2018).

1.6.2 Pressure Effects

Under pressure, nanomaterials demonstrate distinct reaction patterns. Under high-pressure conditions, several nanomaterials transition structurally while their hardness and durability improve through phase transformation behavior. High magnitude pressure transforms CNTs into diamond-like frameworks while enhancing their mechanical resistance. The performance-enhancing properties of pressure-induced effects enable harsh environment applications like space missions and deep-sea operations. Growing pressure intensifies aggregation processes that modify digital, magnetic, and light properties. Under external fields, nanoscale particles with improved magnetic characteristics, including cobalt and iron oxide, develop superparamagnetism since their magnetic domains exhibit an aligned response to external stimuli, yet lose their magnetic behavior after the external field is eliminated. For biomedical imaging and drug delivery needs, precise control of magnetism becomes achievable under various pressures (Neuville 2019).

1.6.3 Chemical Exposure

The environmental chemicals have a substantial impact on the materials' reactivity rate. Titanium dioxide nanoparticles (TiO_2) develop photocatalytic behavior under UV light activation for chemical decomposition against pollutants. Catalytic performance may undergo degradation alongside surface property modifications when exposed to chemicals for extended periods. The nanomaterials list includes BNNTs along with SiC because these two materials demonstrate exceptional resistance to corrosion while remaining chemically inert in hostile chemical conditions. The high resistance of BNNTs to oxidation allows them to perform well in aerospace applications alongside demanding chemical environments requiring extended stability (Gopinath et al. 2020).

1.7 CONCLUSION

The distinctive properties which exceed bulk material capabilities make nanomaterials powerful tools for intelligent manufacturing applications. The unique characteristics of nanomaterials including large surface area-to-volume ratio, quantum confinement, and enhanced reactivity enable their broad utilization across electronics, environmental cleanup techniques, biomedical systems, and energy technologies. This section summarizes the distinct properties that improve manufacturing capabilities and examines the forthcoming examination of nanomaterial creation and implementation in intelligent manufacturing environments.

1.7.1 Summary of How Unique Properties of Nanomaterials Enhance Manufacturing

a. ***Enhanced mechanical strength and durability*:**
Research shows nanomaterials, including CNTs, graphene, and SiC nanocomposites, demonstrate exceptional levels of elasticity and hardness in addition to their tensile strength. The exceptional mechanical properties enable the manufacturing of durable and lightweight materials which serve elevated performance standards in aerospace engineering, automotive production, and construction. Nanomaterials enable manufacturers to deliver components with superior stress resilience, longer service life, and lowered weight through integration into productive processes. Aerospace composite reinforcements incorporating CNTs allow manufacturers to reduce aircraft weight while simultaneously improving structural performance and durability.

b. ***Thermal conductivity and heat dissipation*:**
The semiconductor industry requires efficient heat dissipation, which nanomaterials, specifically graphene and CNTs, enable through exceptional thermal conductivity. Devices with graphene heat-spreading layers achieve

both operational longevity and higher performance through their ability to stabilize operational systems while preventing overheating conditions. Traditional cooling methods prove inadequate for advanced manufacturing of compact high-powered electronics, so this material property proves beneficial for such cases.

c. ***Chemical reactivity and catalytic efficiency*:**
Metal nanoparticles, including gold, silver, and platinum, demonstrate high reactivity which enables the development of advanced, efficient catalytic systems. The materials find applications in environmental remediation, fuel cell development, and chemical sensor production for industry purposes. Nanomaterials demonstrate excellent catalytic properties which lead to faster reaction velocities and reduced activation barriers, so manufacturing operations become more efficient and less harmful to the environment. Studies have shown that gold nanoparticles, particularly at the smallest scales, exhibit excellent catalytic efficiency in oxidizing exhaust emissions, leading to a significant reduction in harmful gases.

d. ***Functionalization for precision and targeted applications*:**
Functionalization of nanomaterials expands their applicability across many manufacturing operations due to their flexible characteristics. Through customized functionalization processes, nanomaterials obtain specific properties which improve their performance in biosensing, drug delivery, and chemical catalysis systems. Functionalized nanomaterials serve as fundamental components for precision manufacturing that requires specific materials in applications such as medical devices and targeted drug delivery technology.

e. ***Optical and magnetic properties for advanced applications*:**
Nanomaterials gain special magnetic and optical properties through quantum confinement which provides important advantages to manufacturing processes. Nanoparticles containing gold and silver demonstrate surface plasmon resonance properties that yield highly sensitive detection methods suitable for biosensors and diagnostics applications. The precise control in magnetic data storage and targeted drug delivery techniques is enabled by the superparamagnetic properties of iron oxide nanoparticles. Unique characteristics discovered within nanomaterials create opportunities for pioneering manufacturing in healthcare, electronics, and environmental monitoring systems.

1.8 FUTURE RESEARCH DIRECTIONS

Research and development must address multiple domains to ensure the complete harnessing of nanomaterials' manufacturing advantages in smart production contexts. Key future directions include:

a. ***Optimizing nanomaterial synthesis for scale-up*:**
One of the main challenges in incorporating nanomaterials into mainstream manufacturing is the scalability of synthesis methods. Current production techniques often require high costs, complex equipment, and strict

environmental controls, making large-scale production challenging. Future research should focus on developing cost-effective, scalable, and environmentally friendly synthesis methods. Techniques such as CVD for graphene and CNTs, as well as green synthesis methods for metal nanoparticles, need to be refined to meet industrial demands. The demand for nanomaterials in smart manufacturing requires research on sustainable synthesis procedures that simultaneously minimize waste and reduce energy consumption.

b. ***Enhancing functionalization techniques for versatile applications*:** The process of functionalization enables targeted nanomaterial applications, yet stability and uniform coverage of complex nanostructures present a major difficulty during functionalization procedures. Future researchers need to develop better methods for functionalizing irregular nanostructures with high aspect ratios, including nanofiber systems, CNTs, and nanowires. The development of standard protocols to functionalize different nanomaterials will generate expanded applications within manufacturing which specifically target biomedical devices, sensors, and catalytic systems.

c. ***Investigating long-term stability and environmental impact*:** The enhanced features from nanomaterials remain partially unknown because researchers still need to understand their basic operational stability and ecological effects. Scientific investigations are necessary to establish how nanomaterials will withstand extended contact with demanding environmental conditions consisting of extreme heat, pressure, and chemical reactivity. Meaningful advancements require extended research into the environmental hazards nanoparticles might present alongside their toxic qualities and biological breakdown behaviors. The success of nanomaterial adoption across industries focused on sustainability will depend on creating nanostructures that endure well and protect the environment.

d. ***Incorporating AI and machine learning in nanomaterial design*:** New efficient opportunities emerge through artificial intelligence (AI) alongside machine learning (ML) technologies for nanomaterial design and synthesis efforts. The combination of data-driven methodologies helps researchers develop predictive capabilities regarding nanomaterial properties with precise control over size, shape, and reactivity to serve manufacturing requirements. Tailored nanomaterial creation speeds up through AI and ML systems that expedite prototype creation and application-driven nanomaterial modification. Processing technologies employ these systems to guarantee product quality by performing continuous sensor-based oversight and feedback operations throughout manufacturing runs.

e. ***Developing hybrid nanomaterials for multifunctional applications*:** Hybrid structures made from multiple nanomaterials show promise for creating systems that deliver multifunctional capabilities. Research demonstrates that hybrid structures formed by joining CNTs to metal nanoparticles produce multifunctional composites which contain both mechanical strength and catalytic activity. The merger of magnetically responsive nanoparticles with plasmonic materials leads to hybrid systems able to detect neither magnetic nor optical indicators for applications including drug targeting and dual-imaging modalities. Science requires a deeper study of nanomaterial interactions in hybrid structures combined with method development for their synthesis. Potential breakthroughs in smart manufacturing become possible because these hybrid materials enable the development of multipurpose components which optimize simultaneous performance.

f. ***Exploring biocompatible nanomaterials for medical manufacturing*:** The development of biocompatible nanomaterials has opened new possibilities for innovation in medical manufacturing. The development of biodegradable and biocompatible nanomaterials, including specific polymers and bioceramics, will drive future progress in medical implants, drug delivery systems, and tissue engineering applications. Moving forward, we need to create safe nanomaterials that degrade naturally while serving biomedical manufacturing needs for medical applications and environmental conservation.

REFERENCES

Ajayan, Pulickel M., Linda S. Schadler, and Paul V. Braun, eds. 2003. *Nanocomposite Science and Technology*. Wiley. https://doi.org/10.1002/3527602127

Ali, Hina Ghulam, Kashif Khan, Muhammad Bilal Hanif, Muhammad Zubair Khan, Iftikhar Hussain, Muhammad Sufyan Javed, Hussein A.Z. AL-bonsrulah, Michał Mosiałek, Maximilian Fichtner, and Martin Motola. 2023. "Advancements in Two-Dimensional Materials as Anodes for Lithium-Ion Batteries: Exploring Composition-Structure-Property Relationships Emerging Trends, and Future Perspective." *Journal of Energy Storage* 73 (December): 108980. https://doi.org/10.1016/j.est.2023.108980

Amin Bhuiyan, Mohammad Ruhul, Hayati Mamur, and Ömer Faruk Dilmaç. 2021. "DA Review on Performance Evaluation of Bi_2Te_3-Based and Some Other Thermoelectric Nanostructured Materials." *Current Nanoscience* 17 (3): 423–46. https://doi.org/10.2174/1573413716999200820144753

Ayanda, Olushola Sunday, Augusta Oluchi Mmuoegbulam, Onyemaechi Okezie, Naseer Inuwa Durumin Iya, Sa'adatu Eri Mohammed, Philip Hegarty James, Abba Bashir Muhammad, et al. 2024. "Recent Progress in Carbon-Based Nanomaterials: Critical Review." *Journal of Nanoparticle Research* 26 (5): 106. https://doi.org/10.1007/s11051-024-06006-2

Balandin, Alexander A. 2011. "Thermal Properties of Graphene and Nanostructured Carbon Materials." *Nature Materials* 10 (8): 569–81. https://doi.org/10.1038/nmat3064

Balasubramanian, Kannan, and Marko Burghard. 2005. "Chemically Functionalized Carbon Nanotubes." *Small* 1 (2): 180–92. https://doi.org/10.1002/smll.200400118

Balaya, Palani. 2008. "Size Effects and Nanostructured Materials for Energy Applications." *Energy & Environmental Science* 1 (6): 645. https://doi.org/10.1039/b809078p

Biderman, Norbert, and Hy Si Bui. 2021. "Atomic Force Microscopy (AFM) as a Surface Characterization Tool for Hair, Skin, and Cosmetic Deposition." In *Surface Science and Adhesion in Cosmetics*, 245–78. Wiley. https://doi.org/10.1002/9781119654926.ch8

Buzea, Cristina, and Ivan Pacheco. 2017. "Nanomaterials and Their Classification." 3–45. https://doi.org/10.1007/978-81-322-3655-9_1

Chen, Dongyang, HanDong Zhang, Guoqi Zhao, Zhiqin Zhu, JingRan Yang, Jie He, JunCheng Li, Zijia Yu, and Zhiqi Zhu. 2024. "Investigating the Corrosion Resistance of Different SiC Crystal Types: From Energy Sectors to Advanced Applications." *Langmuir* 40 (24): 12322–42. https://doi.org/10.1021/acs.langmuir.4c01805

Dubey, Rama, Dhiraj Dutta, Arpan Sarkar, and Pronobesh Chattopadhyay. 2021. "Functionalized Carbon Nanotubes: Synthesis, Properties and Applications in Water Purification, Drug Delivery, and Material and Biomedical Sciences." *Nanoscale Advances* 3 (20): 5722–44. https://doi.org/10.1039/D1NA00293G

Gao, Chengde, Pei Feng, Shuping Peng, and Cijun Shuai. 2017. "Carbon Nanotube, Graphene and Boron Nitride Nanotube Reinforced Bioactive Ceramics for Bone Repair." *Acta Biomaterialia* 61 (October): 1–20. https://doi.org/10.1016/j.actbio.2017.05.020

Golberg, Dmitri, Yoshio Bando, Yang Huang, Takeshi Terao, Masanori Mitome, Chengchun Tang, and Chunyi Zhi. 2010. "Boron Nitride Nanotubes and Nanosheets." *ACS Nano* 4 (6): 2979–93. https://doi.org/10.1021/nn1006495

Gopinath, Kannappan Panchamoorthy, Nagarajan Vikas Madhav, Abhishek Krishnan, Rajagopal Malolan, and Goutham Rangarajan. 2020. "Present Applications of Titanium Dioxide for the Photocatalytic Removal of Pollutants from Water: A Review." *Journal of Environmental Management* 270 (September): 110906. https://doi.org/10.1016/j.jenvman.2020.110906

Guan, Huanqin, Cooro Harris, and Shouheng Sun. 2023. "Metal–Ligand Interactions and Their Roles in Controlling Nanoparticle Formation and Functions." *Accounts of Chemical Research* 56 (12): 1591–1601. https://doi.org/10.1021/acs.accounts.3c00156

Han, Zhidong, and Alberto Fina. 2011. "Thermal Conductivity of Carbon Nanotubes and Their Polymer Nanocomposites: A Review." *Progress in Polymer Science* 36 (7): 914–44. https://doi.org/10.1016/j.progpolymsci.2010.11.004

Hosseini, Hamideh, and Mohammad Ghaffarzadeh. 2022. "Surface Functionalization of Carbon Nanotubes via Plasma Discharge: A Review." *Inorganic Chemistry Communications* 138 (April): 109276. https://doi.org/10.1016/j.inoche.2022.109276

Kumar, Amit, Kamal Sharma, and Amit Rai Dixit. 2021. "A Review on the Mechanical Properties of Polymer Composites Reinforced by Carbon Nanotubes and Graphene." *Carbon Letters* 31 (2): 149–65. https://doi.org/10.1007/s42823-020-00161-x

Liu, Minmin, Ruizhong Zhang, and Wei Chen. 2014. "Graphene-Supported Nanoelectrocatalysts for Fuel Cells: Synthesis, Properties, and Applications." *Chemical Reviews* 114 (10): 5117–60. https://doi.org/10.1021/cr400523y

Long, Nguyen Viet, Toshiharu Teranishi, Yong Yang, Cao Minh Thi, Yanqin Cao, and Masayuki Nogami. 2015. "Iron Oxide Nanoparticles for Next Generation Gas Sensors." *International Journal of Metallurgical & Materials Engineering* 1 (2). https://doi.org/10.15344/2455-2372/2015/119

Maselugbo, Adesewa O., Haley B. Harrison, and Jeffrey R. Alston. 2022. "Boron Nitride Nanotubes: A Review of Recent Progress on Purification Methods and Techniques." *Journal of Materials Research* 37 (24): 4438–58. https://doi.org/10.1557/s43578-022-00672-5

Min, Byoung Koun, and Cynthia M. Friend. 2007. "Heterogeneous Gold-Based Catalysis for Green Chemistry: Low-Temperature CO Oxidation and Propene Oxidation." *Chemical Reviews* 107 (6): 2709–24. https://doi.org/10.1021/cr050954d

Mohammad, S. Noor. 2012. "Concept of Component Seed Vastly Broadens the Understanding of Nanotube Synthesis and Characteristics." *The Journal of Physical Chemistry C* 116 (9): 5312–26. https://doi.org/10.1021/jp208801p

Nandihalli, Nagaraj, Takao Mori, and Holger Kleinke. 2018. "Effect of Addition of SiC and Al_2O_3 Refractories on Kapitza Resistance of Antimonide-Telluride." *AIP Advances* 8 (9). https://doi.org/10.1063/1.5034520

Neuville, Stephane. 2019. "Selective Carbon Material Engineering for Improved MEMS and NEMS." *Micromachines* 10 (8): 539. https://doi.org/10.3390/mi10080539

Ogunbiyi, Olugbenga, and Yingchao Yang. 2024. "Mechanics of Two-Dimensional Materials." In *Mechanics of Flexible and Stretchable Electronics*, 87–138. Wiley. https://doi.org/10.1002/9783527842308.ch4

Pan, Yue, Xuewen Du, Fan Zhao, and Bing Xu. 2012. "Magnetic Nanoparticles for the Manipulation of Proteins and Cells." *Chemical Society Reviews* 41 (7): 2912. https://doi.org/10.1039/c2cs15315g

Pernot, G., M. Stoffel, I. Savic, F. Pezzoli, P. Chen, G. Savelli, A. Jacquot, et al. 2010. "Precise Control of Thermal Conductivity at the Nanoscale through Individual Phonon-Scattering Barriers." *Nature Materials* 9 (6): 491–95. https://doi.org/10.1038/nmat2752

Ren, Xuefeng, Yiran Wang, Anmin Liu, Zhihong Zhang, Qianyuan Lv, and Bihe Liu. 2020. "Current Progress and Performance Improvement of Pt/C Catalysts for Fuel Cells." *Journal of Materials Chemistry A* 8 (46): 24284–306. https://doi.org/10.1039/D0TA08312G

Sankar, Meenakshisundaram, Qian He, Rebecca V. Engel, Mala A. Sainna, Andrew J. Logsdail, Alberto Roldan, David J. Willock, Nishtha Agarwal, Christopher J. Kiely, and Graham J. Hutchings. 2020. "Role of the Support in Gold-Containing Nanoparticles as Heterogeneous Catalysts." *Chemical Reviews* 120 (8): 3890–3938. https://doi.org/10.1021/acs.chemrev.9b00662

Schlexer, Philomena, Anton Bay Andersen, Bela Sebok, Ib Chorkendorff, Jakob Schiøtz, and Thomas W. Hansen. 2019. "Size-Dependence of the Melting Temperature of Individual Au Nanoparticles." *Particle & Particle Systems Characterization* 36 (3). https://doi.org/10.1002/ppsc.201800480

Shaikh, Jasmin S., Navajsharif S. Shaikh, Sawanta S. Mali, Jyoti V. Patil, Krishna K. Pawar, Pongsakorn Kanjanaboos, Chang Kook Hong, J. H. Kim, and Pramod S. Patil. 2018. "Nanoarchitectures in Dye-Sensitized Solar Cells: Metal Oxides, Oxide Perovskites and Carbon-Based Materials." *Nanoscale* 10 (11): 4987–5034. https://doi.org/10.1039/C7NR08350E

Sharma, Navneet, Himanshu Ojha, Ambika Bharadwaj, Dharam Pal Pathak, and Rakesh Kumar Sharma. 2015. "Preparation and Catalytic Applications of Nanomaterials: A Review." *RSC Advances* 5 (66): 53381–403. https://doi.org/10.1039/C5RA06778B

Teo, Boon-Keng., and Joseph C. Calabrese. 1976. "Stereochemical Systematics of Metal Clusters. Crystallographic Evidence for a New Cubane Dblharw. Chair Isomerism in Tetrameric Triphenylphosphine Silver Iodide, $(Ph_3P)_4Ag_4I_4$." *Inorganic Chemistry* 15 (10): 2474–86. https://doi.org/10.1021/ic50164a032

Vysikaylo, Ph. I. 2012. "Cumulative Quantum Mechanics (CQM) Part II. Application of Cumulative Quantum Mechanics in Describing the Vysikaylo Polarization Quantum-Size Effects." *Surface Engineering and Applied Electrochemistry* 48 (5): 395–411. https://doi.org/10.3103/S1068375512050158

Wittstock, Arne, and Marcus Bäumer. 2014. "Catalysis by Unsupported Skeletal Gold Catalysts." *Accounts of Chemical Research* 47 (3): 731–39. https://doi.org/10.1021/ar400202p

Wu, Guohui, Alexander Mikhailovsky, Htet A. Khant, and Joseph A. Zasadzinski. 2009. "Synthesis, Characterization, and Optical Response of Gold Nanoshells Used to Trigger Release from Liposomes." 279–307. https://doi.org/10.1016/S0076-6879(09)64014-3

Wu, Nanhua, Xiaohua Lu, Rong An, and Xiaoyan Ji. 2021. "Thermodynamic Analysis and Modification of Gibbs–Thomson Equation for Melting Point Depression of Metal Nanoparticles." *Chinese Journal of Chemical Engineering* 31 (March): 198–205. https://doi.org/10.1016/j.cjche.2020.11.035

Wu, W., R.F. Giese, and C.J. van Oss. 1999. "Stability versus Flocculation of Particle Suspensions in Water—Correlation with the Extended DLVO Approach for Aqueous Systems, Compared with Classical DLVO Theory." *Colloids and Surfaces B: Biointerfaces* 14 (1–4): 47–55. https://doi.org/10.1016/S0927-7765(99)00023-5

Xu, Mingsheng, Tao Liang, Minmin Shi, and Hongzheng Chen. 2013. "Graphene-Like Two-Dimensional Materials." *Chemical Reviews* 113 (5): 3766–98. https://doi.org/10.1021/cr300263a

Yang, Hailong, Baohai Jia, Lin Xie, Dasha Mao, Junchao Xia, Jianmin Yang, Minhui Yuan, et al. 2024. "Achieving High Thermoelectric Performance through Ultra-Low Lattice Thermal Conductivity Based on Phonon Localization." *Joule* 8 (9): 2667–80. https://doi.org/10.1016/j.joule.2024.06.020

Yi, Zhou, and Shiyu Cui. 2023. "Preparation and Self-Healing Performance of Spherical α-Al_2O_3@$MoSi_2$ Particles Using the Precipitation Method." *Ceramics International* 49 (3): 4331–41. https://doi.org/10.1016/j.ceramint.2022.09.318

Yıldırım, Fatih, Necati Ataberk, and Mürsel Ekrem. 2021. "Mechanical and Thermal Properties of a Nanocomposite Material Which Epoxy Based and Reinforced with Polyvinyl Alcohol Nano Fibers Contained Multiwalled Carbon Nanotube." *Journal of Composite Materials* 55 (10): 1339–47. https://doi.org/10.1177/0021998320969793

Zandieh, Azadeh, Piyapong Buahom, Elnaz Baradaran Shokouhi, Lun Howe Mark, Reza Rahmati, Omid Aghababaei Tafreshi, Mahdi Hamidinejad, Andreas Mandelis, Keun Su Kim, and Chul B. Park. 2024. "Highly Anisotropic Thermally Conductive Dielectric Polymer/Boron Nitride Nanotube Composites for Directional Heat Dissipation." *Small* 20 (48). https://doi.org/10.1002/smll.202404189

Zhang, Lei, and C. Q. Ru. 2019. "A Refined JKR Model for Adhesion of a Rigid Sphere on a Soft Elastic Substrate." *Journal of Applied Mechanics* 86 (5). https://doi.org/10.1115/1.4042574

Zhao, Jun. 2014. "The Use of Ceramic Matrix Composites for Metal Cutting Applications." In *Advances in Ceramic Matrix Composites*, 623–54. Elsevier. https://doi.org/10.1016/B978-0-08-102166-8.00025-6

Zhou, Wen, Xia Gao, Dingbin Liu, and Xiaoyuan Chen. 2015. "Gold Nanoparticles for In Vitro Diagnostics." *Chemical Reviews* 115 (19): 10575–636. https://doi.org/10.1021/acs.chemrev.5b00100

Zoller, Florian, Sebastian Häringer, Daniel Böhm, Jan Luxa, Zdeněk Sofer, and Dina Fattakhova-Rohlfing. 2021. "Carbonaceous Oxygen Evolution Reaction Catalysts: From Defect and Doping-Induced Activity over Hybrid Compounds to Ordered Framework Structures." *Small* 17 (48). https://doi.org/10.1002/smll.202007484

2 Synthetic Techniques in Nanomaterials Manufacture

Rajni, Ankit Mehta and Sneh Lata

2.1 INTRODUCTION

Over the past few decades, nanotechnology has gradually gained a lot of attention from the scientific and professional communities as well as the public. Materials at the nanoscale, often known as "nanomaterials (NMs)," frequently differ significantly from one another in terms of their chemical and physical characteristics. Indeed, compared to their bulk form, NMs are more attractive due to several size-dependent phenomena, which justifies the significance and attention being paid to this area of research. NMs can be obtained by a variety of techniques, most of which are dependent on the material, its form, and its intended uses (Abid et al., 2022; El-Khawaga et al., 2023; Hachem et al., 2022; Mekuye & Abera, 2023; Sajid, 2022). Nowadays, the only physical methods available for precisely achieving epitaxial deposition of thin layers and material clusters, all the way down to atomic levels, on a variety of substrates are specialized vapor deposition methods. When it comes to synthesizing large amounts of NMs at relatively lower prices, other methods that heavily rely on solution chemistry are preferred (Cho et al., 2013; Machado et al., 2015; Nunes et al., 2021; Samrot et al., 2021; B. Yang et al., 2021).

There are two ways to group these various pathways for producing NMs: top-down and bottom-up. The top-down approach produces nanoparticles (NPs) through size reduction, whereas the bottom-up approach uses small units like molecules and atoms to make NPs through clusters (Lai et al., 2019; Liu et al., 2019; Nasrollahzadeh et al., 2019; Pang et al., 2018). The fundamental building blocks of NPs have been mobilized into complex structures using a variety of physical and chemical approaches in a bottom-up manner (Figure 2.1).

The biological environment, where nature uses chemical force to build up all necessary structures for life, serves as an implementation of this bottom-up methodology. To produce NPs, a wide range of bottom-up techniques have been devised, including condensation of atomic vapors and coalescence of liquid atoms. The major objective of this approach is to synthesize NPs at a certain nanoscale while preserving each particle's unique functionality (Baig et al., 2021; Bokov et al., 2021; Getts et al., 2015; Rizvi et al., 2022). On the other hand, the second strategy is top-down, or smaller to larger, as implied by its name. Here, a fabrication technique or a chemical is employed. A chemical process was used to etch the metal using acids to obtain

 DOI: 10.1201/9781003560234-3

the necessary particles (Jiang et al., 2022; Kumari et al., 2023). Furthermore, by using the bottom-up method, nanoshells, ultrafine NPs, and nanotubes can be created, and the deposition parameters can be changed. In addition, a bottom-up strategy is more cost-effective than a top-down one. However, one of the primary issues with the bottom-up approach is that the NPs must be chemically purified immediately before being used. However, the top-down approach does not require the chemical purification of NPs and may be manufactured on a large scale. However, a wide range of sizes (10–1000 nm) and diverse particle geometries or shapes are produced via the top-down approach. Furthermore, the top-down approach involves intricate

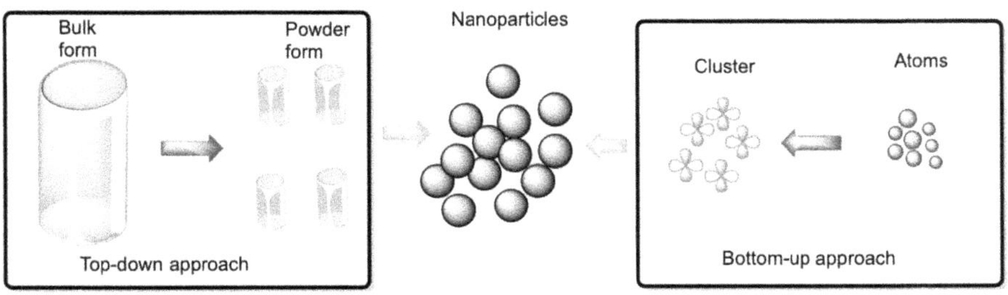

FIGURE 2.1 Schematic representation of the top-down and bottom-up production of NPs.

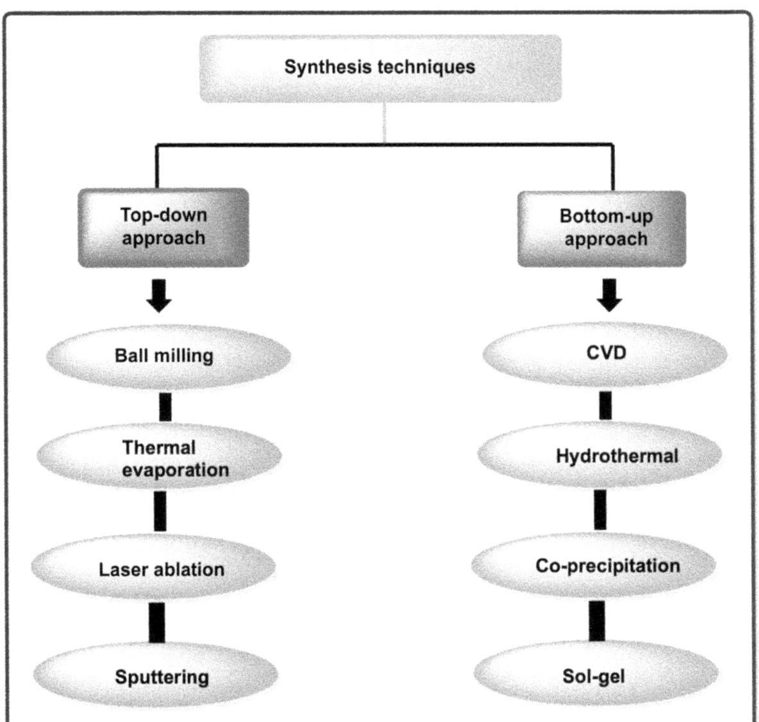

FIGURE 2.2 Classification of top-down and bottom-up approaches.

and expensive procedures (Chan & Kwok, 2011; Harish et al., 2022; Huston et al., 2021; Khan et al., 2022; Niculescu et al., 2021). Broadly top-down and bottom-up approaches are the two major techniques used for NP synthesis. Under each approach, different methods are used; for example, in the case of top-down approach, ball milling, thermal evaporation, laser ablation (LA), and sputtering methods are used, and in the case of the bottom-up approach, chemical vapor deposition (CVD), hydrothermal, co-precipitation, and sol-gel methods are used (Figure 2.2).

2.2 TOP-DOWN APPROACH

The destructive, or top-down, method involves disassembling a substance into its constituent atomic parts. Top-down approaches are well-suited for structures possessing long-range organization and macroscopic links. This technique allows for the breakdown of vast amounts of material into smaller, nanoscale fragments. Top-down approaches are easier to use, but they are not as effective in creating particles with intricate sizes or forms. The difficulty this technology poses in creating particles with the proper size and form is one of its main disadvantages. Typical methods for creating NPs include thermal breakdown, sputtering, LA, and mechanical milling.

2.2.1 Types of Top-Down Approach

2.2.1.1 Ball Milling Process

Ball milling, which generates NPs via the principle of attrition, is the most straightforward and effective mechanical procedure in the top-down method. It is the process by which the material being reduced receives kinetic energy from the grinding medium. Various NPs and metal alloys are produced using this method. In the ball milling method, a requisite amount of bulk material is poured into a milling vial, where it experiences high friction by the interaction with the balls and the walls of the vial vessel, changing its phase at high temperatures and pressure to a fine powder form. It is used to reduce particle size in a variety of materials. A wide range of reactions that are not possible at room temperature can be produced via this method. The milling process could be used for particle size reduction and growth along with its modification. The result is affected due to some factors like ball size, vial volume, total ball mass, mill rotation speed, powder quantity, milling environment, and milling time. It is observed that the efficiency of the process increases when the ball-to-powder weight ratio is larger, and a lesser amount of powder also enhances efficiency due to a greater number of collisions per unit of time.

This process demonstrated many advantages, such as high capacity, low maintenance, and ease of operation. It could also be used for reducing the thickness of conjugated materials. Ball milling also has some disadvantages; it may induce some crystal defects in particles, and it is inconvenient to carry due to its heavy weight. Usage of large amounts of energy was found to be there due to ball grinding and friction between them (Figure 2.3) (Huot et al., 2013; Ravnsbæk et al., 2012; Neikov et al., 2018; Takacs, 2002).

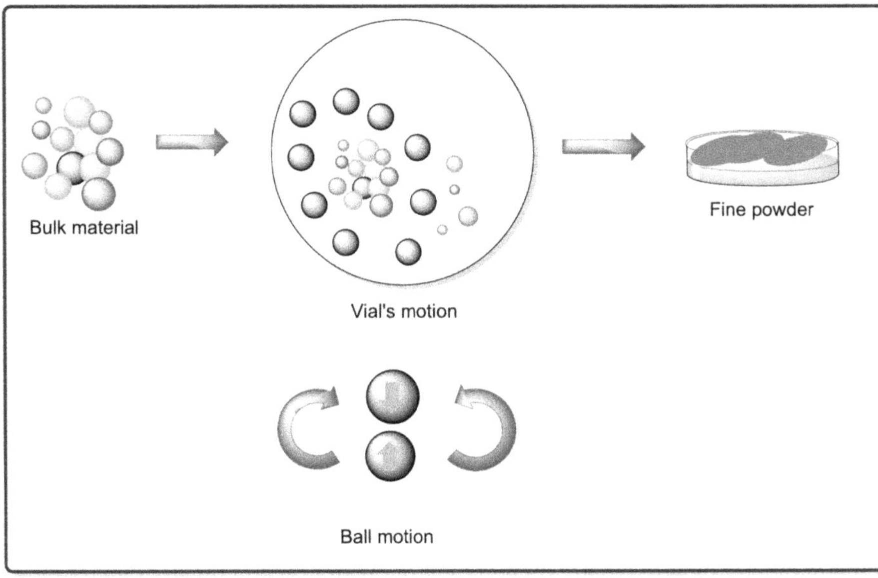

FIGURE 2.3 Schematic representation of ball milling technique.

2.2.1.2 Thermal Evaporation

It is an endothermic process in which heat breaks down chemical substances by breaking chemical bonds. Thermal evaporation is one of the most often used techniques to create stable inorganic NPs in the form of monodisperse suspensions, which have the potential to self-assemble. It is also used to create thin films on a range of substrates. In a typical process, an alumina crucible containing powder is placed in the center of the heating zone, with the powder facing the untreated substrate. The substrate and source are heated using a resistive heating technique. Substrates are cleaned with ethanol and acetone before being put in the evaporator. Temperatures at the source and on a substrate are measured and controlled with the help of a digital display and a PID (proportional integral derivative) controller. A shutter installed at the entrance of the crucible enables control over the deposition time. Before deposition, glass slides are ultrasonically cleaned in acetone and de-ionized water after being cleaned in soap solution.

The thermal evaporation process was found to have many advantages. The process could be carried out in solvent-free conditions, and materials could be deposited uniformly on the substrate surface. This process was found to be suitable for materials with a low boiling point. Monitoring of the process is more convenient as gases or vapors form of NPs could be easily monitored. This method was cost-effective. This process also had some limitations as some alloy compositions could not be deposited uniformly and source material may not be that effective (Figure 2.4) (Guo et al., 2018; Jacob et al., 2019; Keshav & Mahesha, 2021; Zahra et al., 2019).

2.2.1.3 Laser Ablation Method

A laser is a high-intensity electromagnetic radiation produced by the emission of stimulated radiation. It is used to synthesize NPs in various solvents. Lasers have

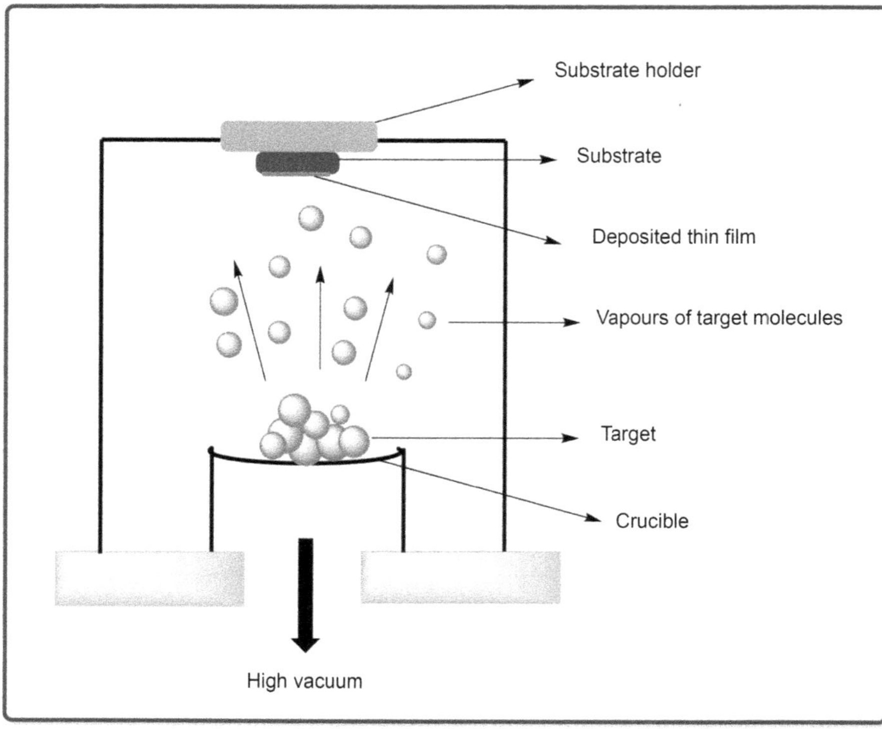

FIGURE 2.4 Thermal evaporation process.

been widely used in the military, industry, information transmission, and medical industries. There are several uses for LA, which uses a pulsed laser for the removal of molecules from the surface of the substrate to form micro/nanostructures in metals, ceramics, glasses, and polymers. LA is a top-down technique that removes a material from a substrate by directing a laser beam above it. Ablation does not happen until a material absorbs enough energy to melt or evaporate. In laser machining applications such as laser cutting, laser beam milling, and high-precision drilling, the general process of LA remains constant. A laser beam is directed at a metal target that is submerged in liquid. This produces vapor, molten metal droplets, or plasma, which then reacts with the liquid medium to form certain compounds and form NPs. A pulsed laser beam is used to irradiate a solid target (usually metal) immersed in a liquid like acetone, ethanol, water, chloroform, isopropanol, etc. to produce NPs.

Ablation rate has been dependent on factors like fluency, wavelength, repetition rate, pulse duration, transmission, the light absorption efficiency of the target material, and the chemical composition of the liquid.

This technique can synthesize ligand-free noble NPs in various solvents with negligible loss of energy. Along with these advantages, this technique has some limitations, such as high energy consumption. It also requires more precise laser sources, as dispersed lasers will not be able to generate efficient NPs on an industrial scale. An increase in ablation reduces this process efficiency duration (Figure 2.5)

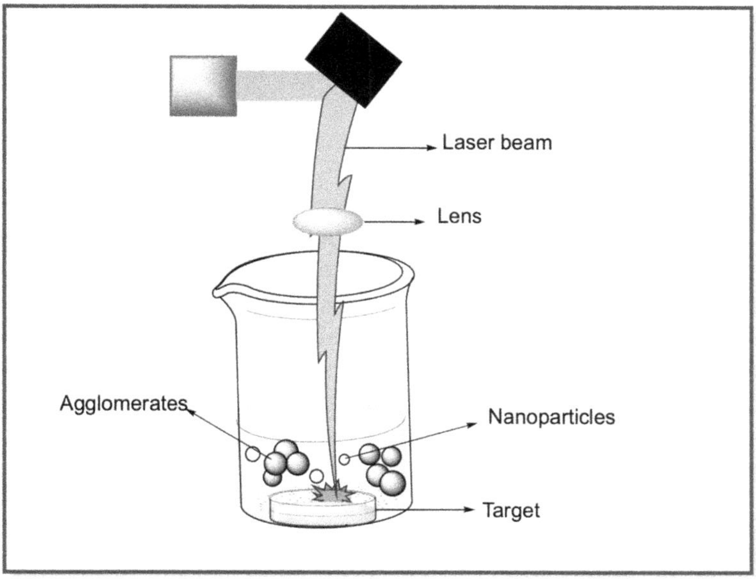

FIGURE 2.5 Schematic representation of LA technique.

(Mintcheva et al., 2018; Mintcheva et al., 2020; Nguyen & Yonezawa, 2018; Oluwatosin Abegunde et al., 2019; ; Ravi-Kumar et al., 2019; Saccomandi et al., 2018; Sportelli et al., 2018; Zhou et al., 2018).

2.2.1.4 Sputtering

Sputtering is a non-thermal vaporization process that makes it more common in bottom-up techniques. Using a vacuum pump, low pressure can be maintained during the sputtering process. Sputtering system components comprise an evacuated chamber equipped with a vacuum pump, sputtering source, gas supply, electrode, and substrate holder. NMs are deposited on a substrate's surface during the sputtering process by ejecting particles through the attack of high-energy ions. The most straightforward source for sputtering is plasma ions, which are formed when gas is ionized by collision among atoms and electronic movements. When Ar^+ ions are present, a negative voltage is supplied to the target during the direct current (DC) sputtering process, releasing the material and producing secondary electrons as a result. These electrons cause more ionization of gaseous particles. Positive ions attack a nonconducting target in the presence of an electric field, which charges a surface. The ion current would vanish. As a result, conducting materials like metals or doped semiconductors are the only ones that can be sputtered with DC. Reactive gases such as nitrogen and oxygen are employed because they react when one of them is introduced to the sputtering chamber. Additionally, when plasma ions strike a target and subsequently revert to their atomic state, they behave as neutralized electric charges. As a result of the neutralization process, positive charges appear on the target spot when an insulator is utilized as the target material. When energetic ions are repulsed by this positive charge, the sputtering process will come to an end. To remove the surface charge in this case, the polarity procedure can be reversed to aggregate the maximum number of electrons from the plasma. By putting a radio-frequency (RF) voltage on the target, an automatic periodic polarity reversal is achieved. Hence, RF sputtering is the term used to describe this kind of sputtering of the magnetron. While using the magnetron sputtering method, the magnetic field is used to deposit a thin film on the substrate.

As a result, there is an increase in argon atom ionization and bombardment ions, which speeds up the deposition process. In comparison to other sputtering methods, this technology has numerous advantages, including a high deposition rate, uniformity, control over thickness, and outstanding adhesion deposition over a relatively large surface. By increasing the deposition rate, increasing plasma ionization, decreasing the dark area in the deposition chamber, and increasing bombardment ion efficiency, the sputtering deposition technique could be improved. The sputtering technique is used for applications of thin-layer deposition, surface coating, and surface etching.

Several factors affect the rate of the sputtering process, such as temperature annealing duration and thickness of the layer, which in turn determine the size and shape of NPs. It is also affected by the angle between substrate and target along with energy, temperature, and mass of bombarding particles.

The sputtering process is preferred over other methods as it could be used at low temperatures. It could sputter alloys, compounds, and elements with large mass uniformly. This technique may show some limitations as its rate is slow in comparison to thermal evaporation, and a lot of energy in this process changes into undesired heat. Products are also produced with less purity (Figure 2.6) (Rane et al., 2018; Sayago et al., 2020; Wang et al., 2020; Zagho et al., 2019; Zhou, 2012).

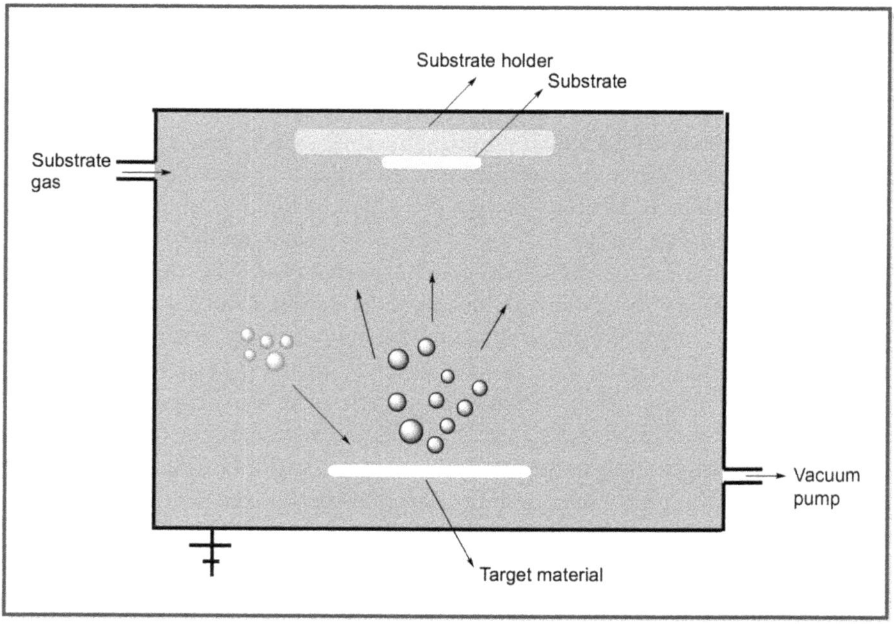

FIGURE 2.6 Sputtering technique for the synthesis of NMs.

2.3 BOTTOM-UP APPROACH

In the bottom-up technique, a material is constructed molecule by molecule, atom by atom, or cluster by cluster, starting from scratch. The strategy that is being discussed is an alternative approach that is more cost-effective and generates less waste. A common method used in the creation of nanoscale materials is the bottom-up technique, which entails building a material from the tiniest units, such as atoms, molecules, or clusters. Various nanoscale materials can be produced with consistent sizes, shapes, and distributions by this procedure.

2.3.1 Types of Bottom-Up Approach

2.3.1.1 Chemical Vapor Deposition (CVD)

CVD is a versatile and effective way to make nanostructures. With this method, NMs are made with comparatively simple ingredients. This method involves covering a substrate with a thin layer of gaseous reactants. A chemical reaction occurs in a reaction chamber when gas meets a heated substrate, and its surface becomes a thin layer which is removed by etching and is usefully employed. The controlled delivery of gas-phase reactants, the availability of an enclosed reaction chamber, the release of gases, the regulation of reaction pressure, the provision of an energy source for chemical reactions, the cleanup of exhaust gases to achieve safe and nontoxic levels, and the automation of process control to increase the stability of the deposition process are the operational requirements.

In this method, the reactor receives the reactant gases first. These reactant gases simply either diffuse directly on the substrate via the boundary layer in both conditions, where intermediate reactants and reactant gases adsorb and diffuse onto the heated substrate surface, or they go through homogeneous reactions in the gas phase to produce intermediate reactants and byproducts of gases. These growing species (intermediate reactants and byproducts) evaporate to get transported by a carrier gas on a substrate where they assemble and continue to grow. Through nucleation, growth, and coalescence, further heterogeneous reactions at the gas–solid interface result in the continuation of thin-film creation as well as the production of reaction byproducts. The reaction zone's surface eventually experiences desorption, allowing undesired gaseous products and undissolved species to be extracted. Gas-phase reactions take place anytime when there is enough temperature or activation energy is supplied in the form of plasma. More importantly, a heterogeneous reaction is needed if the deposition reaction depends on substrate surface catalysis, as it does with some metal surface catalytic growth. In this process, room temperature and pressure are maintained. Several factors, such as gas flow state, substrate and physical characteristics of vapor, deposition time, precursor nature, catalysts, and temperature, greatly affect this process. The maximum rate is observed at low or mild pressure and high temperature. The CVD method shows many advantages, such as the formation of NMs being controlled with excellent efficiency and products being obtained with high purity and crystallinity without any flaws. Orientation, purity, crystal structure, and surface morphology can easily be maintained or measured during this process. Products formed are homogeneous, hard, and robust with good growth rates. It is also

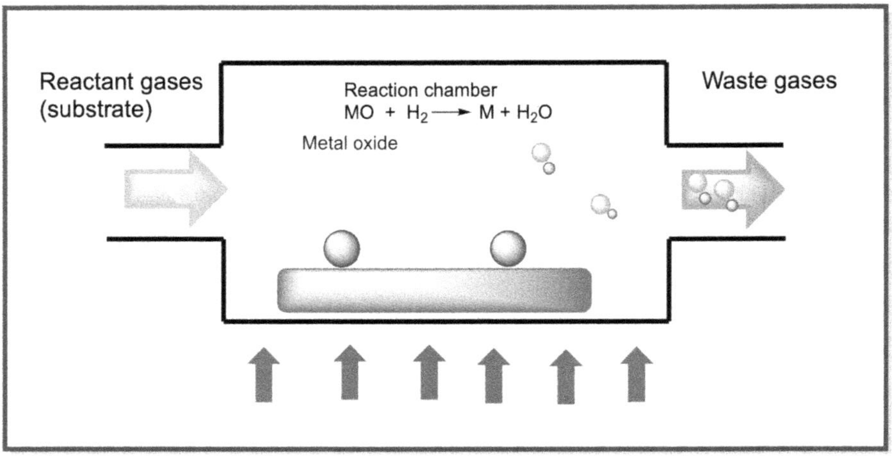

FIGURE 2.7 Schematic illustration of CVD approach.

cost-friendly to operate. The CVD method has some disadvantages like product formed cannot be examination *in situ* as a result product cost increases and some of the chemicals used during the process are toxic, corrosive, flammable, and costly. As high temperature is used in this technique, only selected substrates could be used for NM production and also produce toxic gaseous products (Figure 2.7) (Adachi et al., 2003; Ghasaban et al., 2017; Pottathara et al., 2019; Sun et al., 2021; Yan et al., 2013).

2.3.1.2 Hydrothermal Technique

The process of hydrothermal synthesis involves the reaction of a solid material with an aqueous solution in a reaction vessel at high pressure and temperature, which results in the deposition of tiny particles which upon washing and drying result in the formation of a fine product. A technique based on reactions in water as a solvent is called hydrothermal. In a steel pressure vessel, also known as an autoclave, the hydrothermal process is carried out. Temperatures and pressures can be changed to regulate the processing environment. The water reaches vapor saturation when the temperature is raised above its boiling point. Because of its homogeneous precipitation, low cost, hospitable atmosphere, simple scaling up, and pure final product, the hydrothermal process has made significant contributions to modern science and technology. The hydrothermal process can also be used to prepare functional ceramic powder, treat organic wastes, develop crystals, and synthesize hydrothermally.

In the first step of this process, metal hydroxide is utilized as a mineralizer, and metal alkoxides as a source of metal ions in the concerned solution precursor, which is combined in a solvent. In the second step, hydrothermal treatment is carried out in an autoclave, a sealed reactor in which water and nutrients are fed. The temperature differential between the chamber's upper and lower sections must be maintained. To gather the required crystal, ions are then moved toward the low-temperature area, and seed crystals are deposited at the cooler end.

In the next step, the sample is centrifuged, cleaned, and then dried at various temperatures to get the finished product. Growth conditions and crystal morphology are intimately connected in hydrothermal environments. The morphology study

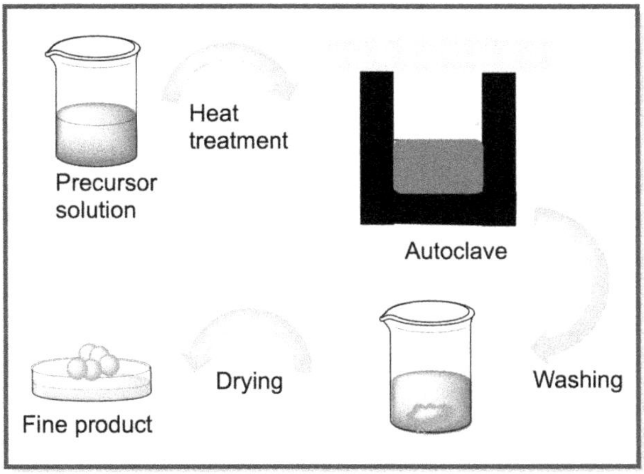

FIGURE 2.8 Schematic representation of hydrothermal technique to make fine particles of NMs.

anticipated the mechanism of crystal formation based on low- or high-pressure state which influences the morphology of the material. Substantial solid preparations such as microporous crystals, superionic conductors, and luminescence phosphors are made via hydrothermal synthesis. It is also a path to condensed matter that is unique, such as gels, thin films, and NPs.

The rate of formation of NPs is influenced by the type of solvent, reaction time, and high pressure and temperature values. This process has the advantage of forming NPs with high vapor pressure and low melting point. Particles produced are found to be in crystalline form tending to show pyrolysis. One of the limitations of this technique is that product formation could not be observed directly and also autoclave used in it is costly (Figure 2.8) (Denkbaş et al., 2016; Kafle, 2020; Wang et al., 2021; Yang & Park, 2019).

2.3.1.3 Co-Precipitation Technique

Co-precipitation is one of the oldest wet chemical methods used in the synthesis of NMs. In this process, impurities precipitate with the product, but they are easily removed by a variety of techniques such as aging, nucleation, and coarsening, or Ostwald ripening, followed by washing and filtration. Using a precipitation reaction to obtain a consistent composition of two or more cations in a homogenous solution is known as the co-precipitation technique. This method has the benefit of directly creating homogenous NMs with small diameters and size distribution through many chemical reactions in solution. To maintain the necessary pH, the solution is mixed either directly or drop by drop with another solution that contains dissolved precipitation agents, such as ammonia, sodium hydroxide, and many more. Oxalates, carbonates, hydroxides, and chlorides are some of the precipitants used in this method. The larger particles created by aging the precipitates are subsequently gathered via centrifugation or filtration. Extra washing with ethanol, distilled water, or other solvents is necessary to get rid of impurities and produce high-purity NPs. Post-treatments such as annealing, sintering, or calcination are used to produce NMs with desirable crystal structures and morphologies. Nucleation, growth, coarsening, and agglomeration must all occur concurrently. The process by which the smallest basic particles of a new thermodynamic phase emerge is known as nucleation, and it is an important stage in the precipitation process. Fast nucleation leads to a large concentration of nuclei, producing nanocrystals of smaller size. Fast nucleation produces much higher concentrations of nuclei and tiny nanocrystals, while slow nucleation produces smaller concentrations and larger particles. The larger particles will absorb the smaller ones to reduce their surface energy during the development phase. The procedure is called coarsening, or Ostwald ripening. To lower surface energy, aggregation may also take place. If coarsening and aggregation are not managed, particles may remain enlarged beyond the nanoscale. It helps to use certain capping or stabilizing agents to limit growth. Agents bonded to the surface of particles induce electrostatic repulsion on their surface by the chemisorption of charged species. If the repelling forces are strong enough, stable particles will form; if not, coagulation will occur. It is used for synthesizing nanotubes, nanorods, and NPs.

The precipitation process is influenced by several factors like temperature, pH, and concentration of ions. All these factors must be controlled for proper size and homogeneous nature of particles by adding different types of surfactants.

The above-mentioned process of making NMs requires low temperatures and is found to be easy to use and prepare. It is also an energy-efficient methodology and prepares homogeneous particles. Some of the disadvantages are also involved with this method like, it will not work if reactants have different precipitation rates, and impurities can also precipitate out which needs to be separated using different techniques (Figure 2.9) (Cuce et al., 2014; Li & Yang, 2019; Su & Chang, 2018).

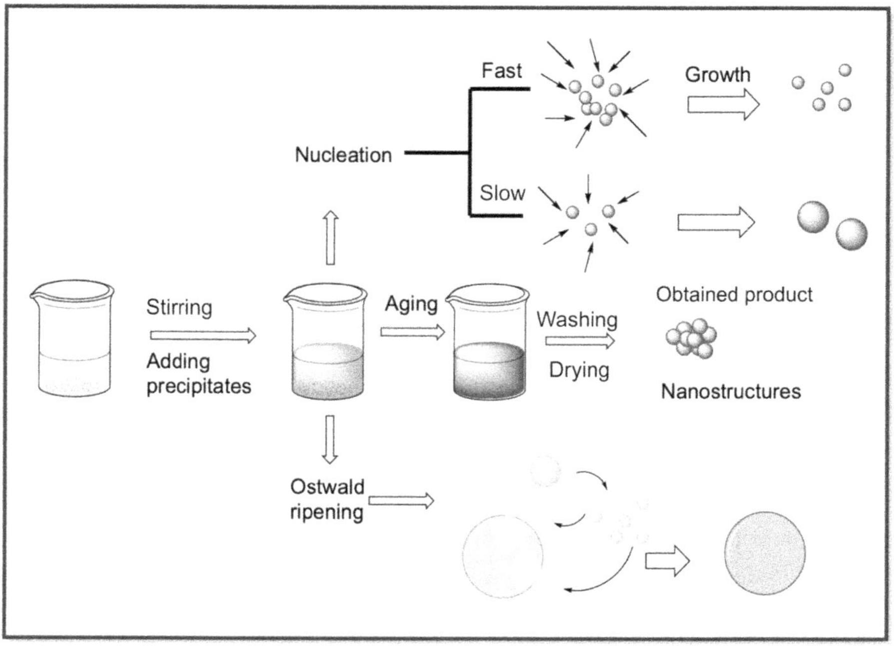

FIGURE 2.9 Schematic diagram of co-precipitation technique.

2.3.1.4 Sol-Gel Process

Sol-gel is the most widely used bottom-up method for synthesizing NMs because of its ease of use. A form of colloidal solution known as sol is composed of solid particles suspended in a liquid. A solid macromolecule called a gel dissolves in a liquid. This process involves hydrolysis, polycondensation, aging, drying, and calcination methods. In the first step, a precursor, such as metal alkoxide (M-OR), is hydrolyzed with hydroxide solution in water or alcohol. It is referred to as an aqueous sol-gel technique if the reaction medium is water, and a nonaqueous sol-gel method if the reaction medium is an organic solvent. In the second step, metal oxide bonds are formed due to the condensation of molecules. In the third step, a porous gel-like structure formed due to polycondensation increases the solvent's viscosity. The next step is supercritical drying, thermal drying, and freeze drying which is collectively called aging to synthesize aerogel, xerogel, and cryogel. In the last step, calcination is done for all three NMs to remove water content.

This technique could be used for synthesizing a wide range of materials, including thin films, NPs, glass, and ceramics. Properties of particles like porosity, microstructure, and crystallinity are greatly affected by temperature, reaction time, pH value, and stirring speed and duration. The heating factor must be there to prepare NMs.

The above-mentioned process is cost-effective and requires low temperatures. It produces compounds with homogeneous microstructure with high purity. This approach is found to be simple for producing composites by doping uniformly. The sol-gel process seems to have some disadvantages as the reaction is time consuming and involves the use of few toxic chemicals. Compounds also require post-purity treatments which are considered a long process with a waste of resources (Figure 2.10) (Alsaiari et al., 2023; Lamy-Mendes et al., 2018; Riffat & Qiu, 2013; Salimian et al., 2018).

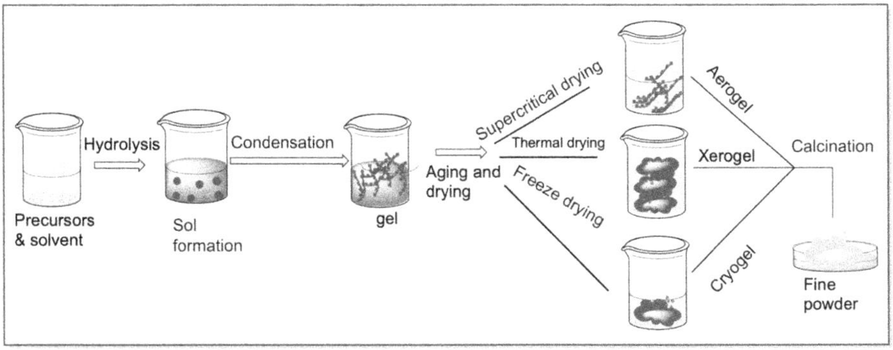

FIGURE 2.10 Sol-gel process presentation for aerogel, xerogel, and Cryogel formation.

2.4 CONCLUSION

Scientists consider nanotechnology to be the breakthrough technology of the twenty-first century. It can be defined as the study, control, and manipulation of materials where traditional physics fails, with diameters ranging from 1 to 100 nm. Worldwide, the application of NMs is growing in fields like chemistry, physics, material science, and healthcare. Several top-down (ball milling, thermal evaporation, LA, sputtering) and bottom-up (CVD, hydrothermal, co-precipitation, sol-gel) methods can synthesize NMs; each has benefits and drawbacks. The selection of synthesis techniques for a range of NMs is contingent upon factors such as the accessibility of raw materials, infrastructure, possible uses, and additional financial and ecological constraints. To choose the best synthesis technique, we must understand the fundamental principles underlying each synthesis process.

REFERENCES

Abegunde, Olayinka Oluwatosin, Esther Titilayo Akinlabi, Oluseyi Philip Oladijo, Stephen Akinlabi, and Albert Uchenna Ude. 2019. "Overview of Thin Film Deposition Techniques." *AIMS Materials Science* 6 (2): 174–199. https://doi.org/10.3934/matersci.2019.2.174

Abid, Namra, Aqib Muhammad Khan, Sara Shujait, Kainat Chaudhary, Muhammad Ikram, Muhammad Imran, Junaid Haider, Maaz Khan, Qasim Khan, and Muhammad Maqbool. 2022. "Synthesis of Nanomaterials Using Various Top-down and Bottom-up Approaches, Influencing Factors, Advantages, and Disadvantages: A Review." *Advances in Colloid and Interface Science* 300: 102597. https://doi.org/10.1016/j.cis.2021.102597

Adachi, Motoaki, Shigeki Tsukui, and Kikuo Okuyama. 2003. "Nanoparticle Synthesis by Ionizing Source Gas in Chemical Vapor Deposition." *Japanese Journal of Applied Physics* 42: L77–79. https://doi.org/10.1143/jjap.42.l77

Alsaiari, Norah Salem, Fatimah Mohammed Alzahrani, Abdelfattah Amari, Haitham Osman, Hamed N. Harharah, Noureddine Elboughdiri, and Mohamed A. Tahoon. 2023. "Plant and Microbial Approaches as Green Methods for the Synthesis of Nanomaterials: Synthesis, Applications, and Future Perspectives." *Molecules* 28 (1): 463. https://doi.org/10.3390/molecules28010463

Baig, Nadeem, Irshad Kammakakam, and Wail Falath. 2021. "Nanomaterials: A Review of Synthesis Methods, Properties, Recent Progress, and Challenges." *Materials Advances* 2 (6): 1821–71. https://doi.org/10.1039/d0ma00807a

Bokov, Dmitry, Abduladheem Turki Jalil, Supat Chupradit, Wanich Suksatan, Mohammad Javed Ansari, Iman H. Shewael, Gabdrakhman H. Valiev, and Ehsan Kianfar. 2021. "Nanomaterial by Sol-gel Method: Synthesis and Application." *Advances in Materials Science and Engineering* 2021 (1). https://doi.org/10.1155/2021/5102014

Chan, Hak-Kim, and Philip Chi Lip Kwok. 2011. "Production Methods for Nanodrug Particles Using the Bottom-Up Approach." *Advanced Drug Delivery Reviews* 63 (6): 406–16. https://doi.org/10.1016/j.addr.2011.03.011

Cho, Eun Jung, Hillary Holback, Karen C. Liu, Sara A. Abouelmagd, Joonyoung Park, and Yoon Yeo. 2013. "Nanoparticle Characterization: State of the Art, Challenges, and Emerging Technologies." *Molecular Pharmaceutics* 10 (6): 2093–2110. https://doi.org/10.1021/mp300697h

Cuce, Erdem, Pinar Mert Cuce, Christopher J. Wood, and Saffa B. Riffat. 2014. "Toward Aerogel Based Thermal Superinsulation in Buildings: A Comprehensive Review." *Renewable and Sustainable Energy Reviews* 34: 273–99. https://doi.org/10.1016/j.rser.2014.03.017

Denkbaş, Emir Baki, Ekin Çelik, Ebru Erdal, Doğa Kavaz, Öznur Akbal, Göknur Kara, and Cem Bayram. 2016. "Magnetically Based Nanocarriers in Drug Delivery." In *Nanobiomaterials in Drug Delivery*. Elsevier.

El-Khawaga, Ahmed M., Alaa Zidan, and Ahmed I. A. Abd El-Mageed. 2023. "Preparation Methods of Different Nanomaterials for Various Potential Applications: A Review." *Journal of Molecular Structure* 1281: 135148. https://doi.org/10.1016/j.molstruc.2023.135148

Getts, Daniel R., Lonnie D. Shea, Stephen D. Miller, and Nicholas J. C. King. 2015. "Harnessing Nanoparticles for Immune Modulation." *Trends in Immunology* 36 (7): 419–27. https://doi.org/10.1016/j.it.2015.05.007

Ghasaban, Samaneh, Mohammad Atai, and Mohammad Imani. 2017. "Simple Mass Production of Zinc Oxide Nanostructures via Low-Temperature Hydrothermal Synthesis." *Materials Research Express* 4 (3): 035010. https://doi.org/10.1088/2053-1591/aa5dcc

Guo, Yifei, Xiuli Fu, and Zhijian Peng. 2018. "Controllable Synthesis of MoS_2 Nanostructures from Monolayer Flakes, Few-Layer Pyramids to Multilayer Blocks by Catalyst-Assisted Thermal Evaporation." *Journal of Materials Science* 53 (11): 8098–8107. https://doi.org/10.1007/s10853-018-2103-0

Hachem, Kadda, Mohammad Javed Ansari, Raed Obaid Saleh, Hamzah H. Kzar, Moaed E. Al-Gazally, Usama S. Altimari, Shaymaa Abed Hussein, Halah T. Mohammed, Ali Thaeer Hammid, and Ehsan Kianfar. 2022. "Methods of Chemical Synthesis in the Synthesis of Nanomaterial and Nanoparticles by the Chemical Deposition Method: A Review." *BioNanoScience* 12 (3): 1032–57. https://doi.org/10.1007/s12668-022-00996-w

Harish, Vancha, Md Mustafiz Ansari, Devesh Tewari, Manish Gaur, Awadh Bihari Yadav, María-Luisa García-Betancourt, Fatehy M. Abdel-Haleem, Mikhael Bechelany, and Ahmed Barhoum. 2022. "Nanoparticle and Nanostructure Synthesis and Controlled Growth Methods." *Nanomaterials* 12 (18): 3226. https://doi.org/10.3390/nano12183226

Huot, J., D. B. Ravnsbæk, J. Zhang, F. Cuevas, M. Latroche, and T. R. Jensen. 2013. "Mechanochemical Synthesis of Hydrogen Storage Materials." *Progress in Materials Science* 58 (1): 30–75. https://doi.org/10.1016/j.pmatsci.2012.07.001

Huston, Matthew, Melissa DeBella, Maria DiBella, and Anisha Gupta. 2021. "Green Synthesis of Nanomaterials." *Nanomaterials* 11 (8): 2130. https://doi.org/10.3390/nano11082130

Jacob, Jolly, R. Wahid, A. Ali, R. Zahra, S. Ikram, N. Amin, A. Ashfa, et al. 2019. "Growth of Cu_2InO_4 Thin Films on Si Substrate by Thermal Evaporation Technique and Enhancement of Thermoelectric Properties by Post-Growth Annealing." *Physica B: Condensed Matter* 562: 59–62. https://doi.org/10.1016/j.physb.2019.03.023

Jiang, Zhiwen, Liwei Li, Hao Huang, Wenbin He, and Wuyi Ming. 2022. "Progress in Laser Ablation and Biological Synthesis Processes: 'Top-down' and 'Bottom-up' Approaches for the Green Synthesis of Au/Ag Nanoparticles." *International Journal of Molecular Sciences* 23 (23): 14658. https://doi.org/10.3390/ijms232314658

Kafle, Bhim Prasad. 2020. "Introduction to Nanomaterials and Application of UV–Visible Spectroscopy for Their Characterization." In *Chemical Analysis and Material Characterization by Spectrophotometry*. Elsevier.

Keshav, Rashmitha, and M. G. Mahesha. 2021. "Investigation on Performance of CdTe Solar Cells with CdS and Bilayer ZnS / CdS Windows Grown by Thermal Evaporation Technique." *International Journal of Energy Research* 45 (5): 7421–35. https://doi.org/10.1002/er.6325

Khan, Yousaf, Haleema Sadia, Syed Zeeshan Ali Shah, Muhammad Naeem Khan, Amjad Ali Shah, Naimat Ullah, Muhammad Farhat Ullah, et al. 2022. "Classification, Synthetic, and Characterization Approaches to Nanoparticles, and Their Applications in Various Fields of Nanotechnology: A Review." *Catalysts* 12 (11): 1386. https://doi.org/10.3390/catal12111386

Kumari, Swati, Sakshi Raturi, Saurabh Kulshrestha, Kartik Chauhan, Sunil Dhingra, Kovács András, Kyaw Thu, Rohit Khargotra, and Tej Singh. 2023. "A Comprehensive Review on Various Techniques Used for Synthesizing Nanoparticles." *Journal of Materials Research and Technology* 27: 1739–63. https://doi.org/10.1016/j.jmrt.2023.09.291

Lai, Xiaochen, Bingyu Lu, Penghao Zhang, Xingguo Zhang, Zhihua Pu, Haixia Yu, and Dachao Li. 2019. "Sticker Microfluidics: A Method for Fabrication of Customized Monolithic Microfluidics." *ACS Biomaterials Science & Engineering* 5 (12): 6801–10. https://doi.org/10.1021/acsbiomaterials.9b00953

Lamy-Mendes, Alyne, Rui F. Silva, and Luisa Durães. 2018. "Advances in Carbon Nanostructure–Silica Aerogel Composites: A Review." *Journal of Materials Chemistry. A, Materials for Energy and Sustainability* 6 (4): 1340–69. https://doi.org/10.1039/c7ta08959g

Li, Huayu, and Bing-Shiou Yang. 2019. "Model Evaluation of Particle Breakage Facilitated Process Intensification for Mixed-Suspension-Mixed-Product-Removal (MSMPR) Crystallization." *Chemical Engineering Science* 207: 1175–86. https://doi.org/10.1016/j.ces.2019.07.030

Liu, Qinqin, Jiyou Shen, Xiaohui Yu, Xiaofei Yang, Wei Liu, Juan Yang, Hua Tang, et al. 2019. "Unveiling the Origin of Boosted Photocatalytic Hydrogen Evolution in Simultaneously (S, P, O)-Codoped and Exfoliated Ultrathin g-C3N4 Nanosheets." *Applied Catalysis B: Environmental* 248: 84–94. https://doi.org/10.1016/j.apcatb.2019.02.020

Machado, S., J. G. Pacheco, H. P. A. Nouws, J. T. Albergaria, and C. Delerue-Matos. 2015. "Characterization of Green Zero-Valent Iron Nanoparticles Produced with Tree Leaf Extracts." *The Science of the Total Environment* 533: 76–81. https://doi.org/10.1016/j.scitotenv.2015.06.091

Mekuye, Bawoke, and Birhanu Abera. 2023. "Nanomaterials: An Overview of Synthesis, Classification, Characterization, and Applications." *Nano Select: Open Access* 4 (8): 486–501. https://doi.org/10.1002/nano.202300038

Mintcheva, Neli, Ali A. Aljulaih, Shusuke Bito, Mitsuhiro Honda, Takahiro Kondo, Satoru Iwamori, and Sergei A. Kulinich. 2018. "Nanomaterials Produced by Laser Beam Ablating Sn-Zn Alloy in Water." *Journal of Alloys and Compounds* 747: 166–75. https://doi.org/10.1016/j.jallcom.2018.02.350

Mintcheva, Neli, Shigeru Yamaguchi, and Sergei A. Kulinich. 2020. "Hybrid TiO2-ZnO Nanomaterials Prepared Using Laser Ablation in Liquid." *Materials* 13 (3): 719. https://doi.org/10.3390/ma13030719

Nasrollahzadeh, Mahmoud, S. Mohammad Sajadi, Mohaddeseh Sajjadi, and Zahra Issaabadi. 2019. "An Introduction to Nanotechnology." In *Interface Science and Technology*. Elsevier.

Neikov, Oleg D., Stanislav Naboychenko, and N. A. Yefimov. 2018. *Handbook of Non-Ferrous Metal Powders: Technologies and Applications*. 2nd ed. Elsevier/The Lancet.

Nguyen, Mai Thanh, and Tetsu Yonezawa. 2018. "Sputtering onto a Liquid: Interesting Physical Preparation Method for Multi-Metallic Nanoparticles." *Science and Technology of Advanced Materials* 19 (1): 883–98. https://doi.org/10.1080/14686996.2018.1542926

Niculescu, Adelina-Gabriela, Cristina Chircov, Alexandra Cătălina Bîrcă, and Alexandru Mihai Grumezescu. 2021. "Nanomaterials Synthesis through Microfluidic Methods: An Updated Overview." *Nanomaterials* 11 (4): 864. https://doi.org/10.3390/nano11040864

Nunes, Daniela, Ana Pimentel, Rita Branquinho, Elvira Fortunato, and Rodrigo Martins. 2021. "Metal Oxide-Based Photocatalytic Paper: A Green Alternative for Environmental Remediation." *Catalysts* 11 (4): 504. https://doi.org/10.3390/catal11040504

Pang, Jinbo, Alicja Bachmatiuk, Yin Yin, Barbara Trzebicka, Liang Zhao, Lei Fu, Rafael G. Mendes, Thomas Gemming, Zhongfan Liu, and Mark H. Rummeli. 2018. "Applications of Phosphorene and Black Phosphorus in Energy Conversion and Storage Devices." *Advanced Energy Materials* 8 (8). https://doi.org/10.1002/aenm.201702093

Pottathara, Yasir Beeran, Yves Grohens, Vanja Kokol, Nandakumar Kalarikkal, and Sabu Thomas. 2019. "Synthesis and Processing of Emerging Two-Dimensional Nanomaterials." In *Nanomaterials Synthesis*. Elsevier.

Rane, Ajay Vasudeo, Krishnan Kanny, V. K. Abitha, and Sabu Thomas. 2018. "Methods for Synthesis of Nanoparticles and Fabrication of Nanocomposites." In *Synthesis of Inorganic Nanomaterials*. Elsevier.

Ravi-Kumar, Sandeep, Benjamin Lies, Hao Lyu, and Hantang Qin. 2019. "Laser Ablation of Polymers: A Review." *Procedia Manufacturing* 34: 316–27. https://doi.org/10.1016/j.promfg.2019.06.155

Ravnsbæk, Dorthe B., Lise H. Sørensen, Yaroslav Filinchuk, Flemming Besenbacher, and Torben R. Jensen. 2012. "Screening of Metal Borohydrides by Mechanochemistry and Diffraction." *Angewandte Chemie* 51 (15): 3582–86. https://doi.org/10.1002/anie.201106661

Riffat, S. B., and G. Qiu. 2013. "A Review of State-of-the-Art Aerogel Applications in Buildings." *International Journal of Low-Carbon Technologies* 8 (1): 1–6. https://doi.org/10.1093/ijlct/cts001

Rizvi, Marziya, Husnu Gerengi, and Preeti Gupta. 2022. "Functionalization of Nanomaterials: Synthesis and Characterization." *ACS Symposium Series*. American Chemical Society.

Saccomandi, Paola, Alfonso Lapergola, Fabio Longo, Emiliano Schena, and Giuseppe Quero. 2018. "Thermal Ablation of Pancreatic Cancer: A Systematic Literature Review of Clinical Practice and Pre-Clinical Studies." *International Journal of Hyperthermia: The Official Journal of European Society for Hyperthermic Oncology, North American Hyperthermia Group* 35 (1): 398–418. https://doi.org/10.1080/02656736.2018.1506165

Sajid, Muhammad. 2022. "Nanomaterials: Types, Properties, Recent Advances, and Toxicity Concerns." *Current Opinion in Environmental Science & Health* 25: 100319. https://doi.org/10.1016/j.coesh.2021.100319

Salimian, S., A. Zadhoush, M. Naeimirad, R. Kotek, and S. Ramakrishna. 2018. "A Review on Aerogel: 3D Nanoporous Structured Fillers in Polymer-based Nanocomposites." *Polymer Composites* 39 (10): 3383–3408. https://doi.org/10.1002/pc.24412

Samrot, Antony V., Chamarthy Sai Sahithya, Jenifer Selvarani, Sajna Keeyari Purayil, and Paulraj Ponnaiah. 2021. "A Review on Synthesis, Characterization and Potential Biological Applications of Superparamagnetic Iron Oxide Nanoparticles." *Current Research in Green and Sustainable Chemistry* 4: 100042. https://doi.org/10.1016/j.crgsc.2020.100042

Sayago, Isabel, Esther Hontañón, and Manuel Aleixandre. 2020. "Preparation of Tin Oxide Nanostructures by Chemical Vapor Deposition." In *Tin Oxide Materials*. Elsevier.

Sportelli, Maria Chiara, Margherita Izzi, Annalisa Volpe, Maurizio Clemente, Rosaria Anna Picca, Antonio Ancona, Pietro Mario Lugarà, Gerardo Palazzo, and Nicola Cioffi. 2018. "The Pros and Cons of the Use of Laser Ablation Synthesis for the Production of Silver Nano-Antimicrobials." *Antibiotics* 7 (3): 67. https://doi.org/10.3390/antibiotics7030067

Su, Shei Sia, and Isaac Chang. 2018. "Review of Production Routes of Nanomaterials." *Commercialization of Nanotechnologies–A Case Study Approach*. Springer International Publishing.

Sun, Luzhao, Guowen Yuan, Libo Gao, Jieun Yang, Manish Chhowalla, Meysam Heydari Gharahcheshmeh, Karen K. Gleason, Yong Seok Choi, Byung Hee Hong, and Zhongfan Liu. 2021. "Chemical Vapour Deposition." *Nature Reviews Methods Primers* 1 (1). https://doi.org/10.1038/s43586-020-00005-y

Takacs, Laszlo. 2002. "Self-Sustaining Reactions Induced by Ball Milling." *Progress in Materials Science* 47 (4): 355–414. https://doi.org/10.1016/s0079-6425(01)00002-0

Wang, Niyou, Jerry Ying Hsi Fuh, S. Thameem Dheen, and A. Senthil Kumar. 2021. "Synthesis Methods of Functionalized Nanoparticles: A Review." *Bio-Design and Manufacturing* 4 (2): 379–404. https://doi.org/10.1007/s42242-020-00106-3

Wang, Yu-Hsiang, Kazi Hasibur Rahman, Chih-Chao Wu, and Kuan-Chung Chen. 2020. "A Review on the Pathways of the Improved Structural Characteristics and Photocatalytic Performance of Titanium Dioxide (TiO2) Thin Films Fabricated by the Magnetron-Sputtering Technique." *Catalysts* 10 (6): 598. https://doi.org/10.3390/catal10060598

Yan, Kai, Lei Fu, Hailin Peng, and Zhongfan Liu. 2013. "Designed CVD Growth of Graphene via Process Engineering." *Accounts of Chemical Research* 46 (10): 2263–74. https://doi.org/10.1021/ar400057n

Yang, Bingjun, Jiangtao Chen, Bao Liu, Yunxia Ding, Yu Tang, and Xingbin Yan. 2021. "One Dimensional Graphene Nanoscroll-Wrapped MnO Nanoparticles for High-Performance Lithium Ion Hybrid Capacitors." *Journal of Materials Chemistry A: Materials for Energy and Sustainability* 9 (10): 6352–60. https://doi.org/10.1039/d1ta00404b

Yang, Guijun, and Soo-Jin Park. 2019. "Conventional and Microwave Hydrothermal Synthesis and Application of Functional Materials: A Review." *Materials* 12 (7): 1177. https://doi.org/10.3390/ma12071177

Zagho, Moustafa M., Hana D. Dawoud, Nasr Bensalah, and T. M. Altahtamouni. 2019. "A Brief Overview of RF Sputtering Deposition of Boron Carbon Nitride (BCN) Thin Films." *Emergent Materials* 2 (1): 79–93. https://doi.org/10.1007/s42247-018-0018-9

Zahra, R., Jolly Jacob, N. Bano, A. Ali, K. Mahmood, S. Ikram, M. I. Arshad, A. Ashfaq, U. Rehman, and Sajad Hussain. 2019. "Effect of Secondary Phases on the Thermoelectric Properties of Zn_2GeO_4 Nano-Crystals Grown by Thermal Evaporation on Au Coated Si Substrate." *Physica. B, Condensed Matter* 564: 143–46. https://doi.org/10.1016/j.physb.2019.02.061

Zhou, A. 2012. "Methods of MAX-Phase Synthesis and Densification – II." In *Advances in Science and Technology of* $Mn+_1AX_n$ *Phases*. Elsevier.

Zhou, Rong, Shi Ding Lin, Yang Ding, Hong Yang, Ka Kei, O. Y. Keng, Meng Hong. 2018. "Enhancement of Laser Ablation via Interacting Spatial Double-Pulse Effect." *Opto-Electronic Advances* 1 (8): 1–6. https://doi.org/10.29026/oea.2018.180014

3 Nanoparticles – Synthesis Techniques of Nanomaterials

Shaman Gupta, Atul Raj, Pushpinder Kumar, Dinesh Deshwal, Sunny Misra and Sandeep Bansal

3.1 INTRODUCTION

Nanomaterials have become a cornerstone of modern science and technology due to their unique properties and versatile applications across various industries. The ability to precisely control their properties and structures is critical for harnessing their full potential. This chapter provides an in-depth exploration of the key methods used to synthesise nanomaterials, offering insights into their principles, processes, and technological advancements. It emphasises the importance of advanced techniques such as chemical vapour deposition (CVD), physical vapour deposition (PVD), the sol-gel process, and electrochemical methods, highlighting their role in producing high-quality nanomaterials. Additionally, the chapter discusses emerging synthesis techniques that are shaping the future of nanomaterial production, presenting both the opportunities and challenges associated with these innovative methods.

3.2 CHEMICAL VAPOUR DEPOSITION (CVD)

3.2.1 Overview of CVD Methods

Chemical vapour deposition (CVD) is considered to be a largely used technique in the case of high-quality nanomaterials, thin films, and coating synthesis (Figure 3.1). CVD's versatility has made this one of the significant procedures within material science, particularly for applications which need controlled deposition in nanoscale [1]. It is involved with volatile precursor's chemical reactions at gas phase which react or decompose on a substrate's surface. It leads to solid material development into a thin layer. Such a method is valued primarily for its capability towards developing uniform films having magnificent structural properties and high purity [2]. This makes them indispensable within industries like photovoltaics, nanotechnology, and manufacturing.

CVD process is categorised broadly depending on the source of energy utilised for driving reaction, deposition mechanism, and operating pressure. Common types involve metal-organic CVD (MOCVD), thermal CVD, atomic layer deposition

DOI: 10.1201/9781003560234-4

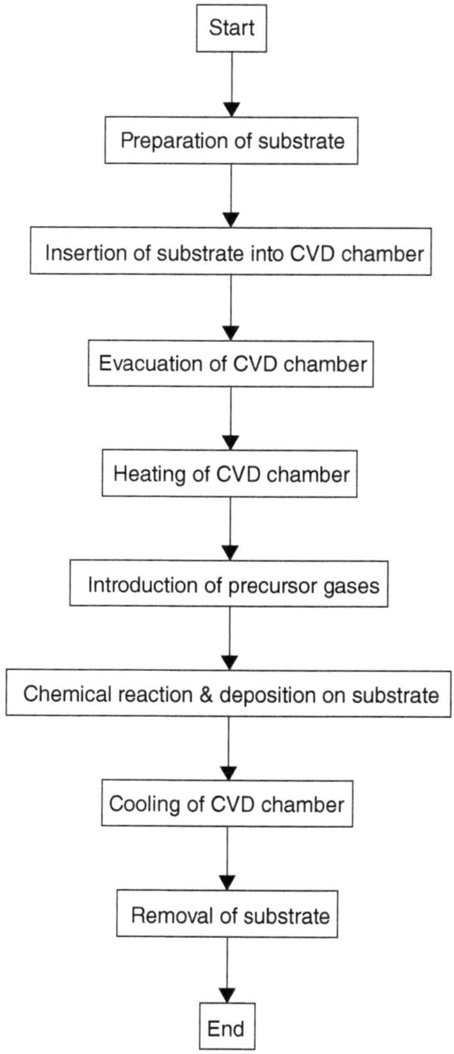

FIGURE 3.1 CVD process overview [2].

(ALD), and plasma-enhanced CVD. Each of them possesses specific features, making them suitable for various applications [2].

3.2.2 Thermal CVD

It is a widely used and simple CVD form that depends upon heat for initiating chemical reaction in precursor gases. This substrate is heated at a temperature that ranges between 500°C and 100°C [3]. This varies based on the material that is being deposited. In this high temperature, precursor gases mainly react or decompose at the surface. This results in solid and thin film formation (Figure 3.2). Thermal CVD is particularly efficient for developing high-performance materials such as

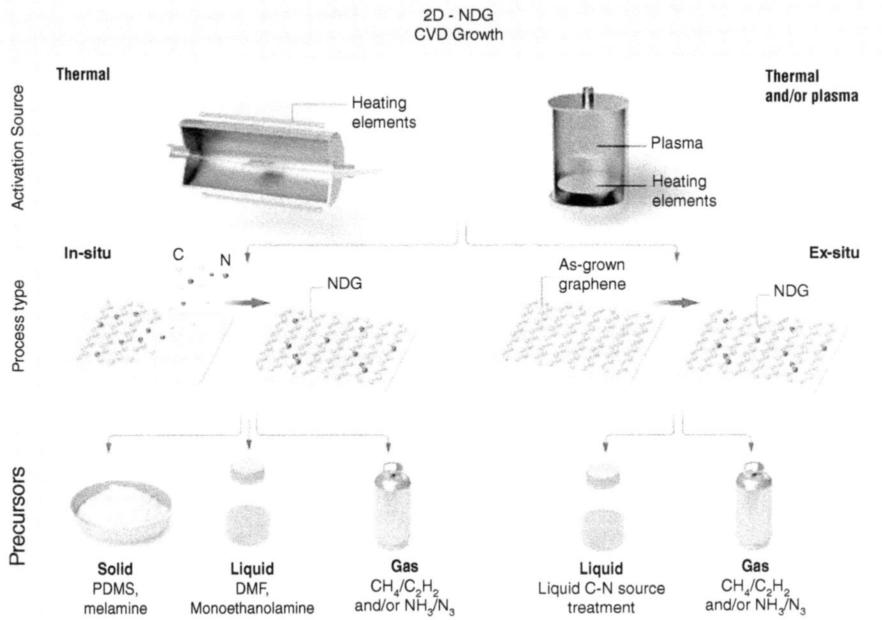

FIGURE 3.2 Tree diagram about NDG (nitrogen-doped graphene) growth through various CVD methods categorised by type [3].

carbon-based nanostructures, including graphene and carbon nanotubes (CNTs), as well as silicon dioxide and silicon. [1]. Although, requirements for higher temperature may limit their utilisation for the substrates that are temperature sensitive. It necessitates the development of an alternative method.

3.2.3 Metal Organic CVD (MOCVD)

This is another significant variation of the CVD procedure. It is specifically associated with compound semiconductors' deposition like indium phosphide and gallium nitride [4]. These are vital in the case of optoelectronic devices such as laser diodes and LEDs. Within MOCVD, precursors are mainly the metal and organic compounds, and in low temperatures, these are volatile in nature [5]. Such compounds are seen to decompose at the surface of the substrate in order to form desirable thin films. This often possesses a higher rate of deposition with better uniformity. Hence, it makes them suitable for bigger industrial production. Although, this procedure may involve complicated chemistry. It also needs adequate control for avoiding contamination while assuring high-quality film.

3.2.4 Plasma-Enhanced CVD (PECVD)

This is thermal CVD's modification which utilises plasma for increasing precursor gases' reactivity [6]. Through plasma generation primarily by microwave excitation or radio-frequency, this reaction may proceed at a lower substrate temperature [1].

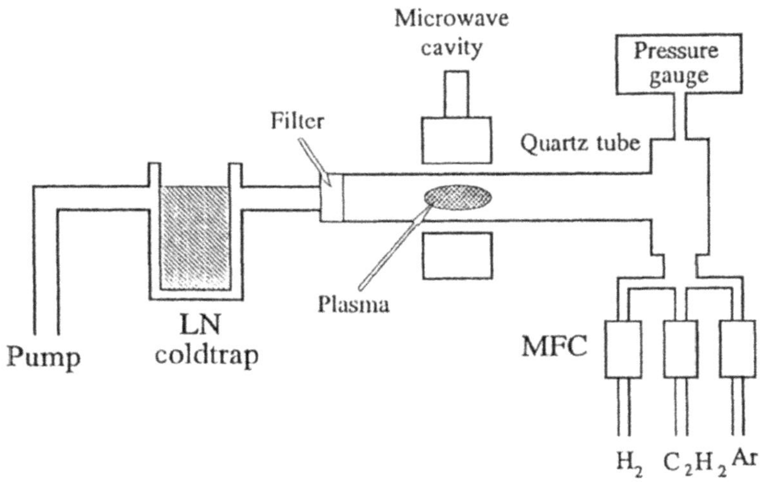

FIGURE 3.3 Diagram of microwave-enhanced CVD chamber [1].

This can even become below 400°C. It makes PECVD feasible in applications that need lower thermal budgets. These are like dielectric film deposition on the materials that are sensitive to temperature, such as flexible substrates or polymers. Utilisation of plasma also facilitates precise control on properties of film involving composition, microstructure, and thickness [7]. It makes PECVD a vital method for optical materials and advanced electronics production (Figure 3.3).

3.2.5 Atomic Layer Deposition (ALD)

ALD is CVD's variant that is mainly distinguished through the procedure of layer-by-layer deposition [8]. In the film thickness, it allows precision at the atomic level. Within ALD, alternating pulses are introduced to precursor gases in a sequential manner where individual pulses result in single ALD on a substrate [9]. Such self-limiting mechanism of reaction assures controlled and uniform films, even upon complicated 3-D surfaces. ALD is valued specifically in the semiconductor sector because of the deposition of barrier and ultrathin dielectric layers [8]. Despite slow deposition rate as compared to different CVD procedures, the control and precision provided by ALD make this effective in applications which require thin films with minimal defects and high uniformity.

3.2.6 Mechanisms of Nanomaterial Synthesis via CVD

Nanomaterial synthesis through CVD is associated with a complicated process. It is mainly influenced by interaction among substrate and precursor gases where chemical reaction takes place, resulting in nanoscale structure formation [10]. CVD's capability towards controlling material deposition at molecular and atomic levels makes it a suitable process to synthesise different nanomaterials. These involve graphene, semiconductor nanowires, and nanotubes (CNTs) [11]. Understanding this nanomaterial synthesis mechanism within the CVD process needs investigation of factors like growth kinetics, process parameters' influence, nucleation, and decomposition of precursor gases.

Nanomaterial synthesis primarily starts with the introduction of volatile precursor gas into the reaction chamber [12]. The precursor, most of the time hydrogen-based or organometallic, reacts or decomposes when it contacts with any heated substrate [8]. Temperature of the substrate is a major factor to initiate chemical reaction or thermal decomposition of precursor gases. It leads to molecular or atomic species formation which diffuses subsequently upon substrate surface. For instance, in case of carbon nanotube synthesis, gases such as acetylene (C_2H_2) or methane (CH_4) are utilised mostly as the source of carbon [13]. Using thermal decomposition, such gases liberate carbon atoms. It, then, acts as the building block for CNT growth.

Nucleation procedure is a major stage for the formation of a nanomaterial. At the time of nucleation, small molecules or atom clusters are developed on the surface of the substrate [14]. Regarding carbon nanotubes, the metal catalysts, like cobalt, nickel, or iron, are deposited on the substrate for influencing the nucleation procedure. Such metal catalysts serve as dissociation sites for the hydrogen gases [3]. It breaks carbon precursors into single carbon atoms. Dissolved carbon atoms are then diffused around or by metal nanoparticles. It, gradually, precipitates on surface, where it forms a CNT cylindrical structure characteristically. Metal nanoparticles' distribution and size impact the length, alignment, and diameter of synthesised nanotubes. This demonstrates the importance of catalyst design in the synthesis of CVD-dependent nanomaterials.

Nanomaterial's growth kinetics, followed by nucleation, is based on a balance of certain factors. These involve precursor gases' concentration, temperature of the substrate, and duration of the deposition process [15]. During the CVD process, gaseous precursors are continuously supplied into the reaction chamber, where they flow over the heated substrate surface, undergo chemical reactions, and subsequently deposit a thin film or nanostructured layer. It assures steady and continuous supply of the reactive species. It is adsorbed on the surface and then migrates to the other side of this. Then, it attaches to growing nanostructures. For example, in graphene synthesis, precursor gases like methane are decomposed at high temperatures, and then they release carbon atoms [16]. These arrange themselves subsequently into a honeycomb lattice, which is 2-D in structure, on the surface of the substrate. Graphene's growth follows a surface-mediated procedure. In this method, carbon atoms stick to certain crystallographic orientations preferentially. This resulted in largely ordered nanostructures.

At the time of CVD method, the crystallinity and morphology of the synthesised nanomaterials are effectively determined through reaction kinetics and surface diffusion [17]. Apart from temperature, other factors like the rate of gas flow and pressure impact such dynamics. For example, the gas flow rate of higher precursors enhances the reaction species' availability. It causes the development of large and potential nanostructure and offers non-uniform growth. The excessive gas flow rates may cause non-uniform growth or amorphous phase formation as a result of incomplete precursor reactions on the substrate surface [18]. . On a similar note, operating this at low pressure is capable of decreasing undesirable gas-phase reactions. It, moreover, offers film production uniformly. This is significant to produce quality nanostructures. The inherent versatility of CVD renders it a highly attractive technique for nanomaterial fabrication, allowing precise control over material properties for targeted applications in photonics, electronics, and advanced energy storage devices.

3.2.7 Overview of PVD Methods

PVD method is associated with encompassing various techniques utilised for depositing thin film upon different substrates by vaporised materials' condensation [19]. Among the widely employed physical vapour deposition (PVD) techniques, cathodic arc deposition, magnetron sputtering, pulsed laser deposition (PLD), and ion-beam assisted deposition (IBAD) are recognised as particularly significant due to their versatility and effectiveness in thin film and nanostructure fabrication [20]. Magnetron sputtering is particularly valuable in thin-film fabrication owing to its high deposition rate and its ability to produce films with accurately controlled stoichiometric composition. It is being operated in AC and DC configurations that allows usage flexibility. The technique is enhanced by plasma assistance which is recognised as plasma-enhanced magnetron sputtering. This helps in improving film properties and efficacy. Deposition of cathodic arcs is featured by their capability due to the larger throughput coating on large areas. It offers films with better density and adhesion, specifically for DC mode [21].

EB-PVD uses high-energy beams of electrons in a focused way for melting and evaporating materials. It enables graded coating production with better adhesion, making this suitable in the application of thermal barriers [22]. PLD is mainly distinguished by its capability regarding transferring stoichiometry from the target towards film. It aids in facilitating complex material deposition at higher rates. IBAD helps in combining the traditional deposition by ion bombardment [23]. It increases intermixing on the interface. This allows adherent coating at low temperatures. Each of these methods is beneficial in different applications, specifically in wear-resistant coating fabrication such as titanium nitride. It has potential in sectors that require wear resistance and durability.

3.2.8 Mechanisms of Nanomaterial Synthesis Through PVD

Nanomaterial synthesis by PVD is associated with a certain mechanism which influences vaporised materials' transformation into nanostructures or thin film [18]. At PVD centre, liquid or solid materials' vaporisation takes place. This takes place under vacuum situations for reducing contamination and facilitating film quality [24]. As mentioned above, different mechanisms like PLD, electron beam evaporation, and magnetron sputtering are engaged in employing varied energy sources for gaining the vaporisation. When they are vaporised, molecules or atoms are associated with travelling through vacuum. After that, these are condensed onto the substrate. In this place, they undergo growth processes and nucleation.

Primary stage of this nucleation is involved with cluster formation from the vapour phase. This may be facilitated by some factors like pressure, substrate characteristics, and temperature [25]. When the clusters are accumulated, they tend to coalesce into bigger entities. This is a procedure which has the potential to lead to nanostructured film development, and mechanism of growth may vary. For example, regarding layer-by-layer development, atoms are noted to deposit sequentially in order to grow distinct layers. In case of 3-D growth, the clusters are aggregated for developing complicated structures [26]. Reactive gases' incorporation in PVD procedure is helpful for compound nanomaterial formation (Figure 3.4). It increases the functional properties and chemical composition of deposited films.

PVD mechanisms allow to get precise control on decomposition parameters involving deposition rate, ambient pressure, and substrate bias. These are crucial for influencing nanomaterial properties and microstructures. Capability for developing thin films having desirable morphologies and uniform thickness makes the PVD a valuable approach in the application areas of biomedical, catalysis, and electronics fields [27]. PVD acts as an effective and versatile process in nanomaterial synthesis. This enables technology advancement in various domains [28].

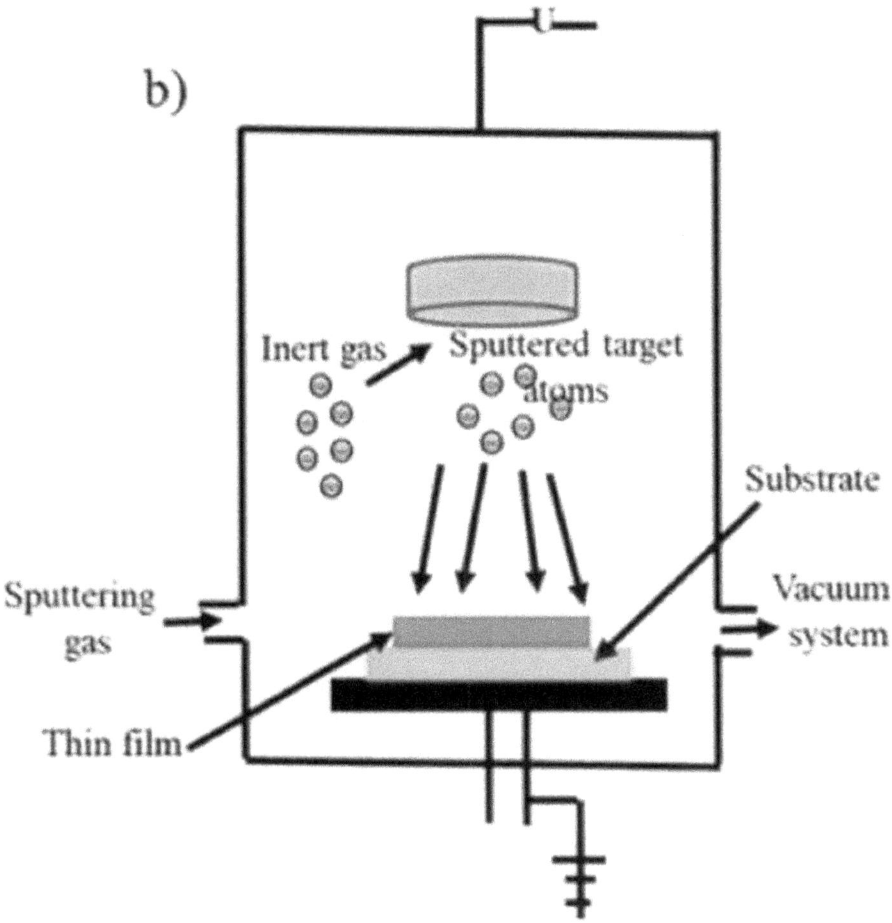

FIGURE 3.4 PVD method [27].

3.3 SOL-GEL PROCESS

3.3.1 About the Process

The sol-gel method is a solution-based process shown in Figure 3.5. It is widely used to produce ceramic particles and allied materials. Others include powders, coatings, films, and monolithic objects. It is typically employed for producing oxides. It incorporates transforming a colloidal suspension into a gel-like network to form materials. The process begins with hydrolysis. It is followed by condensation of metal alkoxides [29]. The process often requires external energy sources like ultrasound. It is necessary for moving the reaction forward. It is crucial for nanomaterial synthesis. It allows for the creation of pure oxide materials as well.

3.3.2 Principles of Sol-Gel Synthesis

The principle includes reaction and precursor. The first is reaction, comprising two parts. The first is hydrolysis and the second is condensation. Water attacks the metalloid atom of the precursor during hydrolysis. It replaces alkoxide groups with hydroxyl groups. In the condensation part, hydroxyl groups react to form an M-O-M network. This releases water and alcohol as byproducts. The reaction is influenced by thermodynamic and kinetic factors. Common precursors for the sol-gel process are metal alkoxides (Figure 3.6). Tetramethoxysilane and tetraethoxysilane are two noteworthy instances [30]. Organic-inorganic hybrid precursors include hydrogen silsesquioxane.

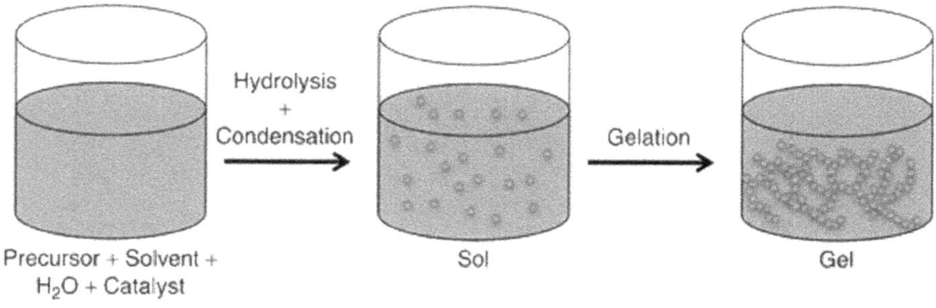

FIGURE 3.5 Schematics of sol-gel process [27].

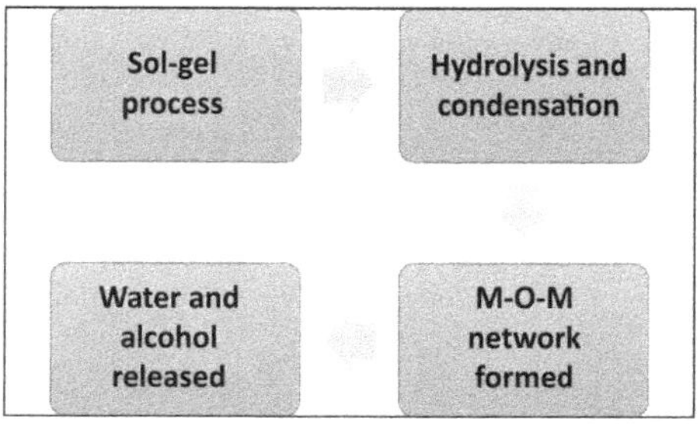

FIGURE 3.6 Sol-gel synthesis reaction process.

3.3.3 Sol-Gel Chemistry for Nanomaterial Fabrication

Metal alkoxides play a crucial role in the sol-gel process. They act as primary precursors for oxide materials. These compounds undergo hydrolysis and condensation reactions. They create a metal-oxygen network when combined. It gives nanomaterials their structural integrity (Figure 3.7). Organic precursors are used to create organic-inorganic hybrids [31]. They provide flexibility in tailoring properties. Examples include mechanical strength and biocompatibility. Morphology in the sol-gel process is a controlled aspect. It is done by adjusting key parameters such as the water-to-alkoxide ratio. There is a need to maintain the catalyst type and concentration. The reaction conditions must be in consideration as well such as pH and temperature.

3.3.4 Synthesis of Metal Oxide Nanoparticles via Sol-Gel

The mentioned process here is widely used to produce metal oxide nanoparticles. Examples include titanium oxide, silica, and alumina. For titanium oxide, titanium alkoxides are used as precursors. They undergo hydrolysis and condensation together. This process is necessary to form TiO_2 nanoparticles. They show value in solar cells due to their high surface area. Silica nanoparticles are typically produced using precursors like tetramethoxysilane [32]. There is just a need to adjust the water-to-alkoxide ratio. After that, the catalyst concentration is rectified as necessitated. It can yield uniform silica nanoparticles. For alumina nanoparticles, aluminium alkoxides are hydrolysed to form Al_2O_3. It is useful in abrasives and coatings.

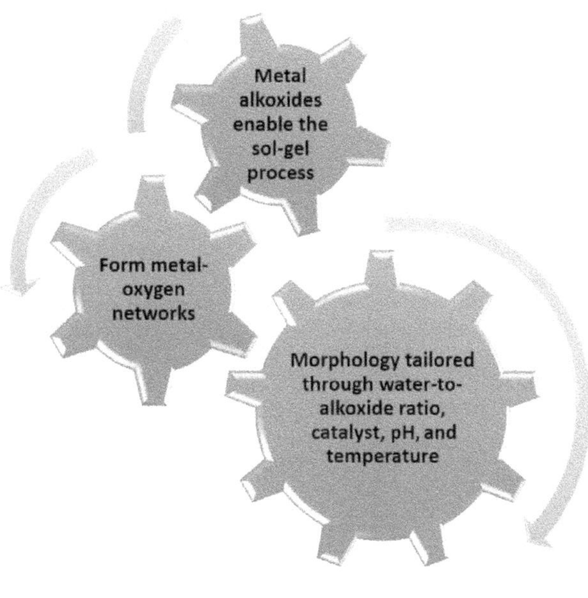

FIGURE 3.7 Nanomaterial fabrication process.

3.3.5 Factors Affecting Sol-Gel Synthesis

The pH of the reaction medium significantly affects the process [33]. Acidic conditions typically slow down hydrolysis. However, they have accelerated condensation that leads to denser structures. Temperature also influences the rate of reaction. Higher temperatures speed up reactions. Lower temperatures allow for better control over particle size. The choice of solvent also affects solubility. They affect the homogeneity of the reaction mixture. Acid or base additives are crucial for controlling the rate as well. Catalysts help modulate reaction kinetics and influence the final structure. Surfactants can control particle growth and prevent aggregation.

3.3.6 Advantages and Limitations

The sol-gel method offers precise control over composition. The production process can control particle size and morphology. It is versatile and allows the creation of many products at the end of the reaction. They include oxides, organic-inorganic hybrids, and bioceramics [34]. The process operates at relatively low temperatures. Hence, it is energy-efficient and suitable for diverse applications (Figure 3.8). There are a few limitations as well. This may happen when dealing with thicker coatings. It also requires careful control of reaction conditions as scaling can affect the reaction in industrial applications.

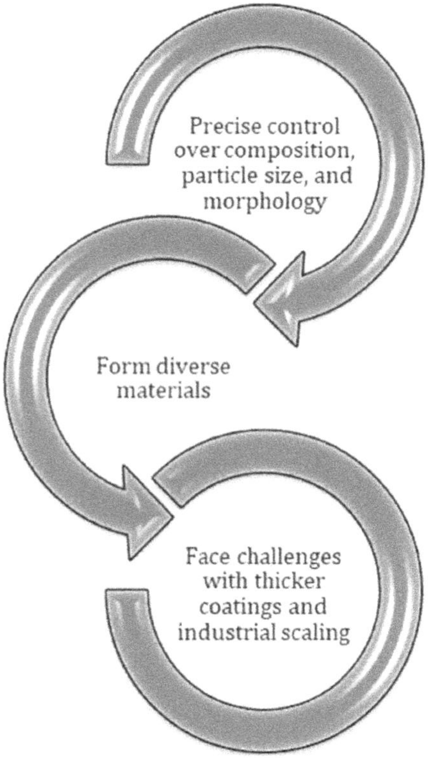

FIGURE 3.8 Advantageous of sol-gel method.

3.3.7 Applications

The process is applicable in various areas. It is used in nanoparticles. In optics, they are used to create coatings. They are used in the formation of waveguides and lenses with high transparency. Their refractive indices are very precise. In catalysis, metal oxide nanoparticles serve as photocatalysts for environmental cleanup. Biomedical applications include drug delivery systems. This is further followed by bioactive coatings and bioceramics. Additionally, sol-gel materials are employed in electronics for sensors [35]. Automotive and aerospace industries use this process to employ the final product as protective coatings. They are helpful in corrosion resistance.

3.3.8 Future Trends

The first area is advances in hybrid organic-inorganic materials. The process is used to create multifunctional nanomaterials with tailored properties for electronics. Green sol-gel processes are gaining traction for sustainable manufacturing. They aim to reduce toxic solvents and lower energy consumption. In the field of 3-D printing, the sol-gel process promises an avenue for creating complex nanostructures [36]. The development of sol-gel techniques for nano-enabled energy storage is effective. Important elements include super-capacitors and batteries.

3.4 ELECTROCHEMICAL METHODS

3.4.1 Background and Fundamentals

The method here is a revolutionary process as it is adaptive. Particle size can be precisely controlled by the procedure. At the electrode-electrolyte contact, electrochemical processes are used [37]. Nanomaterials are created as a result. Usually, they are alloys and metal nanoparticles. It is extensively employed in sensors and catalysis applications. Its capacity to create highly pure nanostructures is the reason behind this. Applying a voltage or current to an electrochemical cell is the fundamental idea. It must have metal ions in solution within it. At the electrode, metal ions are either dissolved or deposited. Through REDOX reactions, this happens. At last, the nanoparticles come into being.

3.4.2 Electrochemical Deposition

Electrodeposition is gaining widespread popularity due to its simplicity. This method involves submerging a conductive substrate in an electrolyte (Figure 3.9). It must have the target material's ions in it. Then a voltage is applied across the solid–electrolyte interface. This controlled voltage drives the charge transfer reaction. The process then results in the deposition of nanomaterial on the substrate. It offers precise control over the material's microstructure. A significant advantage of electrodeposition is the high raw material utility [38]. Energy consumption is low, and waste generation is minimal as well. These features make it commercially attractive. Electronics and energy storage industries always follow this process. However, its ease of use may lead to superficial experimentation.

3.4.3 Electrochemical Synthesis of Metal Nanoparticles

Electroplating is a crucial stage evident in the electrochemical synthesis process. It does so by using controlled reduction of metal ions onto a conductive substrate. Electroplating allows for the formation of highly uniform metal nanoparticles (Figure 3.10). It is done by precisely manipulating the applied potential. This technique is scalable and ideal for producing metal nanoparticles. Then they are applicable in electronics and sensors [40]. Applications for gold nanoparticles in electronics and biology are numerous. It is due to their stability and unique optical properties. Silver nanoparticles are used in coverings and healthcare devices. Copper nanoparticles are commonly employed in electronics and catalysis.

3.4.4 Template-Assisted Electrochemical Method

The process involves using porous membranes. The template guides the growth of nanomaterials. They are made from materials like anodised aluminium oxide (Figure 3.11). They contain nanoscale pores that direct the deposition of metals or other materials. Metal ions are deposited within the pores when immersed in an electrolyte [41]. The following applied potential finally produces nanomaterials having precise shapes. This process is commonly used for synthesising nanowires and

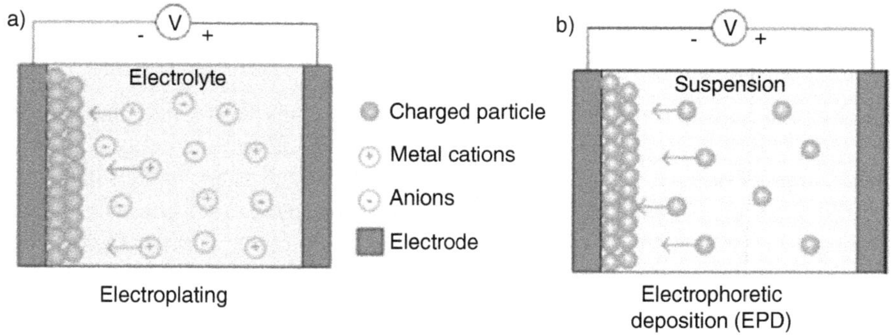

FIGURE 3.9 Electrochemical deposition process [39].

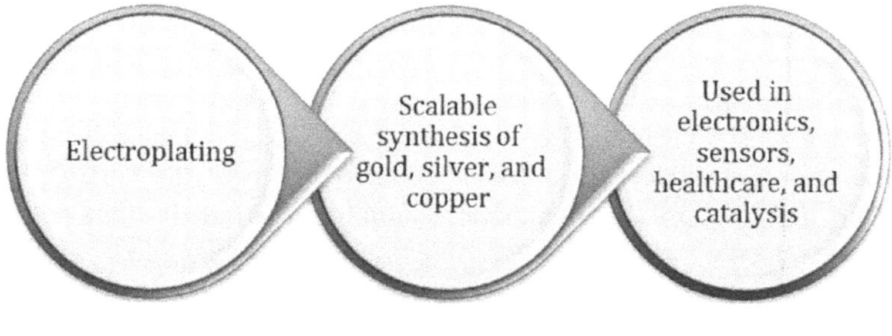

FIGURE 3.10 Metal nanoparticles.

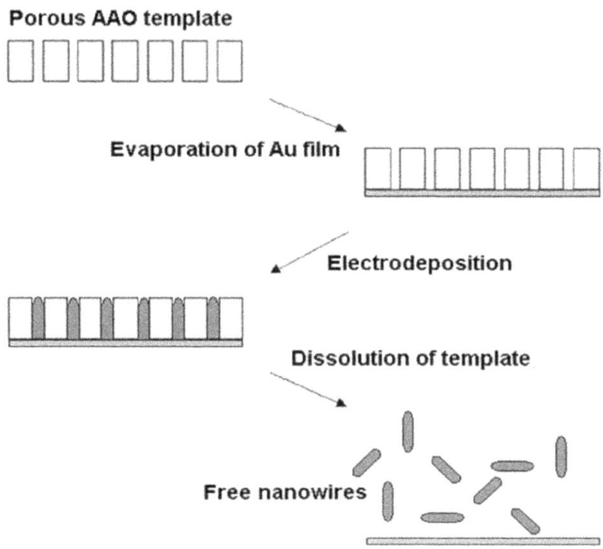

FIGURE 3.11 Template-assisted electrochemical method process [42].

nanotubes. The process can help in producing high-aspect-ratio nanowires. It is done by depositing semiconductors within the pores of a template. Similarly, a hollow template can be used in the formation of nanotubes. They find applications in the electrical and electronics sector.

3.4.5 Benefits, Limitations, and Application for Future

Electrochemical synthesis offers precise control over particle size and composition. It is energy-efficient and scalable for industrial use. It requires careful optimisation of reaction conditions. If not, replication might be challenging. Retaining consistency during the scaling process is essential. Green electrochemical methods are helpful for sustainable nanomaterial production [43]. Advances in template-assisted synthesis will help create more complex nanostructures. Additionally, in situ monitoring of electrochemical processes could lead to better control over reaction dynamics.

3.5 OTHER EMERGING SYNTHESIS METHODS

3.5.1 Ball Milling

It is a top-down mechanical method for synthesising nanoparticles. Here, bulk materials are reduced in size through repeated collisions with ceramic balls. Usually, the procedure involves placing the substance within a cylindrical container [44]. Then it is rotated at higher speeds. As the balls collide, mechanical energy is released. This leads to a reduction in particle size. This method is advantageous for producing ceramics and composites.

3.5.2 Microemulsion Method

It is a versatile technique that uses surfactant-stabilised droplets as nanoreactors. In this approach, two immiscible liquids are combined with surfactants to form tiny droplets on the nanoscale [45]. These droplets can encapsulate precursors. This leads to controlled nucleation and growth of nanoparticles within the confined space. The advantages of this method include high uniformity.

3.5.3 Laser Ablation

It is a physical method that incorporates intense laser pulses. The laser generates localised heating after it is focused on the target. It produces a vapour plume [46]. The target material then gets evaporated. It forms nanoparticles that condense in a liquid medium. It is an advantageous process as it is environmentally friendly. No chemicals are used, and high-purity nanoparticles are formed.

REFERENCES

[1] Y.M. Manawi, Ihsanullah, A. Samara, T. Al-Ansari, M.A. Atieh, A review of carbon nanomaterials' synthesis via the chemical vapor deposition (CVD) method, *Materials (Basel)* 11 (2018) 822. https://doi.org/10.3390/ma11050822

[2] S.P. Kumar, L. Babitha, A.P. Srivastava, R. Kumar, H.A. Ali, S. Lakhanpal, Characterization and modelling of nanomaterials synthesized by chemical vapor deposition, *E3S Web Conf.* 430 (2023) 01122. https://doi.org/10.1051/e3sconf/202343001122

[3] G. Deokar, J. Jin, U. Schwingenschlögl, P.M.F.J. Costa, Chemical vapor deposition-grown nitrogen-doped graphene's synthesis, characterization and applications, *NPJ 2D Mater. Appl.* 6 (2022) 14. https://doi.org/10.1038/s41699-022-00287-8

[4] I. Miccoli, G. Simkus, H. Larhirb, T. Korst, M. Mukinovic, J. Holzwarth, M. Heuken, Enabling MOCVD production on next generation 150 mm Indium Phosphide wafer size, *J. Cryst. Growth* 643 (2024) 127793. https://doi.org/10.1016/j.jcrysgro.2024.127793

[5] T.V. Basova, A. Hassan, N.B. Morozova, Chemistry of gold(I, III) complexes with organic ligands as potential MOCVD precursors for fabrication of thin metallic films and nanoparticles, *Coord. Chem. Rev.* 380 (2019) 58–82. https://doi.org/10.1016/j.ccr.2018.09.005

[6] E. Bertran-Serra, S. Rodriguez-Miguel, Z. Li, Y. Ma, G. Farid, S. Chaitoglou, R. Amade, R. Ospina, J.-L. Andújar, Advancements in plasma-enhanced chemical vapor deposition for producing vertical graphene nanowalls, *Nanomaterials* 13 (2023) 2533. https://doi.org/10.3390/nano13182533

[7] M. Mozetič, A. Vesel, G. Primc, C. Eisenmenger-Sittner, J. Bauer, A. Eder, G.H.S. Schmid, D.N. Ruzic, Z. Ahmed, D. Barker, K.O. Douglass, S. Eckel, J.A. Fedchak, J. Hendricks, N. Klimov, J. Ricker, J. Scherschligt, J. Stone, G. Strouse, I. Capan, M. Buljan, S. Milošević, C. Teichert, S.R. Cohen, A.G. Silva, M. Lehocky, P. Humpoliček, C. Rodriguez, J. Hernandez-Montelongo, D. Mercier, M. Manso-Silván, G. Ceccone, A. Galtayries, K. Stana-Kleinschek, I. Petrov, J.E. Greene, J. Avila, C.Y. Chen, B. Caja-Munoz, H. Yi, A. Boury, S. Lorcy, M.C. Asensio, J. Bredin, T. Gans, D. O'Connell, J. Brendin, F. Reniers, A. Vincze, M. Anderle, L. Montelius, Recent developments in surface science and engineering, thin films, nanoscience, biomaterials, plasma science, and vacuum technology, *Thin Solid Films* 660 (2018) 120–160. https://doi.org/10.1016/j.tsf.2018.05.046

[8] D. Escorcia-Díaz, S. García-Mora, L. Rendón-Castrillón, M. Ramírez-Carmona, C. Ocampo-López, Advancements in nanoparticle deposition techniques for diverse substrates: a review, *Nanomaterials* 13 (2023) 2586. https://doi.org/10.3390/nano13182586

[9] T. Muneshwar, M. Miao, E.R. Borujeny, K. Cadien, Atomic Layer Deposition, in: *Handbook Thin Film Deposition*, Elsevier, 2018: pp. 359–377. https://doi.org/10.1016/B978-0-12-812311-9.00011-6
[10] N. Baig, I. Kammakakam, W. Falath, I. Kammakakam, Nanomaterials: A review of synthesis methods, properties, recent progress, and challenges, *Mater. Adv.* 2 (2021) 1821–1871. https://doi.org/10.1039/d0ma00807a
[11] M. Vir Singh, A. Kumar Tiwari, R. Gupta, Catalytic chemical vapor deposition methodology for carbon nanotubes synthesis, *ChemistrySelect* 8 (2023). https://doi.org/10.1002/slct.202204715
[12] X.-D. Wang, K. Vinodgopal, G.-P. Dai, Synthesis of carbon nanotubes by catalytic chemical vapor deposition, *Perspect. Carbon Nanotub.* (2019) 17. https://doi.org/10.5772/intechopen.86995
[13] K. Hachem, M.J. Ansari, R.O. Saleh, H.H. Kzar, M.E. Al-Gazally, U.S. Altimari, S.A. Hussein, H.T. Mohammed, A.T. Hammid, E. Kianfar, Methods of chemical synthesis in the synthesis of nanomaterial and nanoparticles by the chemical deposition method: a review, *Bionanoscience* 12 (2022) 1032–1057. https://doi.org/10.1007/s12668-022-00996-w
[14] P. Piszczek, A. Radtke, Silver nanoparticles fabricated using chemical vapor deposition and atomic layer deposition techniques: properties, applications and perspectives: review, *Noble Precious Met. - Prop. Nanoscale Eff. Appl.* (2018). https://doi.org/10.5772/intechopen.71571
[15] K.J. Wu, E.C.M. Tse, C. Shang, Z. Guo, Nucleation and growth in solution synthesis of nanostructures – from fundamentals to advanced applications, *Prog. Mater. Sci.* 123 (2022) 100821. https://doi.org/10.1016/j.pmatsci.2021.100821
[16] Y. Yan, F.Z. Nashath, S. Chen, S. Manickam, S.S. Lim, H. Zhao, E. Lester, T. Wu, C.H. Pang, Synthesis of graphene: potential carbon precursors and approaches, *Nanotechnol. Rev.* 9 (2020) 1284–1314. https://doi.org/10.1515/ntrev-2020-0100
[17] J. Zhang, F. Wang, V.B. Shenoy, M. Tang, J. Lou, Towards controlled synthesis of 2D crystals by chemical vapor deposition (CVD), *Mater. Today* 40 (2020) 132–139. https://doi.org/10.1016/j.mattod.2020.06.012
[18] V. Harish, M.M. Ansari, D. Tewari, M. Gaur, A.B. Yadav, M.L. García-Betancourt, F.M. Abdel-Haleem, M. Bechelany, A. Barhoum, Nanoparticle and nanostructure synthesis and controlled growth methods, *Nanomaterials* 12 (2022) 3226. https://doi.org/10.3390/nano12183226
[19] J.T. Gudmundsson, A. Anders, A. von Keudell, Foundations of physical vapor deposition with plasma assistance, *Plasma Sources Sci. Technol.* 31 (2022) 083001. https://doi.org/10.1088/1361-6595/ac7f53
[20] J. Li, G.-K. Ren, J. Chen, X. Chen, W. Wu, Y. Liu, X. Chen, J. Song, Y.-H. Lin, Y. Shi, Facilitating complex thin film deposition by using magnetron sputtering: a review, *JOM* 74 (2022) 3069–3081. https://doi.org/10.1007/s11837-022-05294-0
[21] Y.X. Ou, H.Q. Wang, X. Ouyang, Y.Y. Zhao, Q. Zhou, C.W. Luo, Q.S. Hua, X.P. Ouyang, S. Zhang, Recent advances and strategies for high-performance coatings, *Prog. Mater. Sci.* 136 (2023) 101125. https://doi.org/10.1016/j.pmatsci.2023.101125
[22] K. Mondal, L. Nuñez, C.M. Downey, I.J. van Rooyen, Thermal barrier coatings overview: design, manufacturing, and applications in high-temperature industries, *Ind. Eng. Chem. Res.* 60 (2021) 6061–6077. https://doi.org/10.1021/acs.iecr.1c00788
[23] J. Zhu, Y. Hu, M. Xu, W. Yang, L. Fu, D. Li, L. Zhou, Enhancement of the adhesive strength between Ag films and Mo substrate by Ag implanted via ion beam-assisted deposition, *Materials (Basel)* 11 (2018) 762. https://doi.org/10.3390/ma11050762
[24] A. Baptista, F. Silva, J. Porteiro, J. Míguez, G. Pinto, Sputtering physical vapour deposition (PVD) coatings: a critical review on process improvement and market trend demands, *Coatings* 8 (2018) 402. https://doi.org/10.3390/coatings8110402

[25] M. Muhammed, M. Javidani, T. Ebrahimi Sadrabadi, M. Heidari, T. Levasseur, M. Jahazi, A comprehensive review of cathodic arc evaporation physical vapour deposition (CAE-PVD) coatings for enhanced tribological performance, *Coatings* 14 (2024) 246. https://doi.org/10.3390/coatings14030246
[26] P.O. Oviroh, R. Akbarzadeh, D. Pan, R.A.M. Coetzee, T.C. Jen, New development of atomic layer deposition: processes, methods and applications, *Sci. Technol. Adv. Mater.* 20 (2019) 465–496. https://doi.org/10.1080/14686996.2019.1599694
[27] G. Paramasivam, V.V. Palem, T. Sundaram, V. Sundaram, S.C. Kishore, S. Bellucci, Nanomaterials: synthesis and applications in theranostics, *Nanomaterials* 11 (2021) 3228. https://doi.org/10.3390/nano11123228
[28] M. Fernández-Arias, M. Zimbone, M. Boutinguiza, J. Del Val, A. Riveiro, V. Privitera, M.G. Grimaldi, J. Pou, Synthesis and deposition of Ag nanoparticles by combining laser ablation and electrophoretic deposition techniques, *Coatings* 9 (2019) 571. https://doi.org/10.3390/coatings9090571
[29] M. Poddighe, P. Innocenzi, Hydrophobic thin films from sol–gel processing: a critical review, *Materials (Basel)* 14 (2021) 6799. https://doi.org/10.3390/ma14226799
[30] H. Koreeda, M. Ishijima, K. Kajihara, Cosolvent-free sol–gel synthesis of macroporous silica gels from tetramethoxysilane–tetraethoxysilane mixtures, *J. Sol-Gel Sci. Technol.* 113 (2025) 48–55. https://doi.org/10.1007/s10971-024-06402-z
[31] A. Erigoni, U. Diaz, Porous silica-based organic-inorganic hybrid catalysts: a review, *Catalysts* 11 (2021) 79. https://doi.org/10.3390/catal11010079
[32] A.B.M. Giasuddin, A. Cartwright, D.W. Britt, Silica nanoparticles synthesized from 3,3,3-Propyl(trifluoro)trimethoxysilane or n-Propyltrimethoxysilane for creating superhydrophobic surfaces, *ACS Appl. Nano Mater.* 4 (2021) 4092–4102. https://doi.org/10.1021/acsanm.1c00398
[33] G.B. Veselov, T.M. Karnaukhov, V.O. Stoyanovskii, A.A. Vedyagin, Preparation of the nanostructured Ni-Mg-O oxide system by a sol–gel technique at varied pH, *Nanomaterials* 12 (2022) 952. https://doi.org/10.3390/nano12060952
[34] F. Barrino, Hybrid organic–inorganic materials prepared by sol–gel and sol–gel-coating method for biomedical use: Study and synthetic review of synthesis and properties, *Coatings* 14 (2024) 425. https://doi.org/10.3390/coatings14040425
[35] M.A. Mohammed, Structural, Optical, Electrical and Gas Sensor Properties of ZrO_2 Thin Films prepared by Sol-Gel Technique, *NeuroQuantology* 18 (2020) 22–27. https://doi.org/10.14704/nq.2020.18.3.NQ20146
[36] R. Bento, A. Gaddam, P. Oskoei, H. Oliveira, J.M.F. Ferreira, 3D printing of macro porous sol-gel derived bioactive glass scaffolds and assessment of biological response, *Materials (Basel)* 14 (2021) 5946. https://doi.org/10.3390/ma14205946
[37] E. Fazio, S. Spadaro, C. Corsaro, G. Neri, S.G. Leonardi, F. Neri, N. Lavanya, C. Sekar, N. Donato, G. Neri, Metal-oxide based nanomaterials: synthesis, characterization and their applications in electrical and electrochemical sensors, *Sensors* 21 (2021) 2494. https://doi.org/10.3390/s21072494
[38] D. Tonelli, E. Scavetta, I. Gualandi, Electrochemical Deposition of Nanomaterials for Electrochemical Sensing, *Sensors* 19 (2019) 1186. https://doi.org/10.3390/s19051186
[39] L. Santos, J.P. Neto, A. Crespo, P. Baião, P. Barquinha, L. Pereira, R. Martins, E. Fortunato, Electrodeposition of WO3 Nanoparticles for Sensing Applications, *Electroplat. Nanostructures* (2015). https://doi.org/10.5772/61216
[40] V. M. Varsha, G. Nageswaran, Review—direct electrochemical synthesis of metal organic frameworks, *J. Electrochem. Soc.* 167 (2020) 155527. https://doi.org/10.1149/1945-7111/abc6c6

[41] R.R. Poolakkandy, M.M. Menamparambath, Soft-template-assisted synthesis: a promising approach for the fabrication of transition metal oxides, *Nanoscale Adv.* 2 (2020) 5015–5045. https://doi.org/10.1039/D0NA00599A
[42] I. Sisman, Template-Assisted Electrochemical Synthesis of Semiconductor Nanowires, in: *Nanowires - Implementations Application*, InTech, 2011. https://doi.org/10.5772/20551
[43] M. Huston, M. Debella, M. Dibella, A. Gupta, Green synthesis of nanomaterials, *Nanomaterials* 11 (2021) 2130. https://doi.org/10.3390/nano11082130
[44] M. Kumar, X. Xiong, Z. Wan, Y. Sun, D.C.W. Tsang, J. Gupta, B. Gao, X. Cao, J. Tang, Y.S. Ok, Ball milling as a mechanochemical technology for fabrication of novel biochar nanomaterials, *Bioresour. Technol.* 312 (2020) 123613. https://doi.org/10.1016/j.biortech.2020.123613
[45] A. Salabat, F. Mirhoseini, Polymer-based nanocomposites fabricated by microemulsion method, *Polym. Compos.* 43 (2022) 1282–1294. https://doi.org/10.1002/pc.26504
[46] D. Zhang, Z. Li, K. Sugioka, Laser ablation in liquids for nanomaterial synthesis: diversities of targets and liquids, *J. Phys. Photonics* 3 (2021) 042002. https://doi.org/10.1088/2515-7647/ac0bfd

4 Different Techniques Used for Characterizing Nanomaterials

Sandeep Bansal, Yudhveer Kumar Verma, Pushpinder Kumar, Mayank Choubey and Atul Raj

4.1 INTRODUCTION TO NANOMATERIALS

Nanomaterials have gained considerable attention from researchers and scientists on account of their specialized properties that distinguish them from bulk materials (Kumar et al. 2021). Nanomaterials hold an assemblage of materials whose constituent structural components appear to be smaller than 100 nanometers in at least one dimension. These peculiar properties have opened countless potential applications in medicine, energy, environmental science, and electronics, bringing emphasis on their precise characterization. This chapter delves into the methods used to understand and characterize nanomaterials, giving insight into their unique properties, their behavior, and potential applications in these fields (McIntyre 2012).

4.1.1 Definition

Nanomaterials vary greatly and can be classified in broad categories based on their dimensions. They can be sorted into four classes: zero-dimensional (nanoparticles), one-dimensional (nanowires, nanotubes), two-dimensional (graphene, nanofilms), and three-dimensional (nanocomposites). Because of their small sizes, nanomaterials are characterized by a high surface-to-volume ratio, quantum confinement effects, and unique electronic, mechanical, and optical properties (Li et al. 2017).

4.1.2 Applications

Because of the far-reaching consequences of their various applications, certainly, nanomaterials assume great significance. In medicine, nanomaterials aid in improving targeted drug delivery where drugs are directed specifically to certain cells or tissues, thereby reducing side effects and improving overall treatment efficiency (Deshwal and Narwal 2023a). In medical imaging, nanomaterials are also used to develop more precise and sensitive technologies for diagnosis. Furthermore, they are used as antimicrobial agents offering novel ways of combating drug-resistant

DOI: 10.1201/9781003560234-5

pathogens (Foong et al. 2020). In electronics, nanomaterials serve as a stepping stone into a new era in which electronic devices have truly relied on fast, small, and efficient systems. Their nanoscale properties allow advanced electronic components and materials to be created, thus nurturing innovations and ushering in advancements within the electronics sector (Verma et al. 2024).

Nanomaterials find prime importance in energy applications toward the development of improved solar cells, batteries, and fuel cells. Researchers and engineers strive to improve energy storage and conversion technologies with the help of the unique properties of nanomaterials, contributing to the global shift toward sustainable energy forms (McIntyre 2012).

In environment-related applications, these are utilized for water purification, air filtration, and environmental sensing. Their characteristics such as high surface area, reactivity, and tunable properties make them suitable nanomaterials to tackle environmental issues like water treatment and pollution control. Environmental monitoring may see a revolution of sorts with nanomaterials used as sensors, as these will provide much higher sensitivity and selectivity for detecting different classes of pollution and contamination, thus aiding the conservation of the environment (Deshwal and Narwal 2021).

4.1.3 Challenges in Characterization

Characterizing nanomaterials involves various innumerable challenges due to their extremely small size and the complex interactions occurring at nanoscale. The conventional techniques like microscopy and spectroscopy often require modifications or conjoining with techniques such as X-ray diffraction (XRD) and electron microscopy (EM) to produce accurate and comprehensive information regarding the chemical composition, structure, and properties of nanomaterials (Hong et al. 2015). A critical problem here is attaining spatial resolution sufficient to observe nanoscale features, which often involves the use of advanced imaging techniques or apparatus capable of high magnification and accurate targeting (Himyan et al. 2018).

Sample preparation is another pressing issue; it must be done without interfering with the material properties. Sample preparation requires meticulous attention to detail and special procedures to preserve the samples in their original state. Sectioning, coating, staining, or fixation can be used depending on the sample type and analysis required. Data interpretation is another critical aspect and usually embraces a combination of statistical analysis, image processing, and qualitative observation to derive meaningful conclusions from the information collected (Weckhuysen 2004).

This step is vital for making valid conclusions from any experimental result. Reproducibility helps to ensure that results are valid from one sample and measurement to another. It requires tight quality control, standardized procedures, and careful documentation so that variability may be minimized and the reliability bolstered. To the extent that reproducibility can be achieved, this is a benchmark for sound scientific research, as it is indispensable to confirm results and advance our knowledge into nanomaterials.

In view of the preceding challenges, following are some of the techniques/methods useful for characterization.

4.2 MICROSCOPIC TECHNIQUES

Microscopic methods make the complex structure of nanomaterials viewable and understandable. These advanced techniques allow high-resolution imagery to be obtained by researchers, which provides vital information about the size, structure, and arrangement of nanomaterials at the nanoscale level. To enhance their understanding of the properties and behavior of nanomaterials, various microscopy techniques, such as atomic force microscopy (AFM), scanning electron microscopy (SEM), and transmission electron microscopy (TEM), have been developed. Such advancements can boost various fields, including materials science, nanotechnology, and biomedical research (Jones et al., n.d.).

4.2.1 SEM

A narrow beam of high-energy electrons undertakes the scrutiny of the surface of a sample in an elaborate imaging procedure known as SEM (Ming, Lim, Chen 1993 and Kumar Baghel 2023). The setup is shown in Figure 4.1. While interacting with the surface atoms of the sample, the electrons produce different signals such as secondary electrons, backscattered electrons, and characteristic X-rays.

These signals are detected and then turned into grayscale or color images, providing detailed information about the topography, morphology, and even elemental composition at very high magnifications. Thus, SEM is a very important instrument for both research and industrial applications (Mohammed and Abdullah 2018). SEM

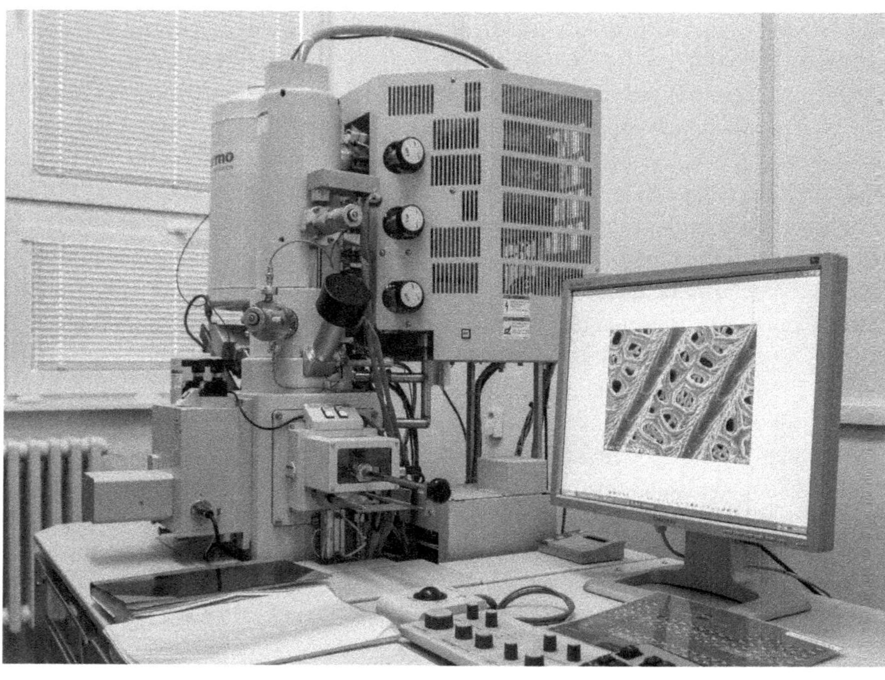

FIGURE 4.1 Actual image of scanning electron microscope ("SEM," n.d.)

is being used in various sciences such as biology, material science, and nanotechnology. It can provide important insights into the properties and behavior of a variety of nanostructures, such as nanoparticles, nanowires, and thin films. Semiautomatic electron microscopy has its own limitations (Hebbale and Srinath 2016). The most prominent is the vacuum environment, which may not be suitable for all types of samples. Other limitations include the resolution being dependent upon the diameter of the electron beam and the particular method providing limited information regarding the internal structure of the sample.

4.2.2 TEM

A stream of electrons is used in TEM, a type of microscopy, to study the ultrastructure of specimens. A tiny sample is positioned in the electron beam's path in a TEM, as shown in Figure 4.2. Some electrons are transmitted, while others are dispersed or absorbed as they go through the sample. The interior structure of the sample is subsequently substantially magnified using the transmitted electrons, providing comprehensive information about its morphology and composition at the nanoscale level. This method is extremely important for many scientific domains of nanotechnology (Tang and Yang 2017).

The transmission electron microscope operates on the principle of removing an extremely thin slice of the sample through an electron beam of high energy. During its passage, the electrons are subjected to electromagnetic lensing to produce a highly magnified image. This is one of the reasons the transmission electron microscope can yield atomic resolution-independent characterization of nanomaterials' internal structure. Its high resolution permits detailed characterization at the nanoscale, especially for thin films, nanoparticles, and nanocomposites. Hence, it helps a researcher visualize and understand the intricate features of these materials, leading to comprehensive analysis and characterization (Xu et al. 2016). However, there are certain drawbacks to the transmission electron microscope. The method requires extensive sample preparation for the thin sections to be made, which may take hours and change the properties of the sample along the way. The operation of the transmission electron microscope also requires a high vacuum environment and is thus limited in size.

4.2.3 AFM

The high-resolution imaging method known as AFM (Figure 4.3) scans a surface of sample using a sharp probe that is attached to a cantilever. The cantilever experiences measured to determine the forces acting on the sample and the probe. By providing comprehensive topography data at the nanoscale, this method offers insights into the characteristics and surface shape of the material. A finely sharpened tip attached to a cantilever is scanned across the sample surface in accordance with the basic idea of AFM. To track the cantilever's minute deflections, a laser beam bounces off its back. AFM operates in various modes—contact, non-contact, and tapping—allowing the creation of intricate topographical maps of the sample with nanometer precision (Himyan et al. 2018). This versatility makes AFM applicable in fields such as materials science, biology, and nanotechnology, where it is used to investigate surface

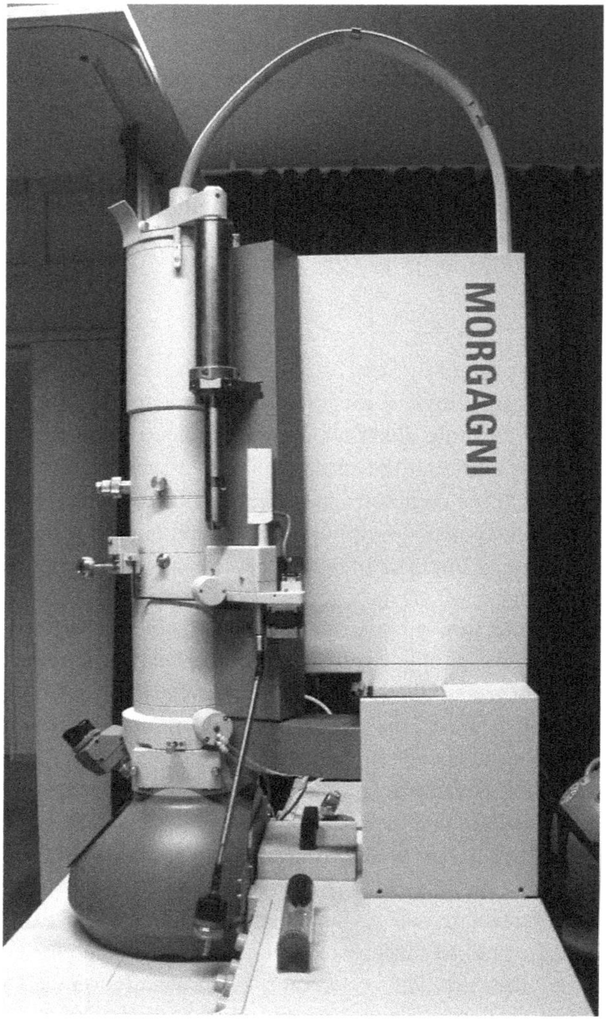

FIGURE 4.2 Actual image of transmission electron microscope ("TEM," n.d.)

morphology, roughness, and mechanical properties of nanomaterials. However, AFM does have limitations. It generally operates at a slower processing speed compared to other microscopic techniques. The quality of the images can be affected by factors such as the condition of the tip and the scanning environment. Additionally, AFM may provide limited information about the internal structure of the sample, focusing primarily on surface characteristics (Giessibl 2003).

4.2.4 Scanning Tunneling Microscopy (STM)

STM is a highly advanced imaging system that relies on the fascinating concept of quantum tunneling, enabling electrons to move through a barrier to create detailed images through precise measurements of the distance between a sharp tip and the

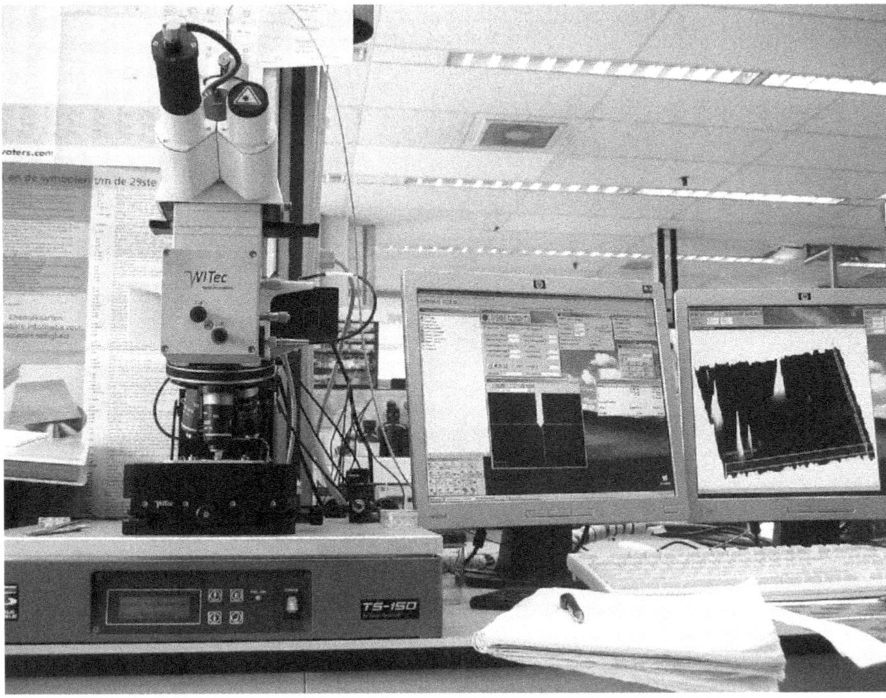

FIGURE 4.3 Actual image of an atomic force microscope ("AFM," n.d.)

surface of the sample. The schematic of STM is shown in Figure 4.4. The key to the STM's operation is to carefully position a sharp metal tip near the surface of the sample. Tunneling current is produced when electrons are able to pass through the vacuum gap by putting a voltage between the tip and the sample. The sensitivity of this current to variations in the distance between the tip and the sample allows for accurate mapping of the sample's surface at atomic resolution. STM has revolutionized the way scientists study the atomic structure and electronic properties of conductive and semi-conductive nanomaterials. This advanced technique is a staple in surface science, materials science, and nanotechnology, providing researchers with the ability to image and manipulate individual atoms and molecules with incredible precision (Giessibl 2003). However, STM is a powerful tool for exploring the atomic-scale world, but its effectiveness is contingent upon the availability of a conductive sample and a surface that is free of contamination and stability issues. Furthermore, the need for a high vacuum or an inert atmosphere can limit the technique's versatility.

4.3 SPECTROSCOPIC TECHNIQUES

Spectroscopic techniques are among the most useful in the study of nanomaterials, enabling scientists to profile the composition, structure, and electronic properties of nanomaterials in a very powerful way. Nanomaterials interact with electromagnetic

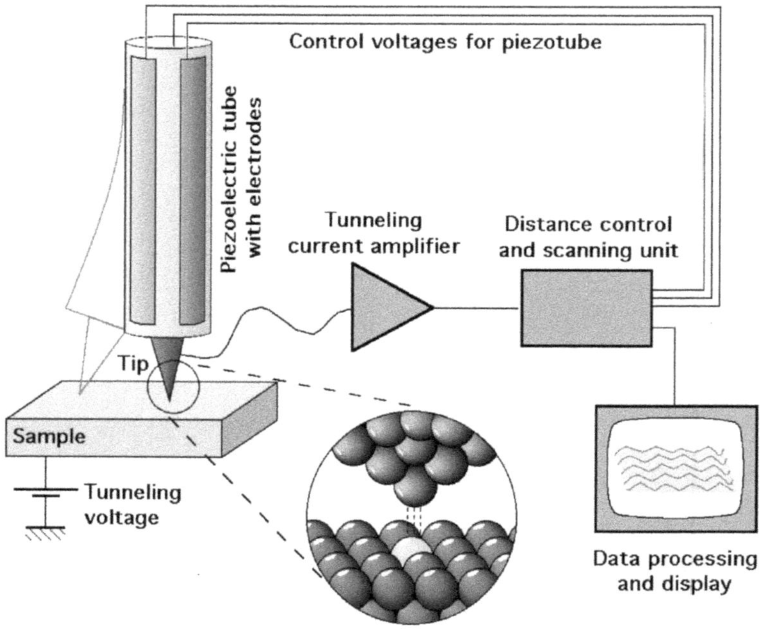

FIGURE 4.4 Schematic of scanning tunnel microscope ("STM," n.d.)

radiation, and from this interaction, properties can be determined, which in turn can aid in the better understanding of such advanced materials (Smith, 2020).

4.3.1 XRD

XRD is one of the latest and most useful techniques for studying nanomaterials, providing important information about phase composition, crystallographic structure, and a number of other significant structural characteristics. In actual, the image of an XRD machine is given in Figure 4.5 (Bansal et al. 2021). Pointing X-rays at a sample for the examination of the intensity and angle of the diffracted beams makes XRD a powerful analytical method. From the diffraction pattern generated by this method, researchers can extract vital information concerning the phase composition, lattice parameters, and crystal structure of the material, which in turn assist in the characterization of the material's properties (Himyan et al. 2018). Due to its widespread applications in characterizing the crystalline structure of numerous materials, XRD serves a number of purposes. These include the identification of phases contained in a sample, orientation of the crystal, and quantification of any possible strain and defects concerning nanomaterials. Although XRD is one of the best techniques for studying the structure of crystalline materials, it does have limitations. Proper sample preparation and an appreciable volume of material are required if the method is to work efficiently. Beyond this, while the structural information obtained from XRD may be very revealing, it cannot represent in full the defects or structural amorphous areas within the sample (Epp 2016).

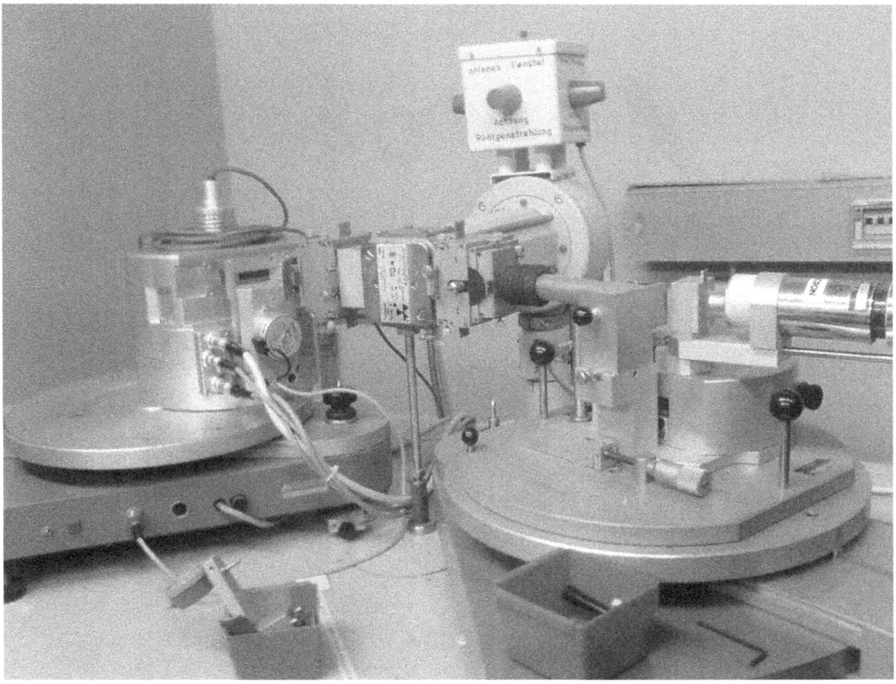

FIGURE 4.5 Actual image of XRD machine ("XRD," n.d.)

4.3.2 Raman Spectroscopy

Owing to its non-destructive nature, Raman spectroscopy is one of the most favored methods for investigating the vibrational modes of molecules, imparting certain information on the molecular content of different materials. The basic principle of Raman spectroscopy is the determination of the chemical composition and arrangement of the molecules (Figure 4.6a and b) of a material based on the scattering of monochromatic laser radiation after an illuminated sample and the measurement of the scattered light. In this way, by observing the frequency shift occurring in the scattering of light, the researchers can obtain useful information regarding the vibrational modes residing in the sample, shedding light on the chemical properties of that sample. Raman spectroscopy stands out as a pivotal tool in the realm of nanomaterial analysis. It is used in many different sectors where it provides useful information about the molecular structure, chemical makeup, and electronic characteristics of materials such as semiconductor nanoparticles, graphene, and carbon nanotubes (Rostron, Gaber, and Gaber 2016).

Despite a powerful analytical technique, Raman spectroscopy has relatively low sensitivity. It might make low-concentration sample identification more difficult, necessitating the use of amplification techniques like surface-enhanced Raman spectroscopy (SERS). Furthermore, the precision of the data can be affected by fluorescence interference, underscoring the significance of careful sample preparation.

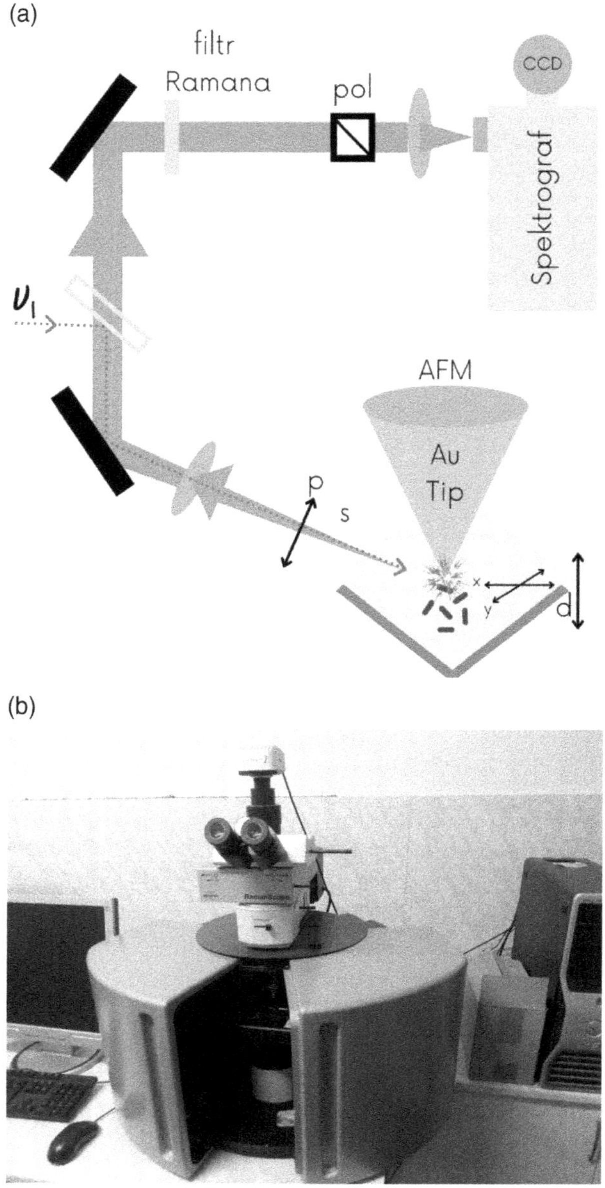

FIGURE 4.6 (a) Schematic and (b) actual image of Raman spectroscopy ("RAMAN," n.d.)

4.3.3 Fourier-Transform Infrared Spectroscopy (FTIR)

An infrared spectrum of absorption or emission from a sample in the form of a solid, liquid, or gas is obtained using the widely used analytical technique known as the FTIR (actual image shown in Figure 4.7), which uses sophisticated mathematical algorithms (Roane and Pepper 2015). Due to its utility in providing valuable information regarding molecular structure and composition, this technique has found its widespread application

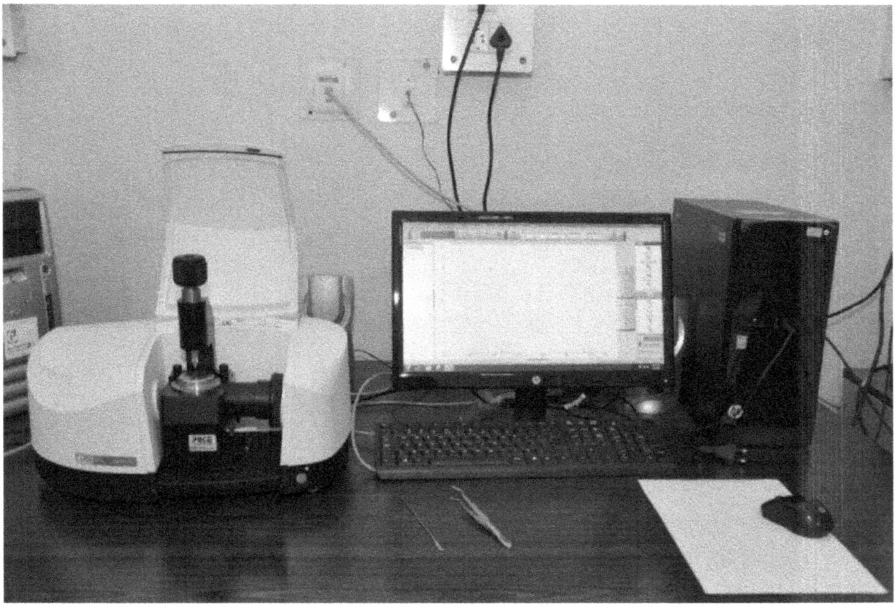

FIGURE 4.7 Actual image of FTIR ("FTIR," n.d.)

in pharmaceuticals, environmental, and material science. FTIR uses infrared radiation, whereby the samples are analyzed according to the wavelengths absorbed as the radiation passes through them. The distinct absorption properties associated with different molecular structures enable the identification of various chemical bonds and functional groups within a material. FTIR studies the chemical composition, functional groups, and bonding in nanomaterials and finds extensive applications for the characterization of organic and inorganic compounds, polymers, and nanocomposites. In view of its rapid and accurate analysis of molecular structures, FTIR can truly be termed an important instrument in research and industry (Dutta 2017). However, FTIR requires careful sample preparation for the accuracy and reproducibility of results. Limitations of the technique include the absence of transparency of the sample for IR radiation and mostly limited structural information of complex mixtures.

Besides these limitations, FTIR provides the advantages of relatively fast sample analysis, little preparative exigencies, and application to solids, liquids, or gaseous samples fairly expediently. The identification of complex mixtures without extensive separation procedures gives it an upper hand in so many analytical applications.

4.3.4 Ultraviolet-Visible Spectroscopy (UV-VIS)

Ultraviolet-visible spectroscopy (UV-Vis) measures a sample's absorbance and reflectance in the ultraviolet and visible light ranges. Information is necessary regarding how UV and visible light interact with materials, which can provide important insights into the chemical species present. This procedure consists of irradiating a sample with UV-visible light and measuring the transmitted or reflected light in turn, as noted

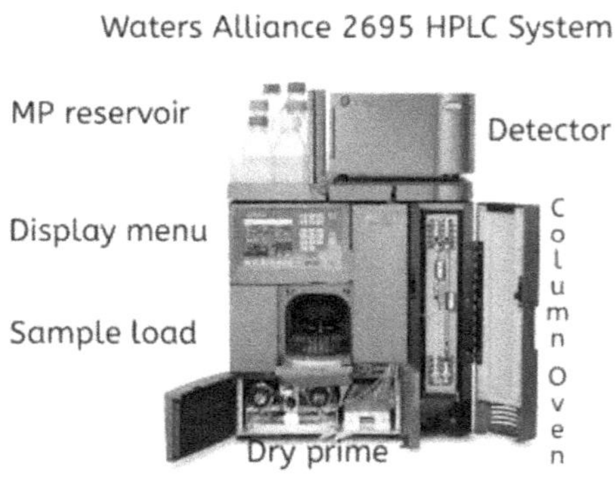

FIGURE 4.8 Actual image of UV-Vis ("UVS," n.d.)

by Cheng et al. 1999. Figure 4.8 shows an actual image. The electronic transitions revealed in the spectrum are linked to the composition and electronic structure of the materials. Light-source ultraviolet and visible radiation is passed through a sample. Light absorbed by the sample at specific wavelengths indicates electronic transitions of molecules in the sample. Such absorbance spectra are useful for indicating the concentration and type of substances present. UV-Vis spectroscopy has various applications in chemistry, biology, and environmental studies. It is used to study optical characteristics, electronic structure, and content involving nanomaterials.

It is also applied for quantitative substance analysis, electronic property evaluation, and environmental pollution detection. It has an equally important role in quality control for the food and pharmaceutical sectors, ensuring product safety and compliance.

It gives limited molecular structural information, requiring complementary methods for detailed investigation. Fast and efficient detection and quantification of substances with minor sample preparation remain the advantages of UV-Vis spectroscopy. Further, being a non-destructive technique, it is applicable in analyzing precious or limited samples.

4.4 SURFACE CHARACTERIZATION TECHNIQUES

Surface characterization techniques give insight into the topography, electrical structure, and chemical composition of the surface of a nanomaterial. The surface characterization techniques are of fundamental importance for the nanomaterials investigation; they help to analyze their chemical composition, electronic structure, and surface topography. Such understanding is important in manipulating the unique properties of nanomaterials and using the same for diverse applications (Aziz and Ismail 2017).

4.4.1 X-Ray Photoelectron Spectroscopy (XPS)

XPS is a surface-sensitive quantitative spectroscopic technique used to analyze the types and chemical states of a material's surface (Figure 4.9). It is done by bombarding the surface with X-ray radiation suitable for the ejection of photoelectrons

FIGURE 4.9 Actual image of XPS ("XPS," n.d.)

from the sample. To establish the elemental concentrations and chemical states pertinent to the surface, the kinetic energy of the emitted electrons is analyzed. In XPS, the binding energy values of the core electrons are measured from kinetic energy values of the photoelectrons, providing information on the elemental composition and chemical states of surface atoms. The core electrons yield information on elemental composition, whereas changes in their binding energy relate to the local chemical environment in which the atom exists. The distinction between different chemical states becomes extremely critical for understanding materials behavior at the nanoscale. XPS has been widely employed for research and quality control in several fields such as electronics, catalysis, energy storage, etc. It describes the oxidation states, chemical bonding, and surface chemistry of nanomaterials (Aziz and Ismail 2017).

It is essential in the technology of energy, for example, in the evaluation of solidity and the cycle of deterioration of materials used in the solid oxide fuel cell and batteries. However, XPS has limitations, including its focus on analyzing only the top few nanometers of a surface and the requirement for a high vacuum environment (Mago et al. 2020). The technique provides average information over the analyzed area and may not reveal details about inhomogeneities or buried interfaces. Additionally, sensitive samples can degrade during the analysis, and careful calibration and data interpretation are necessary to accurately determine the chemical states and composition of complex materials.

4.4.2 Auger Electron Spectroscopy (AES)

A surface-sensitive analytical method called AES is used to look into the chemical states and elemental makeup of a material's surface. The schematic has been shown in Figure 4.10. This technique makes use of the Auger electrons that are emitted as a result of inner-shell ionization to provide intricate details about the surface atoms' chemical surroundings. In AES, a material is exposed to electrons or X-rays, and the kinetic energy of the Auger electrons that are released is measured.

Inner-shell electrons are expelled from the atoms when a high-energy electron beam is directed toward the material's surface. Electrons from higher energy levels then fall into the resulting lower energy vacancies, releasing excess energy as Auger electrons. To identify the elements and their chemical states, the kinetic energy of these released Auger electrons is measured. The technique finds numerous applications in semiconductor fabrication, including assessing surface contamination, analyzing thin film composition, and profiling chemical states in reactive surfaces. This technique requires ultra-high vacuum conditions for optimal performance, complicating sample preparation, and interpreting Auger spectra can be challenging due to overlapping peaks, necessitating careful calibration and spectral analysis (Sardela 2014).

4.4.3 Secondary Ion Mass Spectrometry (SIMS)

A sophisticated surface-sensitive analytical method called SIMS (shown in Figure 4.11) looks at the elemental and isotopic makeup of a material's surface. SIMS produces the ejection of secondary ions by bombarding a sample with intense ion beams; these ions are subsequently measured for their mass-to-charge ratios. Using a main ion beam, atoms and molecules are sputter-deposited onto the sample surface during the procedure (Deshwal and Narwal 2023b). The secondary ions, which can originate from various depths depending on the bombardment intensity,

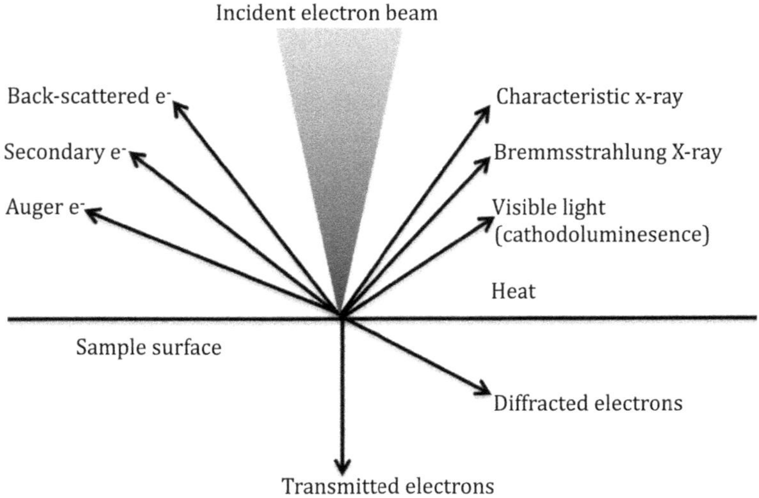

FIGURE 4.10 Schematic of electron beam and sample interaction ("Eb," n.d.)

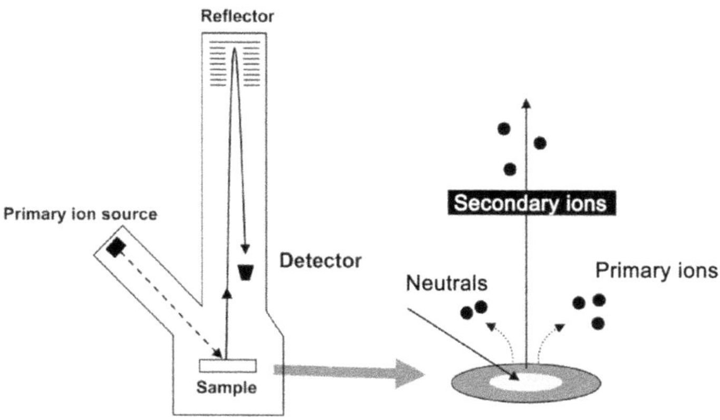

FIGURE 4.11 Schematic of SIMS ("SIMS," n.d.)

are detected and analyzed to provide information about the sample's composition and isotopic ratios. This technique is particularly advantageous for analyzing thin films, coatings, and heterogeneous materials where surface properties are crucial. SIMS is widely applied across multiple disciplines. The technique is utilized in biological research to study the distribution of biomolecules in tissues and cells, highlighting its interdisciplinary relevance (McPhail 2006). SIMS offers excellent sensitivity and has the ability to deliver detailed elemental and isotopic information with high spatial resolution. SIMS faces challenges such as potential sample damage due to high-energy ion bombardment and the complexity of data interpretation, particularly when quantitatively analyzing complex mixtures.

REFERENCES

Mohammed, A., and A. Abdullah. 2018. "Scanning Electron Microscopy (SEM): A Review." *Proceedings of 2018 International Conference on Hydraulics and Pneumatics - HERVEX*, 77–85.

"AFM." n.d. https://www.flickr.com/photos/archeon/91081511

Aziz, M., and A. F. Ismail. 2017. *X-Ray Photoelectron Spectroscopy (XPS). Membrane Characterization*. Elsevier B.V. https://doi.org/10.1016/B978-0-444-63776-5.00005-X

Bansal, Sandeep, Jonty Mago, Dheeraj Gupta, and Vivek Jain. 2021. "Parametric Optimization and Analysis of Cavitation Erosion Behavior of Ni-Based + 10WC Microwave Processed Composite Clad Using Taguchi Approach." *Surface Topography: Metrology and Properties* 9 (1). https://doi.org/10.1088/2051-672X/abda94

Cheng, Yuhang, Yiping Wu, Wenxuan Zhu, Jianguo Chen, Xueliang Qiao, Ying Fan, and Xiaoxia Pan. 1999. "Preparation and Properties of Diamond-CuTi Composite Sintered by Microwave Sintering." *Journal of Materials Science Letters* 18 (23): 1933–35. https://doi.org/10.1023/A:1006641400866

Deshwal, Dinesh, and Anil Kumar Narwal. 2021. "An Extensive Review of Nanotubes-Based Mass Sensors." *Journal of Micro and Nano-Manufacturing* 9 (2). https://doi.org/10.1115/1.4051261

Deshwal, Dinesh, and Anil Kumar Narwal. 2023a. "Analysis of Surface Deviation Impact on Bio-Mass Sensing Application of Boron Nitride Nanotubes." *Results in Engineering* 19 (July): 101282. https://doi.org/10.1016/j.rineng.2023.101282

Deshwal, Dinesh, and Anil Kumar Narwal. 2023b. "Resonant Frequency Analysis of Wavy Boron Nitride Nanotube Based Bio-Mass Sensor." *International Journal on Interactive Design and Manufacturing (IJIDeM)*. https://doi.org/10.1007/s12008-023-01300-0

Dutta, Aastha. 2017. *Fourier Transform Infrared Spectroscopy. Spectroscopic Methods for Nanomaterials Characterization*. Vol. 2. Elsevier Inc. https://doi.org/10.1016/B978-0-323-46140-5.00004-2

"Eb." n.d. https://en.wikipedia.org/wiki/File:Electron_beam_sample_interactions.png

Epp, J. 2016. *X-Ray Diffraction (XRD) Techniques for Materials Characterization. Materials Characterization Using Nondestructive Evaluation (NDE) Methods*. Elsevier Ltd. https://doi.org/10.1016/B978-0-08-100040-3.00004-3

Foong, Loke Kok, Mohammad Mehdi Foroughi, Armita Forutan Mirhosseini, Mohadeseh Safaei, Shohreh Jahani, Maryam Mostafavi, Nasser Ebrahimpoor, Maryam Sharifi, Rajender S. Varma, and Mehrdad Khatami. 2020. "Applications of Nano-Materials in Diverse Dentistry Regimes." *RSC Advances* 10 (26): 15430–60. https://doi.org/10.1039/d0ra00762e

"FTIR." n.d. https://www.iip.res.in/sophisticated-instrumentation-facilities/ftir-spectrometer/

Giessibl, Franz J. 2003. "Advances in Atomic Force Microscopy." *Reviews of Modern Physics* 75 (3): 949–83. https://doi.org/10.1103/RevModPhys.75.949

Hebbale, Ajit M., and M.S. Srinath. 2016. "Microstructure and Experimental Design Analysis of Nickel Based Clad Developed through Microwave Energy." *Perspectives in Science* 8 (September): 257–59. https://doi.org/10.1016/j.pisc.2016.04.044

Himyan, M. A., M. Penchal Reddy, F. Ubaid, R. A. Shakoor, and A. M.A. Mohamed. 2018. "Scanning Electron Microscopic Studies of Microwave Sintered Al-SiC Nanocomposites and Their Properties." *Scanning* 2018. https://doi.org/10.1155/2018/7546573

Hong, Sheng, Yuping Wu, Jianfeng Zhang, Yugui Zheng, Yujiao Qin, and Jinran Lin. 2015. "Ultrasonic Cavitation Erosion of High-Velocity Oxygen-Fuel (HVOF) Sprayed near-Nanostructured WC-10Co-4Cr Coating in NaCl Solution." *Ultrasonics Sonochemistry* 26: 87–92. https://doi.org/10.1016/j.ultsonch.2015.01.012

"IMS." n.d. https://rs-nano.pl/en-us/glossary/time-of-flight-secondary-ion-mass-spectrometry-tof-sims

Jones, Mark I, Maria-Cecilia Valecillos, Kiyoshi Hirao, and Yukihiko Yamauchi. n.d. "Grain Growth in Microwave Sintered Si_3N_4 Ceramics Sintered from Different Starting Powders." www.elsevier.com/locate/jeurceramsoc

Kumar Baghel, Pushp. 2023. "Application of Microwave in Manufacturing Technology: A Review." *Materials Today: Proceedings*. https://doi.org/10.1016/j.matpr.2023.02.008

Kumar, Rajat, Hiralal Bhowmick, Dheeraj Gupta, and Sandeep Bansal. 2021. "Development and Characterization of Multiwalled Carbon Nanotube-Reinforced Microwave Sintered Hybrid Aluminum Metal Matrix Composites: An Experimental Investigation on Mechanical and Tribological Performances." *Proceedings of the Institution of Mechanical Engineers, Part L: Journal of Materials: Design and Applications* 235 (10): 2310–23. https://doi.org/10.1177/14644207211028969

Li, Ruoyu, Feipeng Xiao, Serji Amirkhanian, Zhanping You, and Jia Huang. 2017. "Developments of Nano Materials and Technologies on Asphalt Materials – A Review." *Construction and Building Materials* 143 (July): 633–48. https://doi.org/10.1016/J.CONBUILDMAT.2017.03.158

Mago, Jonty, Sandeep Bansal, Dheeraj Gupta, and Vivek Jain. 2020. "Investigation of Microwave Processing Parameters on Development of Ni-40Cr_3C_2 Composite Clad and Their Characterization." *Metallurgical and Materials Transactions A: Physical Metallurgy and Materials Science* 51 (8): 4288–4300. https://doi.org/10.1007/s11661-020-05832-y

McIntyre, Robin A. 2012. "Common Nano-Materials and Their Use in Real World Applications." *Science Progress* 95 (1): 1–22. https://doi.org/10.3184/003685012X13294715456431

McPhail, D. S. 2006. "Applications of Secondary Ion Mass Spectrometry (SIMS) in Materials Science." *Journal of Materials Science* 41 (3): 873–903. https://doi.org/10.1007/s10853-006-6568-x

Ming, Qian, L. C. Lim, and Z. D. Chen. 1998. "Laser Cladding of Nickel-Based Hardfacing Alloys." *Surface and Coatings Technology* 106 (2–3): 174–82. https://doi.org/10.1016/S0257-8972(98)00524-6

"RAMAN." n.d. https://es.m.wikipedia.org/wiki/Archivo:Raman_Spectroscopy_device_by_Bruker.jpg

Roane, Timberley M., and Ian L. Pepper. 2015. *Microscopic Techniques. Environmental Microbiology*. 3rd Edition. Elsevier Inc. https://doi.org/10.1016/B978-0-12-394626-3.00009-0

Rostron, Paul, Safa Gaber, and Dina Gaber. 2016. "Ijetr042430 (1)." *International Journal of Engineering and Technical Research (IJETR)* 6 (1): 50–64.

Sardela, Mauro. 2014. *Practical Materials Characterization. Practical Materials Characterization*. https://doi.org/10.1007/978-1-4614-9281-8

"SEM." n.d. https://infinitalab.com/material-testing-service/understanding-scanning-electron-microscope/

Smith, Wyatt Tyrone. 2020. "Electron Beam Sample Interactions." Wikimedia Commons, September 14, 2020. https://commons.wikimedia.org/wiki/File:Electron_beam_sample_interactions.png

"STM." n.d. https://en.wikipedia.org/wiki/File:ScanningTunnelingMicroscope_schematic.png

Tang, C. Y., and Z. Yang. 2017. *Transmission Electron Microscopy (TEM). Membrane Characterization*. Elsevier B.V. https://doi.org/10.1016/B978-0-444-63776-5.00008-5

"TEM." n.d. https://commons.wikimedia.org/wiki/File:Transmission_electron_microscope_(Morgagni_268D)_1pl.jpg

"UVS." n.d. https://pharmasciences.in/calibration-of-hplc/

Verma, Yudhveer Kumar, Kanwarjeet Singh, and Gaurav Arora. 2024. "Innovation and Possibility of Various Coating Materials with Microwave Hybrid Heating." https://doi.org/10.1177/25165984241281008

Weckhuysen, Bert M. 2004. "CHAPTER 12 Ultraviolet-Visible Spectroscopy." In B. M. Weckhuysen (ed.), *In-Situ Spectroscopy of Catalysts*, 255–70. San Diego: American Scientific Publishers.

"XPS." n.d. https://www.cfi.lu.lv/en/services/all-services/full-equipment-list/x-ray-photoelectron-spectrometer-xps/

"XRD." n.d. https://commons.wikimedia.org/wiki/File:X_Ray_Diffractometer.JPG

Xu, Peng, Cheng Zhang, Wei Wang, and Lin Liu. 2016. "Pitting Mechanism in a Stainless Steel-Reinforced Fe-Based Amorphous Coating." *Electrochimica Acta* 206: 61–69. https://doi.org/10.1016/j.electacta.2016.04.130

5 Green Synthesis Approaches for Nanomaterials

Recent Advances and Applications

Mayank Choubey, Shweta Mishra, Ekta Verma, Sandeep Bansal, Sunny Misra and Prem Narayan Ahirwar

5.1 INTRODUCTION

Nanotechnology became a significant scientific topic following major public interest in recent years. The measurement scale for nanoparticles spans one nanometer to a hundred nanometers. Nanoscale metals reveal greater surface area compared to traditional metals. The absence of microscopic size effects and surface boundaries, along with quantum characteristics, renders non-nanometals unable to display distinctive physical and chemical characteristics found in nanometals. These materials have grown in importance and excitement over the past decade because of their vast applications in medicine and engineering [Zhao et al. 2016]. At the nanoscale, these substances have a more excellent condensed structure of atoms and molecules, resulting in unique phenomena and improved mechanical [Wu et al. 2020], electrical [Min et al. 2019], magnetic [Martinez-Mera et al. 2007], optical [Kelly et al. 2003], catalytic [Zeng et al. 2010], and antibacterial [Alvarez-Chimal et al. 2021; Rayees-Carmona et al. 2023] properties that are distinct from those of larger molecules [Baig et al. 2021].

The nanomaterials, nanoparticles, nanosheets, nanotubes, and quantum dots are some examples that have found widespread use in the various fields of pharmaceuticals, agriculture, and electronics. Most metal nanoparticles, including silver, gold, iron oxide, selenium, cadmium sulfide, and lead sulfide, possess useful features in various applications. Various methods can be utilized to produce nano-sized materials; nevertheless, two methodologies typically are used in their creation: top-down synthesis and bottom-up synthesis, as shown in Figure 5.1. Nanomaterials are often synthesized using one of two approaches: conventional or green synthesis methods. The conventional techniques offer many attractive advantages, including wide scalability, excellent control over nanoparticle morphology, power storage/conservation

DOI: 10.1201/9781003560234-6

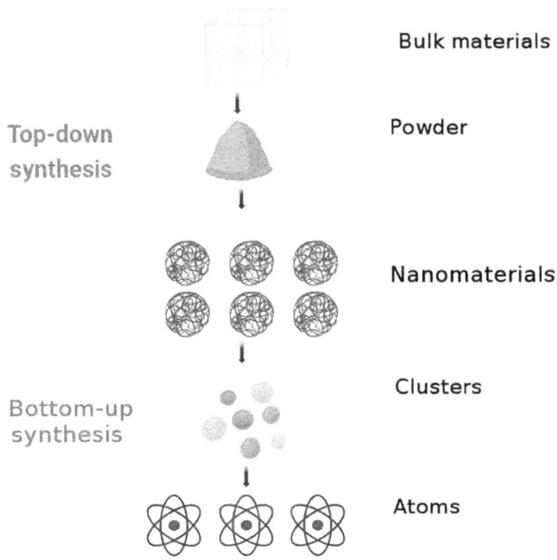

FIGURE 5.1 Demonstration of top-down and bottom-up synthesis.

applications, electronic applications, targeted illness treatment, and innovative battery conduction. Despite the advantages, there are several unavoidable shortcomings to these methods. For example, most conventional procedures include using organic solvents, which have neurobehavioral and reproductive consequences during nanomaterial synthesis. Also, these procedures involve high pressure, and high temperatures can lead to dangerous working circumstances and the greenhouse effect due to excessive carbon dioxide generation.

Green synthesis approaches are progressively substituting physical and chemical techniques owing to the abuse of energy, producing poisonous and dangerous compounds [Horwat et al. 2011, Alsammarraie et al. 2018, Hoag et al. 2009], and sophisticated equipment and synthesis conditions [Baruwati et al. 2009; Ahmed and Ikram 2015]. Physical approaches encompass aerosols, UV light, and temperature breakdown, all needing elevated pressure and temperature levels [Smirniotis et al. 2018; Wojnarowicz et al. 2018; Ahmed and Ikram 2016]. As an illustration, a flame temperature of 2400 K is generally required to generate droplets of atomized aerosol and create metallic nanoparticles. The production of PdO nanoparticles using plasma-enhanced physical vapor deposition requires three heat cycles ranging from 250 °C to 800 °C, resulting in significant energy consumption. Chemical treatments often involve sodium borohydride, which is both costly and poisonous, in addition to dispersion stabilizers and organic solvents [Hussain et al. 2016]. On the other hand, green synthesis uses natural, environmentally beneficial ingredients, such as reducing agents. The possible risks outweigh the advantages of using conventional methods for synthesizing nanomaterials. Green synthesis has gained traction due to the disuse of conventional synthesis techniques.

The term "green" refers to "eco-friendly," as in "clean," "harmless," "cost-effective," and "ecologically friendly" methods of synthesizing nanomaterials.

The production of ecologically friendly, chemically, and physically directed nanoscale metals is the subject of increasing research [Wang et al. 2007; Horwat et al. 2011]. Certain environmentally friendly materials can function both as end-capping agents and as dispersants [Devi et al. 2019]. This decreases energy usage and eliminates the need for poisonous and dangerous substances. Currently, green synthesis primarily depends on microorganisms like fungi, bacteria, and algae [Subramaniyam et al. 2015; Arsiya et al. 2017], as well as extracts derived from various sections of plants, including leaves [Chahardoli et al. 2018; Leili et al. 2018], flowers [Thovhogi et al. 2016, Thovhogi et al. 2015; Sone et al. 2020] roots, peels, fruits, and seeds [Ehrampoush et al. 2015; Kumar et al. 2017; Dhand et al. 2016; Gao et al. 2016].

The green materials comprise polyphenols and proteins, which can serve as alternative reducing agents for chemical reactions. These chemicals can reduce the valence of metal ions [Can 2020]. Metal nanoparticles can be formed when green materials are present and certain conditions, such as temperature, concentration, and ambient air, are satisfied. Under certain conditions, metal nanoparticles produced using green technology can surpass in quality those made by traditional chemical methods. For example, Fe_3O_4 nanoparticles created using the green synthesis process range in size from 2 to 80 nm, which is a substantially smaller size than that of particles synthesized using the wet chemical method, which ranges in size from 87 to 400 nm [Gokila et al. 2021]. Studies have shown that green synthesis delivers advantages over conventional chemical synthesis while highlighting promising future applications for this approach [Ren 2019].

Green synthesis approaches for nanomaterials offer numerous advantages over classical chemical methods and traditional physical methods. Green synthesis stands out because of its harmless character [Subramaniyam et al. 2015] while also eliminating harmful releases [Alsammarraie et al. 2018] and being environmentally friendly [Nasrollahzadeh and Mohammad Sajadi 2016] with affordable operational costs [Kataria and Garg 2018]. Recommendations about raw material exploitation demand further exploration and concerns for extended production times and end product quality. The synthesis process takes an extended time, while essential ingredients demonstrate restricted availability [Turunc et al. 2017] and produced materials display fixed sizes of particles [Gao et al. 2017]. However, Subramaniyam et al. [2015] found that the synthesis approach is time-consuming. A brief review of green nanomaterial synthesis techniques and their associated constraints is presented in this chapter. The authors of this study focus on recognizing key barriers within green nanoscale metallic nanoparticle synthesis methodologies while proposing new research directions. A detailed exploration of green synthesis techniques follows a broader analysis, which shows potential paths forward for future sustainable research.

5.2 A BRIEF SUMMARY OF TRADITIONAL METHODS OF SYNTHESIS OF NANOMATERIALS

The production of nanomaterials benefits from multiple established synthesis procedures. The frequently used traditional techniques include the following:

5.2.1 Chemical Reduction

In this method, the chemical substances like borohydride or sodium hydroxide reduce the metal ions to create nanoparticles [Manikam et al. 2011].

5.2.2 Soft Template Method

The production of nanostructured materials through this technique represents a basic routine method. Soft templates work effectively for their ease of adoption in addition to requiring gentle experimental conditions while supporting various material topologies [Liu et al. 2013]. The soft templating technique entails synthesizing nanoporous materials using a range of soft templates, including block copolymers, flexible organic molecules, and surfactants with anionic, cationic, and non-ionic surfactants [Li and Zhao 2013].

5.2.3 Hard Template Method

The complex template method, often known as nano-casting, refers to a specific technique. Highly designed solid materials act as templates, with holes filled with precursor chemicals to form nanostructures customized for specific applications [Poolakkandy and Menamparambath 2020].

5.2.4 Co-Precipitation

The synthesis process entails the combination of multiple solutions that include metal ions. Metal ions undergo precipitation from the solution upon mixing and forming nanoparticles [Nam and Luong 2019].

5.2.5 Sol-Gel Method

This is a wet chemical technique widely employed for developing nanoparticles. The method unites metal salts with gelling agents and solvents to create the nanomaterials. The dissolved substances undergo evaporation and leave behind the gel substance. Subsequently, the exposure of the material to heat leads to solidification and produces nanoparticles [Bokoy et al. 2021].

5.2.6 Microemulsion Technique

Surfactants serve as essential components of this procedure because they dissolve in water and oil and combine liquid-soluble and oil-soluble materials. During the emulsification process, the solution forms droplets containing metal ions that become tiny discrete elements. The increase in droplet temperature leads to metal ion precipitation in solutions which forms nanoparticles [Cid 2018].

5.2.7 Solvothermal/Hydrothermal Synthesis

The hydrothermal process maintains its status as the preferred and widely utilized method for creating nanostructured materials [Wu et al. 2011; Cao et al. 2016].

The process requires pressure without saturating either water or organic solvent while maintaining metal ion solution exposure. The increased pressure and temperature in the solution initiate metal ion precipitation into nanoparticles [Nalwana et al. 2021].

5.2.8 Sonochemical/Electrochemical Synthesis

The technique converts metal salts into nanoparticles through electrical currents or ultrasonography waves [Ashassi-Sorkhabi et al. 2015]. Mechanical nanoparticle synthesis methods rely on tools that physically apply laser-blasting, ball milling, and sputtering to create nanostructures [Krishnia et al. 2022]. The selection of a suitable production method depends on the desired characteristics of generated nanoparticles, including morphology and dimension, and the available chemicals and apparatus [Mat Isa et al. 2022].

5.2.9 Challenges and Limitations of Conventional Nanoparticle Synthesis Methods

It is crucial to acknowledge the widespread usage of harmful and environmentally damaging substances in the manufacturing procedures of synthetic nanoparticles. Even with the fundamental advancements, nanotechnology still has a big obstacle. Converting scientific results from academic journals into industrial technology applications remains a significant problem. Furthermore, to substitute current materials with nanoparticles possessing enhanced functionalities, there is a requirement for large-scale manufacturing of nanomaterials. The industry is only willing to invest significantly in developing new large-scale conventional methods for nanomaterial manufacture if a substantial profit is guaranteed. Some of the critical limitations of conventional methods are discussed below:

- Traditional techniques for producing metal nanoparticles utilize hazardous chemical-reducing reagents such as hydrazine, sodium citrate, and sodium borohydride [Mat Isa et al. 2022; Sengani et al. 2017]. Similarly, using particular abrasive physical processes, such as pyrolysis and attrition, in synthesizing diverse nanomaterials results in instant and lasting environmental consequences [Soares and Soares 2021].
- The utilization of organic compounds can have an adverse impact on the environment, as well as on the well-being of humans and animals, leading to many ailments, including harm to the liver [Ray et al. 2009].
- The process of synthesizing nanoparticles might produce effluent that may contain hazardous substances [Khoshnamvand et al. 2020].
- Another drawback is the low yield, as only a small proportion of the initial components are transformed into nanoparticles, producing trash from unused raw materials. The applications of these materials can be limited due to factors such as the expensive starting ingredients, tools, labor, the lengthy synthesis process, and the impairment in regulating dimensions and shape [Rahimi and Doostmohammadi 2019; Jamkhande et al. 2019].

5.3 RECENT GREEN SYNTHETIC TECHNIQUES FOR NANOMATERIALS

In the fast-growing field of nanoscience, there is an urgent need for green strategies for synthesizing nanomaterials to minimize associated potential risks. Different facets of nanomaterials were synthesized using the basic twelve principles of green chemistry to reduce the number of harmful materials emitted into the natural surroundings and the amount of exposure that humans have to these chemicals (Figure 5.2). These approaches often involve using natural sources such as plant extracts, microorganisms, and compostable materials to create nanomaterials with a minor environmental impact. This section will explore several crucial green synthesis techniques [Fahimirad et al. 2019].

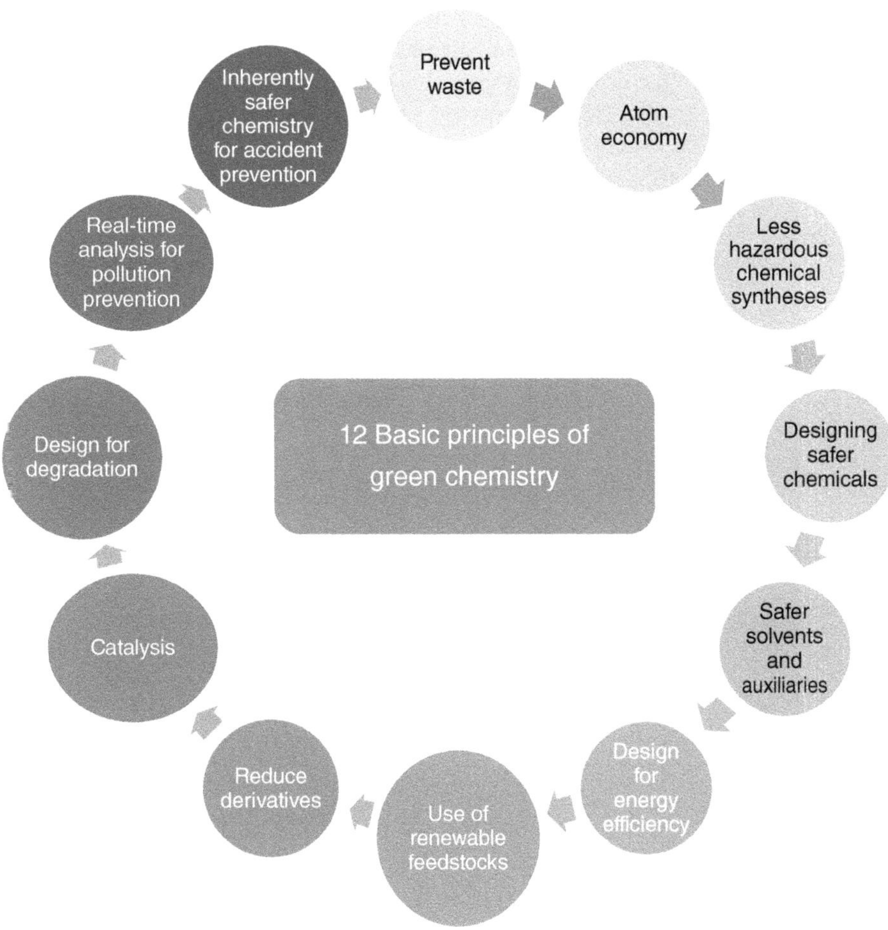

FIGURE 5.2 Basic principles of green chemistry.

5.3.1 Plant-Mediated Nanomaterial Synthesis

Plants and their extracts are rich in bioactive compounds like alkaloids, phenolics, terpenoids, flavonoids, and proteins, which are important for synthesizing nanoparticles, capping, and stabilizing. These phytochemicals assist in reducing metal ions to their elemental form, leading to the formation of nanoparticles. Most of the green nanomaterials are synthesized using the strongest reducing agents such as amino acids, aldehydes, citric acid, tartaric acid, flavonoids, NADP reductase, secondary metabolism products, and other related compounds [Kasthuri et al. 2009]. The representative process of nanoparticle synthesis is given in Figure 5.3. Additionally, it serves as a capping agent, adhering to the nanoparticle's surface via chemical or molecular interaction to stop aggregation and aid in stabilizing the nanoparticles with the desired molecules [De Lima et al. 2012].

Nanomaterials may be synthesized using various plant sections, such as leaves, roots, branches, seeds, and flowers, with the selection depending on the type and concentration of bioactive compounds. Various nanomaterials were synthesized using plant extracts, as illustrated in Table 5.1. The technique entails combining plant extract with a metal salt solution and stirring the reaction mixture under specific conditions based on the desired nanoparticles. The resulting nanoparticles are then characterized using UV-VIS spectroscopy, X-ray diffraction, scanning EM, and transmission EM [Velusamy et al. 2016]. Several factors affect nanoparticles' size, shape, stability, and general qualities, including pH, temperature, stirring time, light

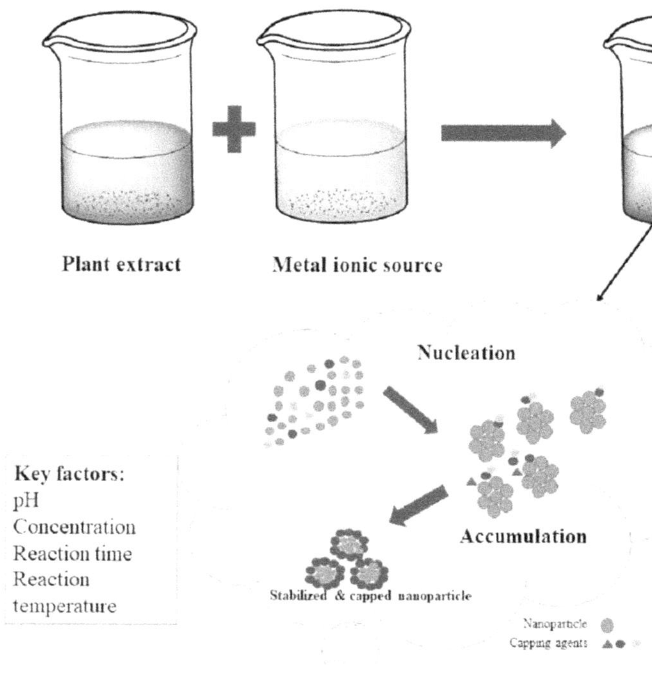

FIGURE 5.3 Representative process of nanoparticle synthesis.

TABLE 5.1
Some Important Nanoparticles Produced by Plant Extracts

S. No	Extracts	Part	Particle Size	Activity	References
	Silver Nanoparticles				
1	*Conocarpus lancifolius*	Fruits	21–173 Nm	Antibacterial, anticancer	[Oves et al. 2022]
2	*Alhagi graecorum*	Leaves	22–36 Nm	Anticancer, antifungal	[Hawar et al. 2022]
3	*Hagenia abyssinica*	Leaf	22.2 Nm	Antibacterial	[Melkamu and Bitew 2021]
4	*Piper chaba*	Stem	19 Nm	Catalytic reduction	[Mahiuddin et al. 2020]
5	*Annona squamosa*	Leaf and fruit	15–90 Nm	Antibacterial, anticancer, antidiabetic	[Malik et al. 2022]
6	*Astragalus tribuloides*	Roots	34.2 ± 8.0 Nm	Antioxidant, antibacterial, and anti-inflammatory	[Sharifi-Rad et al. 2020]
7	*Citrus limon*	Zest	535.5 Nm	Antioxidant, antibacterial	[Khane et al. 2013]
	Gold Nanoparticles				
1	*Terminalia arjuna*	*Leaf*	*15–30 Nm In*	Antibacterial	[Dudhane et al. 2019]
2	*Annona muricata*	*Leaf*	25.5 Nm	Antimicrobial	[Folorunso et al. 2019]
3	*Nerium oleander*	Leaf	2–10 Nm	Antioxidant	[Tahir et al. 2015]
4	*Nepenthes khasiana*	Leaf	50–80 Nm	Antimicrobial	[Bhau et al. 2015]
5	*Abelmoschus esculentu*	Seed	45–75 Nm.	Antifungal	[Jayaseelan et al. 2013]
	Zinc Nanoparticles				
1	*Passiflora caerulea*	Leaf	300–500 Nm.	UTI	[Santoshkumar et al. 2017]
2	*Parthenium hysterophorus*	Leaf	16–45	Antibacterial	[Datta et al. 2017]
3	*Hibiscus subdariffa* leaf extract		12–46 Nm	Antibacterial and antidiabetic activity	[Bala et al. 2015]
4	*German chamomilla*	Flower	40.5–124.0 Nm	Antibacterial activity	[Qgunyemi et al. 2019]
	European olive	Leave			
	Lycopersicon esculentum M.	Fruit			
5	*Artemisia pallens*	Whole	50–100 Nm	Antimicrobial	[Gomathi and Suhana 2021]

exposure, duration, the solvent used, reagent purity, and the concentrations of both the extract of the plant and metallic salt [Nadagouda et al. 2010; Shou et al. 2011].

Although considerable research is being done worldwide, the precise mechanism underlying the creation of nanoparticles is yet unknown. The majority of studies show that biomolecules act as reducing and capping agents during the formation of nanoparticles, either independently or together. This process occurs in three stages: (1) the activation phase, (2) the growth phase, and (3) the termination phase. During the activation phase, reduced metal ions self-organize (nucleate) to create new structures. To increase the thermodynamic stability of the nanoparticles, these newly created structures expand during the development phase and further reduce metal ions. In the final stage, the nanoparticles assume their final shape, influencing the resulting particle's stability [Devatha and Thalla 2018].

5.3.2 Microorganisms-Mediated Nanomaterial Synthesis

Microbial-mediated nanomaterial synthesis is an eco-friendly and sustainable method that leverages microorganisms such as fungi, algae, bacteria, and yeasts to generate nanomaterials. This biogenic technique harnesses the natural metabolic processes of these microorganisms to reduce metal ions and produces nanomaterials with defined sizes, shapes, and functionalities [83].

5.3.3 Eco-Friendly Bacterial Synthesis of Nanoparticle

Various species of bacteria can be employed to produce nanoparticles of metals such as silver, gold, copper, selenium, iron, titanium, and manganese. These nanoparticles have diverse applications, including the development of biosensors, biofilm creation, cell labeling and imaging, sensor devices, and diagnostics [Ahmed et al. 2016]. Additionally, these nanoparticles have shown potential as anticancer, antibacterial, antioxidant, anticoagulant, antimigration, and antiproliferative agents. For instance, certain bacteria like *Pseudomonas stutzeri* (a facultative aerobe) and *P. aeruginosa* (a facultative anaerobe) can survive and thrive in atmospheres with high concentrations of heavy metals and produce silver, gold, selenium, zinc oxide nanoparticles [Huston et al. 2021; Alsaiari et al. 2023].

5.3.4 Eco-Friendly Synthesis of Nanoparticles Using Fungi

Fungal biomass has gained recognition for nanomaterials synthesis due to its straightforward handling procedure and affordable cost structure. The tolerance of fungi toward heavy metal ions that bind to their cell walls enables these microorganisms to synthesize higher amounts of nanoparticles than bacterial nanoparticle synthesis. Increased fungal protein production is a leading factor that boosts nanoparticle production scale. Recently, extensive research has concentrated on the extracellular synthesis of nanoparticles utilizing fungi, as they efficiently secrete extracellular enzymes. For example, *Fusarium oxysporum* strains have been employed to produce extracellular silver nanoparticles with sizes ranging from 20 to 50 nm. In a separate study, gold nanoparticles were synthesized intracellularly within the biomass of *Verticillium* species [Sandhu and Goel 2023; Huston et al. 2021; Ahmed et al. 2016].

5.3.5 Biosynthesis of Nanoparticles by Algae

Algae are some photosynthetic microorganisms classified into two types: 1. unicellular (*Chlorella*) and 2. multicellular (brown algae). Like other microorganisms, algae have been shown to play an important part in biologically creating nanomaterials. Algae have the ability to reduce various accumulated heavy metals, making them increasingly popular for nanoparticle production. For instance, *Cystoseira trinodis* was used to biochemically synthesize CuO nanoparticles approximately 7 nm in size, demonstrating enhanced antioxidant and antibacterial properties and the potential for methylene blue degradation. Additionally, the marine alga *Sargassum wightii Greville* was observed to rapidly produce extracellular gold nanoparticles, ranging from 8 to 12 nm in size [Sandhu and Goel 2023; Huston et al. 2021]. Microorganisms may convert inorganic compounds into nanoparticles using enzymes found both within and outside the cell. In the intracellular procedure, positively charged metal ions become entrapped within the negatively charged cell wall of microbial cells. Inside the cells, these metal ions are converted into nanoparticles by cellular enzymes, forming nanoclusters. Eventually, nanoparticles eventually diffuse out of the cell wall and into the fluid. In the extracellular process, metal ions adhere to the cell surface, where cellular enzymes reduce them to form nanoparticles.

Microbes are cultivated in the appropriate medium for the creation of nanoparticles through an extracellular process. Once the microbial cells have formed, the broth is centrifuged, and the supernatant containing microbial enzymes is collected to produce nanoparticles. In a separate jar, the supernatant containing the reductase enzyme is mixed with metal ions. The bioreduction of metal ions in cell-free supernatant results in the formation of nanoparticles. In the intracellular process, microbial cultures are grown in an appropriate liquid medium before being centrifuged and washed with sterile distilled water to create biomass. Following this biomass exposure to an aqueous metal solution, the combination is incubated under certain circumstances until a noticeable color shift that suggests the creation of nanoparticles is seen. For example, the production of zinc and manganese nanoparticles is indicated by a whitish-yellow to yellow color; the synthesis of gold nanoparticles is indicated by a light yellow to pinkish blue; and the synthesis of silver nanoparticles is marked by a pale yellow to brownish color [Koul et al. 2021; Plucinski et al. 2021].

5.3.6 Biopolymer-Mediated Nanomaterial Synthesis

Natural biopolymers consist mainly of polysaccharides; structurally, they can be linear or branched, unlike proteins and nucleic acids. Polysaccharides are classified according to their primary functions, such as participating in the structural elements of cells or storing energy. For example, cellulose is a structural polysaccharide, whereas starch is largely used to store energy. Polysaccharides are able to coordinate metal ions, making them effective stabilizers. Furthermore, they have reducing capabilities that help synthesize metal nanostructures. This dual functionality enables the creation of metal nanoparticles with better characteristics and usefulness, including varied dimensions and shapes, hydrophilicity, biocompatibility, precision, and non-toxicity. Due to their non-toxic nature, polysaccharides are recognized as environmentally friendly reducing agents in synthesizing nanostructures. The hydroxyl groups on their structure give them a high reducing ability and water solubility. Other natural polysaccharides, such as chitosan, have also received interest for the production of nanoparticles [Mavila et al. 2016]. Several techniques have been devised to create nanoparticles from natural polysaccharides. The choice of an optimal synthesis technique is determined by parameters such as particle size, repeatability, toxicity, and endurance of actives and finished products.

5.3.6.1 Cross-Linking Process

Cross-linking chemistry is critical for the production of nanoparticles. The primary mechanisms of cross-linking include covalent and ionic cross-linking, polyelectrolyte complexation, and self-assembly. Covalent cross-linking is particularly stable due to its irreversible cross-linking points, making it resistant to various solvents and a broad pH level. Dialdehydes, which make acetal connections, and multifunctional carboxylic acids, which form ester or amide bonds, are two common polysaccharide cross-linkers [Palacios et al. 2023]. Ionic cross-linking is another form of cross-linking in which two tiny, different-charged ionic molecules with different functions are added to the polymer. Similarly, coacervation or polyelectrolyte complexation replaces oppositely charged small molecules with polysaccharides/polyelectrolytes. This method may also be used to create layer-by-layer assemblies. Compared to covalent cross-linking, ionic cross-linking and polyelectrolyte complexation have poorer chemical and physical stability, making them more susceptible to pH, strength of ions, and temperature [Qiu et al. 2023].

Self-assembly is a technique for building biopolymeric nano-systems using non-covalent linkages. Hydrophilic polysaccharides are commonly modified with hydrophobic groups (such as bile, fatty, and cholesterol), which combine in an aqueous environment to create micellar structures. This produces nanoparticles with a repellent core and a hydrophilic polysaccharide exterior. This repellent core can serve as a reservoir for hydrophobic drugs and other substances. The features of micellar nanoparticles can be modified by varying the molecular weight of the polysaccharide, increasing the ratio of hydrophobic to hydrophilic components, and changing the manufacturing method, such as micelle formation kinetics or the solvent system used [Kucuk et al. 2023].

5.3.6.2 Nanoprecipitation Process

Two compatible phases are required for the nanoprecipitation method: water-based and organic or oil-based. Normally, hydrophobic substances are initially mixed in a solvent that is miscible in water and where they exhibit good solubility. Then, this mixture is combined with a significant surplus of a solvent inhibitor, like water or buffer mixtures. Because hydrophobic solutes have reduced solubility in the aqueous solution, they form particles by precipitating out. The particles are gathered by evaporating, dialyzing, or freeze-drying to remove extra solvent. Barresi et al.'s research indicates that the size of nanoparticles is significantly affected by the mixing of phases, as better mixing leads to smaller nanoparticles and poor mixing leads to larger nanoparticles. The formation of nanoparticles is influenced by the disparity in surface tension between the aqueous and organic phases. Continuous solvent vortices at the interface cause polymers, such as polysaccharides, to gradually precipitate onto the organic surface, facilitating nanocapsule production [Plucinski et al. 2021].

5.3.6.3 The Emulsification Method

Emulsions are generally classified according to whether the dispersed phase is oil or water, resulting in oil-in-water (O/W) or water-in-oil (W/O) emulsions, with water and oil as the continuous phases, respectively. The other types of complex emulsions include oil-in-water-in-oil (O/W/O) and water-in-oil-in-water (W/O/W). These emulsions can be created by dispersing O/W or W/O emulsions in oil or water using specific methods. In addition, based on the size of the droplets, emulsions are broadly categorized as microemulsions, miniemulsions, or macroemulsions. The droplet sizes of miniemulsions and macroemulsions are more significant than 1 μm and between 100 nm and 1 μm, respectively, and they are both thermodynamically unstable. Nanoparticles of polysaccharides form through emulsification techniques that entail solvent dissolution of polysaccharides in deionized water before emulsion production. A solution containing polysaccharides dissolves in deionized water or alternative solvents. The water-based solution combines with an oil phase under stirring or ultrasonic action to generate W/O or O/W emulsions. Methods including solvent evaporation, solvent diffusion, and reverse salting-out produce the synthesized polysaccharide nanoparticles [Kucuk et al. 2023].

5.3.7 Complex Coacervation Method

The synthesis of microcapsules in textiles exploits complex coacervation as its foundational process. The synthesis approach requires water solutions of two separate polymers with distinct charges which creates nanomaterials through this approach while avoiding organic solvents. The mixture of solution containing polycation and polyanion polyelectrolytes forms a denser coacervate-rich phase along with a less concentrated solution phase. The method excels by allowing scientists to manipulate nanoparticle net charge through altering the polyelectrolyte concentrations above or below each other while producing core-shell structures where the excess component becomes the outer shell. The physical characteristics of polyelectrolyte complexes formed via coacervation respond to variations in fluid pH, temperature, ion strength, charge density, and polymer molar mass [Nitta and Numata 2013]. The researchers at Delair established that maintaining a charge ratio ($R = n+/n-$) less than 0.6 enhanced the equilibrium formation of dextran sulfate and chitosan polyelectrolyte complexes [Palacios et al. 2023].

5.3.7.1 Electrospray Technique

The electrospray method represents an effective technique which generates nanoparticles or nanospheres with high performance. A common device setup features a combination of a polymer solution pumping system from a syringe and a powerful electrical source serving as an active electrode. The system utilizes a metal foil collector which functions as the grounding electrode. Parameters of discharge and potential in electro-spraying systems depend on solution composition. The applied high electrical power through the nozzle-operated capillary transforms polymer solutions into microscopic aerosolized droplets before the solvent evaporates and leaves behind nanoparticles. The properties of needle gauge diameter, voltage application, discharge, and working distance are adjusted according to what is required for drug delivery. High voltage electro-sprayers eject polymer particles from their syringe tips into nanoparticles [Sridhar and Ramakrishna 2013].

5.3.8 Carbon-Based Synthetic Technique for Nanomaterials

Carbon-based nanomaterials span numerous intriguing materials, including graphene, carbon nanotubes, and carbon dots of various types. Synthesizing these elements occurs through both traditional chemical procedures and eco-friendly chemical methods. Organic materials from plant leaves, together with agricultural residues, serve as raw materials in these synthesis methods.

5.3.8.1 Carbon Dots Preparation

Different synthesis methods used to make carbon dots involve electrochemical approaches combined with chemical oxidation and hydrothermal/solvothermal procedures, microplasma-based methods, and microwave systems. Carbon dots are generally classified into three categories on the basis of their structural characteristics, optical characteristics, and synthesis techniques: Carbon dots and carbonized polymer dots together form one type of carbon quantum-dot alongside graphene dots. Targeted dots can be produced through the manipulation of reaction parameters such as temperature, voltage, and time duration. The prepared products undergo additional purification steps that commonly include centrifugation, dialysis, and filtration techniques. Multi-step purification strategies utilizing cataphoresis together with silica gel column-based chromatography and advanced liquid chromatography systems help in their attainment. The hydrothermal/solvothermal synthesis method represents the leading environmentally friendly technique for carbon dot production. When conducting a standard hydrothermal synthesis reaction, the method employs autoclaved containers under pressure with heating levels between 120 °C and 240 °C throughout durations of 3–12 hours [Tuerhong et al. 2017].

5.3.8.2 Carbon Nanotubes Preparation

The scientific community has lately shown heightened interest in carbon nanotube synthesis because these nanomaterials excel in mechanical strength while offering electrical conductivity and superior thermal properties. The production of carbon nanotubes requires multiple synthesis methods such as plasma discharge, laser destruction, and chemical vapor deposition. The chemical vapor deposition process, known for producing high-purity nanomaterials, involves two main steps: (1) transporting gas-phase materials into the chamber and facilitating the gas-phase reaction, and (2) forming and depositing the final nanomaterial onto a substrate [Carlsson and Martin 2010]. Additionally, spray pyrolysis has become a viable substitute for industrial carbon nanotube manufacturing. This method is appealing because of its simplest form, use of inexpensive equipment, and lack of need for high vacuum or reducing agents, making it suitable for scaled-up production. Spray pyrolysis becomes even more attractive when using renewable biomass as a starting material. For example, incorporating wood sawdust in spray pyrolysis produced high-quality carbon nanotubes and fibers. Notably, this process operates at 750°C temperature, compared to the 1200°C typically used in chemical vapor deposition, producing carbon nanotubes with a diameter of 50 nm [Verma et al. 2022; Qasim et al. 2023]. Both single- and multi-wall carbon nanotubes have been successfully produced by a spray pyrolysis technique using different oils. For instance, multi-wall carbon nanotubes with consistent alignment were formed by the pyrolysis of neem oil. In contrast, eucalyptus and turpentine oils were used to produce single-wall carbon nanotubes with a thickness varying from 0.79 to 1.71 nm. Carbon nanotubes derived from turpentine oil, developed at 700 °C within 60 minutes, were shorter than those from eucalyptus oil and showed a high degree of graphitization with well-defined concentric shells. These turpentine oil-based carbon nanotubes also had fewer defects and demonstrated superior electrical density and high electron emission strength.

5.3.8.3 Graphene Nanosheets Preparation

Graphite is a three-dimensional sp^2 carbon atom composed of millions of graphene layers. Upon oxidation with strong oxidizing agents, oxygen-containing functional groups are introduced into graphite, leading to an increased interlayer spacing and allowing it to disperse in water. There are numerous methods to functionalize graphene, including hydrothermal procedures, microemulsion, sol-gel processes, oxidation, deposition, and electrochemical reduction. Through the addition of functional groups, the functions of carbon nanostructures are improved. Graphene oxide has a variety of applications due to its electrical, thermal, and mechanical capabilities, as well as its ease of manufacture and high hydrophilic properties. Because of its feasibility and potential applications, graphene oxide nanosheet is employed in fields such as sensors, nanoelectronics, and nanocomposite materials. As an illustration, Hummers and colleagues created high-grade graphene oxide by treating graphite with a mixture of $KMnO_4$, $NaNO_3$, conc H_2SO_4, and pomegranate juice, which acts as a natural reducing agent, thereby transforming graphite oxide into graphene nanosheets. Due to its hydrophilic nature, the easy exfoliation of graphene oxide facilitates the formation of stable colloidal suspensions of thin sheets in water. The chemical method for preparing graphene oxides is advantageous because of its simplicity, relatively short duration, and high safety [Tavakoli et al. 2015; Liu et al. 2022].

5.4 CONCLUSION

The use of nanomaterials in nanotechnology has been the subject of extensive research in recent years. Due to significant advancements in many areas, including biomedical, electronics and telecommunication, medicine, skin care products, food engineering, and agriculture, nanotechnology is developing quickly. Conventional nanomaterial production technologies cause harm to the environment due to hazardous chemicals and high energy needs. In this scenario, green chemistry is rapidly advancing nanomaterial synthesis technologies. Green synthesis methods are as effective as traditional methods while causing less harm to the environment and humans involved in their manufacture. This chapter discusses green synthesis approaches for producing nanomaterials using natural plant and microorganism-based active substrates to facilitate the synthesis. However, in order to effectively produce and use green-synthesized nanomaterials, several challenges need to be addressed. These challenges include low yield, inconsistent size of particles, sophisticated extraction processes, and the availability of basic ingredients that vary seasonally and regionally. Future research should focus on using cost-effective raw materials, increasing the production of nanoscale metal particles, and implementing basic energy-saving technologies. In summary, the development of green nanomaterial synthesis must continue to advance as the world adapts to climate change, aiming to protect energy resources, the environment, and the integrity of scientific research.

REFERENCES

Ahmed, Shakeel, and Saiqa Ikram. "Biosynthesis of gold nanoparticles: A green approach." *Journal of Photochemistry and Photobiology B: Biology* 161 (2016): 141–153.

Ahmed, Shakeel, and Saiqa Ikram. "Silver nanoparticles: One pot green synthesis using Terminalia arjuna extract for biological application." *Journal of Nanomedicine & Nanotechnology* 6, no. 4 (2015): 1–6.

Ahmed, Shakeel, Mudasir Ahmad, Babu Lal Swami, and Saiqa Ikram. "Green synthesis of silver nanoparticles using Azadirachta indica aqueous leaf extract." *Journal of Radiation Research and Applied Sciences* 9, no. 1 (2016): 1–7.

Alsaiari, Norah Salem, Fatimah Mohammed Alzahrani, Abdelfattah Amari, Haitham Osman, Hamed N. Harharah, Noureddine Elboughdiri, and Mohamed A. Tahoon. "Plant and microbial approaches as green methods for the synthesis of nanomaterials: Synthesis, applications, and future perspectives." *Molecules* 28, no. 1 (2023): 463.

Alsammarraie, Fouad K., Wei Wang, Peng Zhou, Azlin Mustapha, and Mengshi Lin. "Green synthesis of silver nanoparticles using turmeric extracts and investigation of their anti-bacterial activities." *Colloids and Surfaces B: Biointerfaces* 171 (2018): 398–405.

Alvarez-Chimal, Rafael, Víctor Irahuen García-Pérez, Marco Antonio Álvarez-Pérez, and Jesús Ángel Arenas-Alatorre. "Green synthesis of ZnO nanoparticles using a Dysphania ambrosioides extract. Structural characterization and antibacterial properties." *Materials Science and Engineering: C* 118 (2021): 111540.

Arsiya, Farzaneh, Mohammad Hossein Sayadi, and Sara Sobhani. "Green synthesis of palladium nanoparticles using Chlorella vulgaris." *Materials Letters* 186 (2017): 113–115.

Ashassi-Sorkhabi, Habib, Babak Rezaei-Moghadam, Robabeh Bagheri, Leila Abdoli, and Elnaz Asghari. "Synthesis of Au nanoparticles by thermal, sonochemical and electrochemical methods: Optimization and characterization." *Physical Chemistry Research* 3, no. 1 (2015): 24–34.

Baig, Nadeem, Irshad Kammakakam, and Wail Falath. "Nanomaterials: A review of synthesis methods, properties, recent progress, and challenges." *Materials Advances* 2, no. 6 (2021): 1821–1871.

Bala, Niranjan, S. Saha, M. Chakraborty, M. Maiti, S. Das, R. Basu, and P. Nandy. "Green synthesis of zinc oxide nanoparticles using Hibiscus subdariffa leaf extract: Effect of temperature on synthesis, anti-bacterial activity and anti-diabetic activity." *RSC Advances* 5, no. 7 (2015): 4993–5003.

Baruwati, Babita, Vivek Polshettiwar, and Rajender S. Varma. "Glutathione promoted expeditious green synthesis of silver nanoparticles in water using microwaves." *Green chemistry* 11, no. 7 (2009): 926–930.

Bhau, B. S., Sneha Ghosh, Sangeeta Puri, B. Borah, D. K. Sarmah, and Raju Khan. "Green synthesis of gold nanoparticles from the leaf extract of Nepenthes khasiana and antimicrobial assay." *Advanced Materials Letters* 6, no. 1 (2015): 55–58.

Bokov, Dmitry, Abduladheem Turki Jalil, Supat Chupradit, Wanich Suksatan, Mohammad Javed Ansari, Iman H. Shewael, Gabdrakhman H. Valiev, and Ehsan Kianfar. "Nanomaterial by sol-gel method: Synthesis and application." *Advances in Materials Science and Engineering* 2021, no. 1 (2021): 5102014.

Can, Mustafa. "Green gold nanoparticles from plant-derived materials: An overview of the reaction synthesis types, conditions, and applications." *Reviews in Chemical Engineering* 36, no. 7 (2020): 859–877.

Cao, Shixiu, Cong Zhao, Tao Han, and Lingling Peng. "Hydrothermal synthesis, characterization and gas sensing properties of the WO_3 nanofibers." *Materials Letters* 169 (2016): 17–20.

Carlsson, Jan-Otto, and Peter M. Martin. "Chemical vapor deposition." In *Handbook of Deposition Technologies for Films and Coatings*, pp. 314–363. William Andrew Publishing, 2010.

Chahardoli, Azam, Naser Karimi, and Ali Fattahi. "Nigella arvensis leaf extract mediated green synthesis of silver nanoparticles: Their characteristic properties and biological efficacy." *Advanced Powder Technology* 29, no. 1 (2018): 202–210.

Cid, Antonio. "Synthesis of NPs by microemulsion method." In *Microemulsion-A Chemical Nanoreactor*, 2018. IntechOpen.

Datta, Arinjoy, Chandi Patra, Himani Bharadwaj, Sukhjeet Kaur, Neha Dimri, and Robinka Khajuria. "Green synthesis of zinc oxide nanoparticles using parthenium hysterophorus leaf extract and evaluation of their antibacterial properties." *Journal of Biotechnology and Biomaterials* 7, no. 3 (2017): 271–276.

De Lima, Renata, Amedea B. Seabra, and Nelson Durán. "Silver nanoparticles: A brief review of cytotoxicity and genotoxicity of chemically and biogenically synthesized nanoparticles." *Journal of Applied Toxicology* 32, no. 11 (2012): 867–879.

Devatha, Chella Purushothaman, and Arun K. Thalla. "Green synthesis of nanomaterials." In *Synthesis of Inorganic Nanomaterials*, pp. 169–184. Woodhead Publishing, 2018.

Devi, Henam Sylvia, Muzaffar Ahmad Boda, Mohammad Ashraf Shah, Shazia Parveen, and Abdul Hamid Wani. "Green synthesis of iron oxide nanoparticles using Platanus orientalis leaf extract for antifungal activity." *Green Processing and Synthesis* 8, no. 1 (2019): 38–45.

Dhand, Vivek, L. Soumya, Somayajula Bharadwaj, Shilpa Chakra, Deepika Bhatt, and B. J. M. S. Sreedhar. "Green synthesis of silver nanoparticles using Coffea arabica seed extract and its antibacterial activity." *Materials Science and Engineering: C* 58 (2016): 36–43.

Dudhane, Amol A., Samadhan R. Waghmode, Laxmikant B. Dama, Vaibhav P. Mhaindarkar, Anup Sonawane, and Santosh Katariya. "Synthesis and characterization of gold nanoparticles using plant extract of Terminalia arjuna with antibacterial activity." *International Journal of Nanoscience and Nanotechnology* 15, no. 2 (2019): 75–82.

Ehrampoush, Mohammad Hassan, Mohammad Miria, Mohammad Hossien Salmani, and Amir Hossein Mahvi. "Cadmium removal from aqueous solution by green synthesis iron oxide nanoparticles with tangerine peel extract." *Journal of Environmental Health Science and Engineering* 13 (2015): 1–7.

Fahimirad, Shohreh, Fatemeh Ajalloueian, and Mansour Ghorbanpour. "Synthesis and therapeutic potential of silver nanomaterials derived from plant extracts." *Ecotoxicology and Environmental Safety* 168 (2019): 260–278.

Folorunso, Aderonke, Sunday Akintelu, Abel Kolawole Oyebamiji, Samuel Ajayi, Babawale Abiola, Ibrahim Abdusalam, and Adetoun Morakinyo. "Biosynthesis, characterization and antimicrobial activity of gold nanoparticles from leaf extracts of Annona muricata." *Journal of Nanostructure in Chemistry* 9 (2019): 111–117.

Gao, Jing-Feng, Hong-Yu Li, Kai-Ling Pan, and Chun-Ying Si. "Green synthesis of nanoscale zero-valent iron using a grape seed extract as a stabilizing agent and the application for quick decolorization of azo and anthraquinone dyes." *RSC Advances* 6, no. 27 (2016): 22526–22537.

Gokila, V., V. T. Perarasu, and R. Rufina. "Qualitative comparison of chemical and green synthesized Fe_3O_4 nanoparticles." *Advances in Nano Research* 10, no. 1 (2021): 71–76.

Gomathi, R., and Harindran Suhana. "Green synthesis, characterization and antimicrobial activity of zinc oxide nanoparticles using Artemisia pallens plant extract." *Inorganic and Nano-Metal Chemistry* 51, no. 12 (2021): 1663–1672.

Hawar, Sumaiya N., Hanady S. Al-Shmgani, Zainb A. Al-Kubaisi, Ghassan M. Sulaiman, Yaser H. Dewir, and Jesamine J. Rikisahedew. "Green synthesis of silver nanoparticles from Alhagi graecorum leaf extract and evaluation of their cytotoxicity and antifungal activity." *Journal of Nanomaterials* 2022, no. 1 (2022): 1058119.

Hoag, George E., John B. Collins, Jennifer L. Holcomb, Jessica R. Hoag, Mallikarjuna N. Nadagouda, and Rajender S. Varma. "Degradation of bromothymol blue by 'greener' nano-scale zero-valent iron synthesized using tea polyphenols." *Journal of Materials Chemistry* 19, no. 45 (2009): 8671–8677.

Horwat, D., D. I. Zakharov, J. L. Endrino, F. Soldera, A. Anders, S. Migot, R. Karoum, Ph Vernoux, and J. F. Pierson. "Chemistry, phase formation, and catalytic activity of thin palladium-containing oxide films synthesized by plasma-assisted physical vapor deposition." *Surface and Coatings Technology* 205 (2011): S171–S177.

Hussain, Imtiyaz, N. B. Singh, Ajey Singh, Himani Singh, and S. C. Singh. "Green synthesis of nanoparticles and its potential application." *Biotechnology letters* 38 (2016): 545–560.

Huston, Matthew, Melissa DeBella, Maria DiBella, and Anisha Gupta. "Green synthesis of nanomaterials." *Nanomaterials* 11, no. 8 (2021): 2130.

Jamkhande, Prasad Govindrao, Namrata W. Ghule, Abdul Haque Bamer, and Mohan G. Kalaskar. "Metal nanoparticles synthesis: An overview on methods of preparation, advantages and disadvantages, and applications." *Journal of Drug Delivery Science and Technology* 53 (2019): 101174.

Jayaseelan, Chidambaram, Rajendiran Ramkumar, Abdul Abdul Rahuman, and Pachiappan Perumal. "Green synthesis of gold nanoparticles using seed aqueous extract of Abelmoschus esculentus and its antifungal activity." *Industrial Crops and Products* 45 (2013): 423–429.

Gao, J.F., H.Y. Li, K.L. Pan, and C.Y. Si, "Green synthesis of silver and palladium nanoparticles using Lithodora hispidula (Sm.) Griseb. (Boraginaceae) and application to the electrocatalytic reduction of hydrogen peroxide." *Materials Chemistry and Physics* 202 (2017): 310–319.

Kasthuri, J., S. Veerapandian, and N. Rajendiran. "Biological synthesis of silver and gold nanoparticles using apiin as reducing agent." *Colloids and Surfaces B: Biointerfaces* 68, no. 1 (2009): 55–60.

Kataria, Navish, and V. K. Garg. "Green synthesis of Fe_3O_4 nanoparticles loaded sawdust carbon for cadmium (II) removal from water: Regeneration and mechanism." *Chemosphere* 208 (2018): 818–828.

Kelly, K. Lance, Eduardo Coronado, Lin Lin Zhao, and George C. Schatz. "The optical properties of metal nanoparticles: The influence of size, shape, and dielectric environment." *The Journal of Physical Chemistry B* 107, no. 3 (2003): 668–677.

Khane, Yasmina, Khedidja Benouis, Salim Albukhaty, Ghassan M. Sulaiman, Mosleh M. Abomughaid, Amer Al Ali, Djaber Aouf et al. "Green synthesis of silver nanoparticles using aqueous Citrus limon zest extract: Characterization and evaluation of their antioxidant and antimicrobial properties." *Nanomaterials* 12, no. 12 (2022): 2013.

Khoshnamvand, Mehdi, Zhineng Hao, Oluniyi O. Fadare, Parichehr Hanachi, Yongsheng Chen, and Jingfu Liu. "Toxicity of biosynthesized silver nanoparticles to aquatic organisms of different trophic levels." *Chemosphere* 258 (2020): 127346.

Koul, Bhupendra, Anil Kumar Poonia, Dhananjay Yadav, and Jun-O. Jin. "Microbe-mediated biosynthesis of nanoparticles: Applications and future prospects." *Biomolecules* 11, no. 6 (2021): 886.

Krishnia, Lucky, Preeti Thakur, and Atul Thakur. "Synthesis of nanoparticles by physical route." In *Synthesis and Applications of Nanoparticles*, pp. 45–59. Singapore: Springer Nature Singapore, 2022.

Kucuk, Nika, Mateja Primozic, Zeljko Knez, and Maja Leitgeb. "Sustainable biodegradable biopolymer-based nanoparticles for healthcare applications." *International Journal of Molecular Sciences* 24, no. 4 (2023): 3188.

Kumar, Brajesh, Kumari Smita, Luis Cumbal, Alexis Debut, and Yolanda Angulo. "Biofabrication of copper oxide nanoparticles using Andean blackberry (Rubus glaucus Benth.) fruit and leaf." *Journal of Saudi Chemical Society* 21 (2017): S475–S480.

Leili, Mostafa, Mehdi Fazlzadeh, and Amit Bhatnagar. "Green synthesis of nano-zero-valent iron from Nettle and Thyme leaf extracts and their application for the removal of cephalexin antibiotic from aqueous solutions." *Environmental Technology* 39, no. 9 (2018): 1158–1172.

Li, Wei, and Dongyuan Zhao. "An overview of the synthesis of ordered mesoporous materials." *Chemical Communications* 49, no. 10 (2013): 943–946.

Liu, Yiding, James Goebl, and Yadong Yin. "Themed issue: Chemistry of functional nanomaterials." *Chemical Society Reviews* 42 (2013): 2610–2653.

Liu, Zhixin, Qian Ling, Yawen Cai, Linfeng Xu, Jiahao Su, Kuai Yu, Xinyi Wu, Jiayi Xu, Baowei Hu, and Xiangke Wang. "Synthesis of carbon-based nanomaterials and their application in pollution management." *Nanoscale Advances* 4, no. 5 (2022): 1246–1262.

Mahiuddin, Md, Prianka Saha, and Bungo Ochiai. "Green synthesis and catalytic activity of silver nanoparticles based on Piper chaba stem extracts." *Nanomaterials* 10, no. 9 (2020): 1777.

Malik, Maria, Muhammad Aamir Iqbal, Mariam Malik, Muhammad Akram Raza, Wajeehah Shahid, Jeong Ryeol Choi, and Phuong V. Pham. "Biosynthesis and characterizations of silver nanoparticles from Annona squamosa leaf and fruit extracts for size-dependent biomedical applications." *Nanomaterials* 12, no. 4 (2022): 616.

Manikam, Vemal Raja, Kuan Yew Cheong, and Khairunisak Abdul Razak. "Chemical reduction methods for synthesizing Ag and Al nanoparticles and their respective nanoalloys." *Materials Science and Engineering: B* 176, no. 3 (2011): 187–203.

Martínez-Mera, I., M. E. Espinosa-Pesqueira, Raúl Pérez-Hernández, and J. Arenas-Alatorre. "Synthesis of magnetite (Fe_3O_4) nanoparticles without surfactants at room temperature." *Materials Letters* 61, no. 23–24 (2007): 4447–4451.

Mat Isa, Siti Zaleha, Rafidah Zainon, and Mahbubunnabi Tamal. "State of the art in gold nanoparticle synthesisation via pulsed laser ablation in liquid and its characterisation for molecular imaging: A review." *Materials* 15, no. 3 (2022): 875.

Mavila, Sudheendran, Or Eivgi, Inbal Berkovich, and N. Gabriel Lemcoff. "Intramolecular cross-linking methodologies for the synthesis of polymer nanoparticles." *Chemical Reviews* 116, no. 3 (2016): 878–961.

Melkamu, Walelign Wubet, and Legesse Terefe Bitew. "Green synthesis of silver nanoparticles using Hagenia abyssinica (Bruce) JF Gmel plant leaf extract and their antibacterial and anti-oxidant activities." *Heliyon* 7, no. 11 (2021).

Min, Soo-Hong, Tae Hun Lee, Sangwook Lee, Ji-Hyeon Song, Gil-Yong Lee, Daniel Zontar, Christian Brecher, and Sung-Hoon Ahn. "Simulation of electrical conductivity for nanoparticles and nanotubes composite sensor according to geometrical properties of nanomaterials." *Composites Part B: Engineering* 174 (2019): 107003.

Nadagouda, Mallikarjuna N., Alicia B. Castle, Richard C. Murdock, Saber M. Hussain, and Rajender S. Varma. "In vitro biocompatibility of nanoscale zerovalent iron particles (NZVI) synthesized using tea polyphenols." *Green Chemistry* 12, no. 1 (2010): 114–122.

Nam, Nguyen Hoang, and Nguyen Hoang Luong. "Nanoparticles: Synthesis and applications." In *Materials for Biomedical Engineering*, pp. 211–240. Elsevier, 2019.

Nasrollahzadeh, Mahmoud, and S. Mohammad Sajadi. "Pd nanoparticles synthesized in situ with the use of Euphorbia granulate leaf extract: Catalytic properties of the resulting particles." *Journal of Colloid and Interface Science* 462 (2016): 243–251.

Ndlwana, Lwazi, Naledi Raleie, Kgogobi M. Dimpe, Hezron F. Ogutu, Ekemena O. Oseghe, Mxolisi M. Motsa, Titus A. M. Msagati, and Bhekie B. Mamba. "Sustainable hydrothermal and solvothermal synthesis of advanced carbon materials in multidimensional applications: A review." *Materials* 14, no. 17 (2021): 5094.

Nitta, Sachiko Kaihara, and Keiji Numata. "Biopolymer-based nanoparticles for drug/gene delivery and tissue engineering." *International journal of molecular sciences* 14, no. 1 (2013): 1629–1654.

Ogunyemi, Solabomi Olaitan, Yasmine Abdallah, Muchen Zhang, Hatem Fouad, Xianxian Hong, Ezzeldin Ibrahim, Md Mahidul Islam Masum, Afsana Hossain, Jianchu Mo, and Bin Li. "Green synthesis of zinc oxide nanoparticles using different plant extracts and their antibacterial activity against Xanthomonas oryzae pv. oryzae." *Artificial Cells, Nanomedicine, and Biotechnology* 47, no. 1 (2019): 341–352.

Oves, Mohammad, Mohd Ahmar Rauf, Mohammad Aslam, Huda A. Qari, Hana Sonbol, Irfan Ahmad, Gaffar Sarwar Zaman, and Mohd Saeed. "Green synthesis of silver nanoparticles by Conocarpus Lancifolius plant extract and their antimicrobial and anticancer activities." *Saudi Journal of Biological Sciences* 29, no. 1 (2022): 460–471.

Palacios, A., M. E. Navarro-Rivero, B. Zou, Z. Jiang, M. T. Harrison, and Y. Ding. "A perspective on phase change material encapsulation: Guidance for encapsulation design methodology from low to high-temperature thermal energy storage applications." *Journal of Energy Storage* 72 (2023): 108597.

Plucinski, Alexander, Zan Lyu, and Bernhard V.K.J. Schmidt. "Polysaccharide nanoparticles: From fabrication to applications." *Journal of Materials Chemistry B* 9, no. 35 (2021): 7030–7062.

Poolakkandy, Rasha Rahman, and Mini Mol Menamparambath. "Soft-template-assisted synthesis: A promising approach for the fabrication of transition metal oxides." *Nanoscale Advances* 2, no. 11 (2020): 5015–5045.

Qasim, Muhammad, Andrew N. Clarkson, and Simon FR Hinkley. "Green synthesis of carbon nanoparticles (CNPs) from biomass for biomedical applications." *International Journal of Molecular Sciences* 24, no. 2 (2023): 1023.

Qiu, Liang, Xinyue Han, Chengfen Xing, and Ulrich Glebe. "Polymerization-induced self-assembly: An emerging tool for generating polymer-based biohybrid nanostructures." *Small* 19, no. 18 (2023): 2207457.

Rahimi, Hamid-Reza, and Mohsen Doostmohammadi. "Nanoparticle synthesis, applications, and toxicity." *Applications of Nanobiotechnology* 10 (2019).

Ray, Paresh Chandra, Hongtao Yu, and Peter P. Fu. "Toxicity and environmental risks of nanomaterials: Challenges and future needs." *Journal of Environmental Science and Health Part C* 27, no. 1 (2009): 1–35.

Ren, Jingyi. "Research on green synthetic iron nanoparticles." *IOP Conference Series: Materials Science and Engineering* 612, no. 2 (2019): 022025.

Reyes-Carmona, Lorena, Enrique Camps, Enrique Campos-González, Gabriela Mercado-Celis, Alejandra Cervantes-Garduño, Ezequiel A. Pérez-Ibarra, Rafael Álvarez-Chimal, Sandra E. Rodil, and Argelia Almaguer-Flores. "Antimicrobial evaluation of bismuth subsalicylate nanoparticles synthesized by laser ablation against clinical oral microorganisms." *Optics & Laser Technology* 158 (2023): 108930.

Sandhu, Aashi, and Anjana Goel. "Biosynthesis of nanoparticles by micro-organisms and its applications." *Journal of Young Pharmacists* 15, no. 3 (2023): 430–440.

Santhoshkumar, J., S. Venkat Kumar, and S. Rajeshkumar. "Synthesis of zinc oxide nanoparticles using plant leaf extract against urinary tract infection pathogen." *Resource-Efficient Technologies* 3, no. 4 (2017): 459–465.

Sengani, Manimegalai, Alexandru Mihai Grumezescu, and V. Devi Rajeswari. "Recent trends and methodologies in gold nanoparticle synthesis–A prospective review on drug delivery aspect." *OpenNano* 2 (2017): 37–46.

Sharifi-Rad, Majid, Pawel Pohl, Francesco Epifano, and José M. Álvarez-Suarez. "Green synthesis of silver nanoparticles using Astragalus tribuloides delile. root extract: Characterization, antioxidant, antibacterial, and anti-inflammatory activities." *Nanomaterials* 10, no. 12 (2020): 2383.

Shou, Qinghui, Chen Guo, Liangrong Yang, Lianwei Jia, Chunzhao Liu, and Huizhou Liu. "Effect of pH on the single-step synthesis of gold nanoparticles using PEO–PPO–PEO triblock copolymers in aqueous media." *Journal of Colloid and Interface Science* 363, no. 2 (2011): 481–489.

Smirniotis, Panagiotis G., Thirupathi Boningari, and Siva Nagi Reddy Inturi. "Single-step synthesis of N-doped TiO_2 by flame aerosol method and the effect of synthesis parameters." *Aerosol Science and Technology* 52, no. 8 (2018): 913–922.

Soares, Eduardo V., and Helena M. V. M. Soares. "Harmful effects of metal (loid) oxide nanoparticles." *Applied Microbiology and Biotechnology* 105 (2021): 1379–1394.

Sone, B. T., A. Diallo, X. G. Fuku, A. Gurib-Fakim, and M. Maaza. "Biosynthesized CuO nano-platelets: Physical properties & enhanced thermal conductivity nanofluidics." *Arabian Journal of Chemistry* 13, no. 1 (2020): 160–170.

Sridhar, Radhakrishnan, and Seeram Ramakrishna. "Electrosprayed nanoparticles for drug delivery and pharmaceutical applications." *Biomatter* 3, no. 3 (2013): e24281.

Subramaniyam, Vidhyasri, Suresh Ramraj Subashchandrabose, Palanisami Thavamani, Mallavarapu Megharaj, Zuliang Chen, and Ravi Naidu. "Chlorococcum sp. MM11—a novel phyco-nanofactory for the synthesis of iron nanoparticles." *Journal of Applied Phycology* 27 (2015): 1861–1869.

Tahir, Kamran, Sadia Nazir, Baoshan Li, Arif Ullah Khan, Zia Ul Haq Khan, Peng Yu Gong, Shahab Ullah Khan, and Aftab Ahmad. "Nerium oleander leaves extract mediated synthesis of gold nanoparticles and its antioxidant activity." *Materials Letters* 156 (2015): 198–201.

Tavakoli, Farnosh, Masoud Salavati-Niasari, and Fatemeh Mohandes. "Green synthesis and characterization of graphene nanosheets." *Materials Research Bulletin* 63 (2015): 51–57.

Thovhogi, N., A. Diallo, A. Gurib-Fakim, and M. Maaza. "Nanoparticles green synthesis by Hibiscus sabdariffa flower extract: Main physical properties." *Journal of Alloys and Compounds* 647 (2015): 392–396.

Thovhogi, N., E. Park, E. Manikandan, M. Maaza, and A. Gurib-Fakim. "Physical properties of CdO nanoparticles synthesized by green chemistry via Hibiscus Sabdariffa flower extract." *Journal of Alloys and Compounds* 655 (2016): 314–320.

Tuerhong, Mhetaer, X. U. Yang, and Y. I. N. Xue-Bo. "Review on carbon dots and their applications." *Chinese Journal of Analytical Chemistry* 45, no. 1 (2017): 139–150.

Turunc, Ersan, Riza Binzet, Ilkay Gumus, Gun Binzet, and Hakan Arslan. "Green synthesis of silver and palladium nanoparticles using Lithodora hispidula (Sm.) Griseb. (Boraginaceae) and application to the electrocatalytic reduction of hydrogen peroxide." *Materials Chemistry and Physics* 202 (2017): 310–319.

Velusamy, Palaniyandi, Govindarajan Venkat Kumar, Venkadapathi Jeyanthi, Jayabrata Das, and Raman Pachaiappan. "Bio-inspired green nanoparticles: Synthesis, mechanism, and antibacterial application." *Toxicological Research* 32, no. 2 (2016): 95–102.

Verma, Sandeep Kumar, Ashok Kumar Das, Saikat Gantait, Yogita Panwar, Vinay Kumar, and Marian Brestic. "Green synthesis of carbon-based nanomaterials and their applications in various sectors: A topical review." *Carbon Letters* 32, no. 2 (2022): 365–393.

Wang, Yong, Sean Maksimuk, Rui Shen, and Hong Yang. "Synthesis of iron oxide nanoparticles using a freshly-made or recycled imidazolium-based ionic liquid." *Green Chemistry* 9, no. 10 (2007): 1051–1056.

Wojnarowicz, Jacek, Tadeusz Chudoba, Stanisław Gierlotka, and Witold Lojkowski. "Effect of microwave radiation power on the size of aggregates of ZnO NPs prepared using microwave solvothermal synthesis." *Nanomaterials* 8, no. 5 (2018): 343.

Wu, Qiong, Wei-shou Miao, Yi-du Zhang, Han-jun Gao, and David Hui. "Mechanical properties of nanomaterials: A review." *Nanotechnology Reviews* 9, no. 1 (2020): 259–273.

Wu, Xia, Gao Qing Max Lu, and Lianzhou Wang. "Shell-in-shell TiO_2 hollow spheres synthesized by one-pot hydrothermal method for dye-sensitized solar cell application." *Energy & Environmental Science* 4, no. 9 (2011): 3565–3572.

Zeng, Jie, Qiang Zhang, Jingyi Chen, and Younan Xia. "A comparison study of the catalytic properties of Au-based nanocages, nanoboxes, and nanoparticles." *Nano Letters* 10, no. 1 (2010): 30–35.

Zhao, Xiao, Wen Liu, Zhengqing Cai, Bing Han, Tianwei Qian, and Dongye Zhao. "An overview of preparation and applications of stabilized zero-valent iron nanoparticles for soil and groundwater remediation." *Water Research* 100 (2016): 245–266.

Part 2

Advanced Manufacturing and AI Integration

6 Smart Manufacturing at the Nanoscale

Exploring AI-Enhanced Magnetorheological Finishing

Manpreet Singh

6.1 INTRODUCTION

Magnetorheological (MR) finishing (MRF) is a type of optical manufacturing used to improve the surface quality of optical components. This process is based on an MR fluid, which is defined as the suspension of micron-sized magnetic particles in a carrier liquid [1–5]. By changing the viscosity of the fluid owing to the applied magnetic field, the finishing tool also changes; hence, the MRF is implemented. There are several ways artificial intelligence (AI) can be taken advantage of within the MRF process, thence improving its efficiency, accuracy, and other measures [6]. Therefore, the purpose of the chapter is to highlight the areas where AI can be implemented and the improvements AI brings.

An in-depth analysis of using AI to enhance the MRF process will be covered in each section. The advantages offered by such applications and possible challenges will also be addressed in detail. Specifically, some AI components could be machine learning (ML) algorithms to optimize some process parameters such as wheel speed, fluid viscosity, and magnetic field strength [7–10]. The AI system could study the data available from the previous finishing runs to determine workpiece material, shape, and the needed surface requirements [11–15]. This will allow for improved and accurate finishing as it will reduce the time and cost. Some ML algorithms to optimize MRF process parameters include neural networks and genetic methods.

An example of such AI technology is the artificial neural networks. These are ML algorithms, which process vast amounts of data obtained from the previous finishing runs and find patterns and relationships between the process parameters and the surface quality achieved. Using those patterns, an AI system can predict what would be the process parameters, using which this quality can be reached. Artificial neural networks are built of interconnected nodes, which are also called neurons [16, 17]. They take an input dataset, process it, and produce the output. In the case of the MRF

DOI: 10.1201/9781003560234-8

process parameter optimization, the dataset is the historical database of the finishing run reports. The output is the recommendation on workpiece.

Once the network has received training, it can be refined and updated when new data are acquired. As a result, the network can reinforce its ability to forecast with time [18–22]. Genetic algorithms are one of the other ML algorithms that can be employed in parametric optimization of the process. Genetic algorithms are algorithms that generate and evaluate solution space by natural selection of the species. Genetic algorithms can generate and evaluate how different pairs of parameters might perform in the case of MRF. Further, the best parents generated can become the daughter or make babies that will be used to test next. By optimizing process parameters, the MRF process can improve, which translates to reduced costs of the process as a finishing technique and time spent finishing the workpiece [23–28].

Furthermore, AI-based parameter optimization can be used to increase the accuracy of the process and guarantee that the requisite surface quality is always achieved [29–32]. Because the outcome of the process is based on prior experiments and results, the AI system can examine historical data to determine the best parameters to achieve the target surface quality. Additionally, optimized process parameters can further reduce waste in the MRF process, lowering the number of trials necessary to obtain the needed surface quality. This results in time and cost savings, as well as a reduction in the environmental impact associated with the process. The utility of AI-based parameter optimization is contingent on the availability of quality historical data [33]. Figure 6.1 shows the architecture of the AI that is implemented on various processes.

Because ML algorithms may be complicated, implementation and upkeep of such systems may need significant computational resources and ability. For best results, it is necessary to use the appropriate algorithm for the MRF process requirements [34–38].

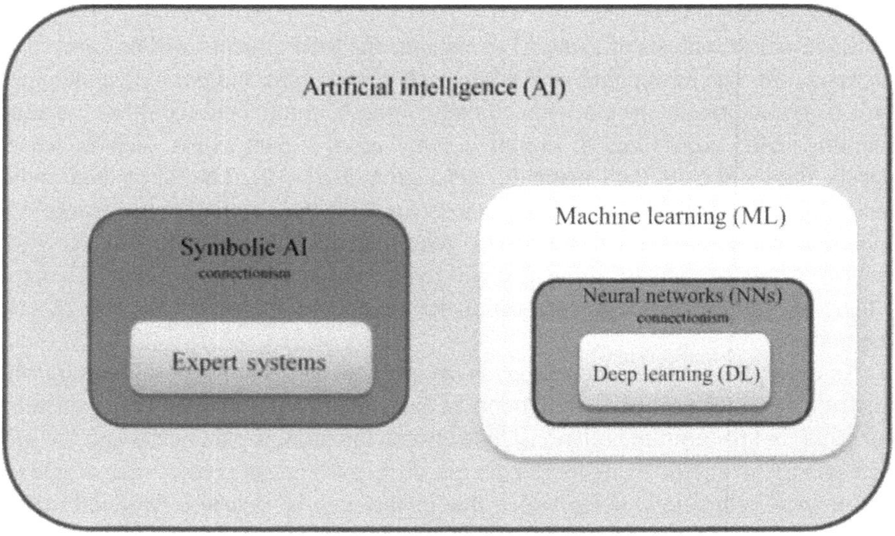

FIGURE 6.1 Architecture of AI.

Furthermore, to function while maintaining a level of volume which the AI is operating in the Pareto optimal domain, AI-based parameter optimization systems should continuously and autonomously adjust based on which new information they get due to the execution of the process. In practice, this entails regularly accumulating new data, analyzing it, and then changing the system, which can be quite time-consuming and laborious [39]. In this manner, AI can also be used to control real-time monitoring and detect unplanned variations from the finishing terms. Data capturing may be achieved by sensory monitoring the workpiece surface quality, including hydrostatic pressure levels and sensor displacement [40–42].

6.2 AI TECHNIQUES IN MRF

The incorporation of AI into the MRF regime has revolutionized the approach to high-precision surface finishing, ushering in a new epoch of optimization, flexibility, and accuracy [43–45]. MRF is a dedicated approach to polishing that utilizes MR fluids' unique properties. In this regard, MRF is a sound tool for reaching excellent surface quality and accuracy. However, MRF's numerous operational facets of superimposed parameters as well as the requirement for constant observation and control have always been two of the most significant hindrances [46].

As in other fields, AI's influence on MRF can be traced back to the technologies and algorithms that have been implemented to enhance the process's proficiency. More specifically, AI-powered solutions utilize a range of techniques, each specializing in improving a certain activity linked to the MRF, be it the analysis of process data in real time, predictive modeling, or decision-making. The most common technique in MRF is ML [47–52]. ML using algorithms such as neural networks, support vector machines, and decision trees played a crucial part in the creation of dependable predictive systems that are capable of reflecting the interaction of the MR fluid, workpiece, and the magnetic field.

Moreover, computer vision and image processing techniques have also significantly contributed to the development of MRF [53–58]. Modern AI-based MRF systems use advanced sensors and high-resolution cameras to monitor the surface topography of the workpiece, which, in turn, helps detect defects, measure roughness, and track the polishing process progress. Visual data obtained in this manner are further processed through ML approaches, such as convolutional neural networks or support vector machines, to provide real-time feedback to the system and enable dynamic adjustments of the process. Figure 6.2 shows the general ML pipeline for supervised learning. Besides applying another AI approach of reinforcement learning, MR artists tame AI as if they were teaching a child. This method, which mimics the way an animal learns by experience, behavior adaptation, and decision-making using the trial-and-error approach, has been used in AI to solve some of the problems in the process control [41].

By continuously amending the process performance in line with the desired outcomes, these algorithms can alter the MR fluid properties, magnetic strength of the field, and other parameters to remain accountable for the productivity and surface quality [12]. The contribution of expert systems, a particular field of AI centered on human-like decision-making, is likewise of great importance in the scope of

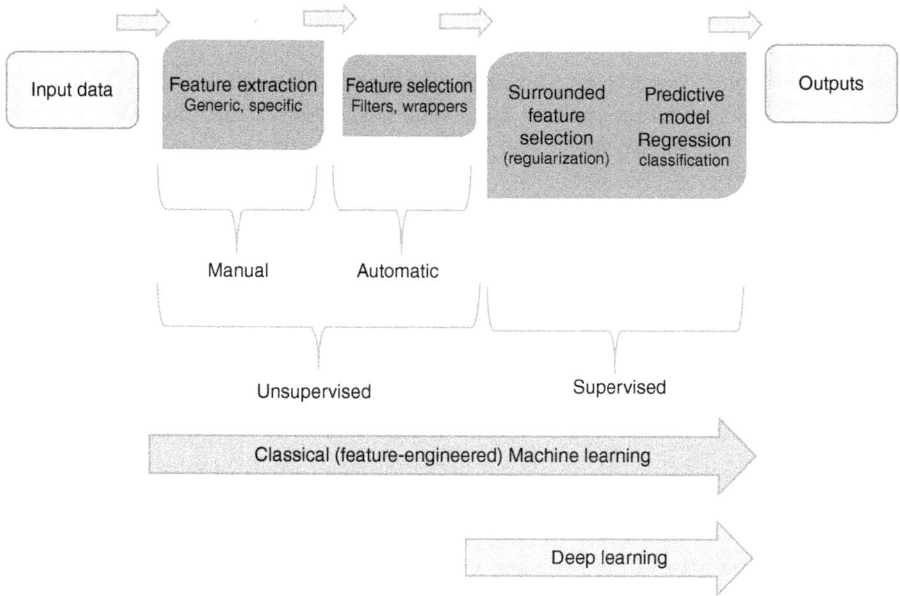

FIGURE 6.2 General ML pipeline for supervised learning.

enhancing the MR final product. Utilizing knowledge bases, decision rules, and inference engines, these systems are able to make diagnostic and prognostic decisions, including detection of malfunctions, identification of potential cases, and prediction of remaining work life for critical parts including lapping fluids and polishing tools. Fuzzy logic, a method of reasoning where an outcome is derived from a set of raw data that may be incomplete or ambiguous, is used in other areas related to MRF.

The fuzzy logic–based system effectively models the uncertain and subjective relationship between erythrocyte parameters and resulting outcomes. It enables decision-making and adaptive adjustments that account for the variability and imprecision inherently present in the MRF process [11–14]. The MRF sector has also witnessed a concomitant development of the new AI approaches such as deep learning and hybrid modeling besides the well-known AI techniques. Deep learning is a suitable tool for building predictive models that could simulate the nonlinear and multiscale phenomena essential for MRF, as illustrated in [16]. These models, provided with extensive datasets, can achieve outcomes previously impossible and visualize intricate relationships between process parameters and the desired result, leading to more exact control and fine-tuning [30-32]. Hybrid modeling which combined the force of different AI methods such as ML and physics-based simulations has enjoyed increased popularity in the simulation of the MRF process. These hybrid models can utilize the data-driven competence of ML to improve the precision and power of regular computational fluid dynamics (CFD) and finite element analysis (FEA) simulations which in the end bring about stronger credibility gains. These credibility gains are related to material removal rates, surface topography, and other key performance indicators [31].

With new technologies being developed frequently and possibilities of deep reinforcement learning, federated learning, and meta-learning, the prospects for MRF to utilize more of the potential will become even more comprehensive. Several industries are in the process of adopting these transformative technologies, including optics and electronics, aerospace, biomedical, and so on, since they have a keen eye on the competitive advantages brought about by the technologies in terms of increased productivity, better product quality, and lower manufacturing costs. Moving forward then, the adoption of AI in MRF will in fact be a catalyst to technological development, whereby smart and highly agile manufacturing systems can be built more easily and can accommodate all the changing dynamics and needs of the modern industrial landscape [29]. While the researchers and engineers simply state "the sky is the limit," the AI robot itself, with its ability to continuously adapt to new applications and solutions, stands at the apex of the evolution that is revolutionizing high-precision surface finishing as well as making some new horizons available. Figure 6.3 shows the 3D model of MRF process with sensor-based real-time monitoring.

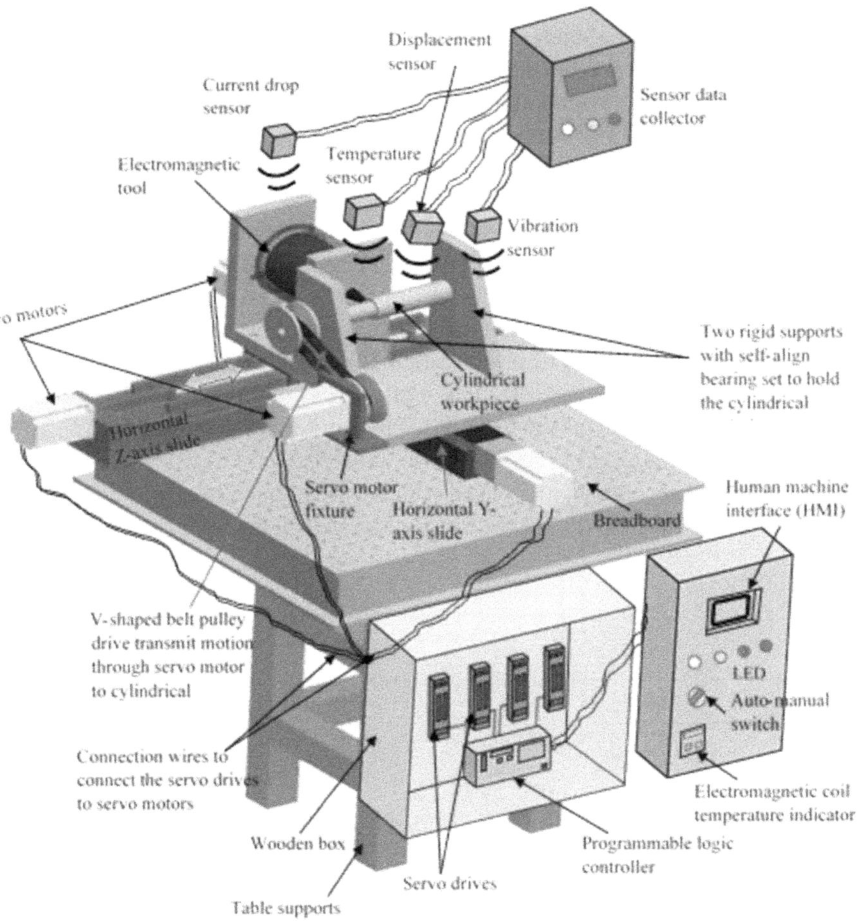

FIGURE 6.3 3D model of MRF process with sensor based real-time monitoring.

6.3 BENEFITS AND ADVANTAGES OF AI IN MRF

The highly digitized, automated, and accurate implementation of AI in MRF processing regimes has provided a paradigm shift in the way the finishing of high-precision surfaces was traditionally handled. MRF refers to a sophisticated polishing procedure which is unified with MR fluids, well acknowledged for having remarkable properties for achieving high-performance surfaces along with great smoothness. Although numerous factors influence the quality of MR-finished products and the growing need for online control and real-time monitoring has increased the complexity of the process, the significant advantages and benefits of MR-finished products continue to motivate both manufacturers and consumers to adopt and rely on this technology. Implementation of AI-powered systems has been already showing some tangible results as they have overcome the challenges of conventional MRF, the era of which has brought profound improvements in efficiency, adaptability, precision, and many more [12]. The most important feature of the AI in MRF—its ability to change the parameters in real time—has to be highlighted. Pollution in traditional MRF processes mainly arises from poor understanding of the mechanism, lack of standardization, and trial-and-error optimization of parameters like media composition, magnetic field strength, and polishing time. However, when tested, these methods proved unsuitable for adapting to environmental variations, differing material properties, and changing customer requirements.

In contrast, AI-driven MRF techniques utilize advanced control algorithms, machine learning, feed-forward models, intelligent sensors, and data analytics tools to continuously monitor, analyze, and optimize the process in real time. These systems can analyze information from different sources like sensors and computer vision to identify the patterns of the MRF process, as well as detect abnormalities and quickly intervene to stay within the given requirement on the quality and performance. AI has been applied to several automatic ML areas.

Figure 6.4 is the schematic illustration of a force control system architecture for a gripper, and the end-effector can be replaced by any kind of manipulator [59]. Through this architecture, the smart control can assure that the end-effector's interactions with the environment are correctly regulated with the programmed force levels. The framework has the capacity to regulate the forces generated by the end-effector and precisely

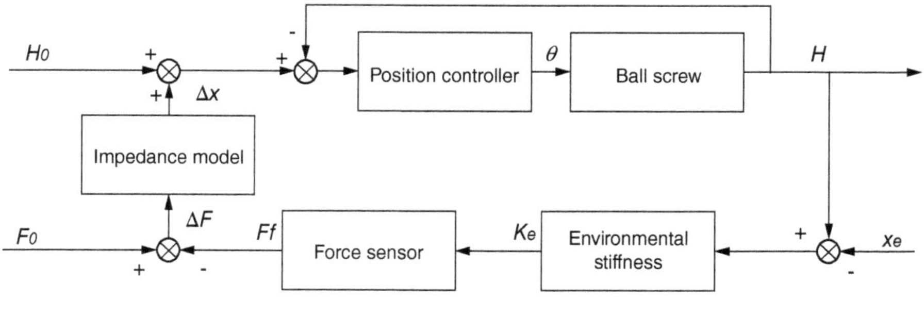

FIGURE 6.4 Schematic representation of the force control system architecture for an end-effector, enabling precise force regulation during interactions with the environment [59].

apply forces on interfaces or objects against which end-effector is in touch inside the workspace. In various applications—such as assembly, material handling, and object manipulation—human performance is often constrained by the need to apply controlled force. Exceeding these limits can lead to damage or failure of the operation.

Figure 6.5 helps to visualize a constructional structure of a constant force control system, where the active disturbance rejection control (ADRC) methodology is applied [59]. This method utilizes the act of active rejection for a regime that can regulate the strength of the force and reject any disturbance which can impinge on the system's performance. ADRC technique is used as an effective method throughout the entire force channel to minimize uncertainties, external disturbances, and model inaccuracies that can have very bad effects on the force control performance. Implementing this technology will assist in the system's control of the force output and consequent maintenance of the rhythm resulting from constant and repeated output in various scenarios. Therefore, AI systems are able to continue updating their predictive models and control strategies. Consequently, AI-based MRF systems can adapt to material property modifications, tool wear, environmental perturbations, and other disturbances. This assures that surface finishing results are highly reliable and consistent. This flexibility feature extends the lifespan of machines and turns them into smart devices; therefore, the efficiency of the production process will be increased by reducing the amount of manual intervention and the downtime [17]. The further value of AI in MRF is its capability of optimizing the material removal and surface finish properties. In conventional MRF methods, achieving the desired surface quality often required compromising the material removal rate to meet customer expectations. [8]. AI-based machines, on the contrary, may make the best use of predictive model and control algorithm, in order to find a way that will ensure the highest material quality, with minimum energy and time consumption.

With the power to selectively alter the magnet flux, the fluid flow, etc., AI-based MR systems can introduce microscopic alterations to the surface of the workpiece, minimizing surface imperfections, sag, and rough edges, hence ensuring high productivity [4–6]. This high level of control is highly valuable to such fabrication processes as optical components, semiconductor devices, and the specialty products that would serve in many areas and where the surface finish is critically important in determining such aspects as the final product's performance and reliability.In addition to real-time process optimization, the AI-enhanced MRF system also enables comprehensive

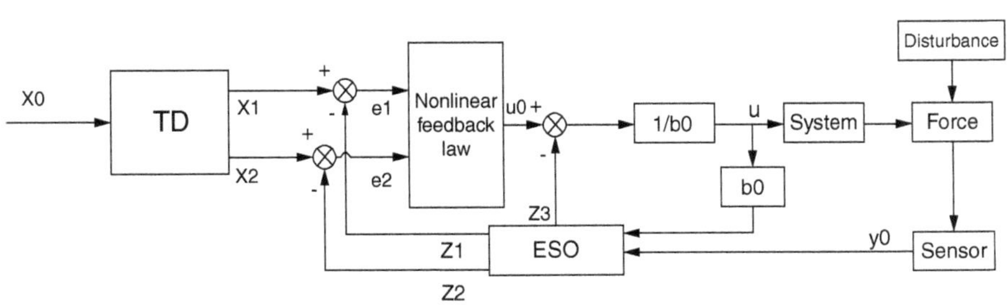

FIGURE 6.5 Architectural diagram illustrating a constant force control system employing an ADRC methodology for robust force regulation [59].

process modeling and simulation. By utilizing machine learning algorithms and CFD simulations, these systems create predictive models that accurately capture the complex interactions between the MR fluid, the workpiece, and the magnetic field [5–11]. All these models could be used to maximize process parameters, material removal rates prediction, and even analogize surface topography evolution, with less prototypes and experimentations [3]. AI in MRF shows an increasing direction of identifying the disease in the early stages and predicting the outcome. Through regular monitoring of the process conditions, sensor data, and machine parameters, AI is powerful enough to detect any anomalies, perform diagnoses on potential problems, and predict the remaining useful life of the wearing parts, like the MR fluid and the polishing tools [53–55]. This maintenance predictive feature will be critical in preventing or shortening downtime, reducing the maintenance costs, and ensuring the reliability of the MRF process. In addition, the unique character of AI makes it possible to create modern systems of control and decision-making for MRF processes.

These systems are able to regulate automatically the process parameters, to respond adaptively to the environment changes, as well as to take decisions based on their own experience to improve the surface finishing using the manufacture workflow and therefore requiring less human intervention and increasing the overall efficiency of the production cycle. The superiority of AI in transitional finishing goes far beyond this by allowing these features of real-time optimization and control adaption. Similarly, the activation of these technologies has been instrumental in the design and implementation of intelligent manufacturing systems, where the MRF process is incorporated with other avant-garde technologies including additive manufacturing, CAD (computer-aided design), and CAM (computer-aided manufacturing). Such tightly knitted systems can utilize AI-driven algorithms to find the best possible combination of all production phases, including product design and surface finishing, and thereby guarantee the generation of components with reliable performance properties [6].

Apart from that, AI has helped to identify new pathways for investigations and improvement. AI provides a unique, excellent analysis data tool that enables researchers to look into new material, fluid composition, and polishing techniques, which consequently has seen a speeding up of growth in technology. The capability of instantly building and testing various scenarios, complemented by the real-time attribute of the AI-powered systems, has markedly improved the capacity to innovate and to create the next generation of superior MR coagulation therapy solutions [3]. The scope of AI is becoming wider, and the technological world is advancing by leaps and bounds, outdoing itself constantly, boosting the potential of AI in MRF. These advanced manufacturing technologies are already getting adopted by a large number of manufacturing industries, such as optics, electronics, aerospace, and biomedical industries, which recognize their benefits that make competition easier by improving productivity, augmenting product quality, and cutting production operation costs.

6.4 AI-BASED CONTROL SYSTEMS FOR MRF

MRF is an innovative polishing technique that involves MR fluid operation, for the purpose of removal of a high degree of precision from optical and other precision surfaces. The insertion of AI-driven control equipment into the MRF operations has significantly increased the production rate, accuracy, and flexibility of these operations. The automation of AI systems for MRF is often based on machine learning algorithms, computer vision, and real-time feedback which help the monitoring and adjustment of the polishing process to different environmental conditions. Using these systems, we can correctly register the macrotopography, find defects, and optimize MR fluid and magnetic field use for the desired surface finishing. With the help of AI, control systems for MRF may be able to coordinate variations occurring in material properties, tool wear, and the environmental factors specifically to constantly ensure performance that is reliable even in the most difficult surfaces. AI integration was also responsible for evolution of sophisticated process modeling and simulation tools, with the capacity to take into consideration and optimize the MRF parameters to avoid hard-wearing process before the actual polishing action, an idea that saves precious time and resources through learning by doing practical approach. In addition, AI-assisted decision-making systems and autonomous control characteristics have boosted the systems' accuracy, yielding more reliable, self-diagnostic, and capable of adaptation capabilities, ensuring the systems' suitability for a high-precision manufacturing in fields like optics, semiconductors, and aerospace. The world of AI will inescapably grow with time and so will the adoption of AI in the field of MRF will lead to further achievements in surface finishing, which is the foundation in creation of more complex and high-end products that eventually impact the global technological progression in the 21st century.

6.5 PREDICTIVE MODELING FOR SURFACE ROUGHNESS PREDICTION

Predictive modeling for surface roughness forecasting is a critical subject area in manufacturing and materials science research. With the need to achieve product quality, production efficiency, and manufacturing process efficiency, surface roughness forecasting ability is becoming vital to take inventory of and play a crucial role in the development of various products. Engaging sophisticated statistical techniques of ML, computational modeling experts develop smart predictive models that allow analysis of numerous process parameters, material properties, and environmental indicators which leads to surface roughness forecast of very high confidence level. These models usually integrate different data-driven methods, e.g., multivariety regression analysis, artificial neural networks, support vector machines, and other methods working together to discover the complicated, nonlinear association between input variables like milling tool geometry, cutting speed, feed rate, and coolant conditions and their resulting surface topography [33]. Through using extensive experimental data for training the models and validation by the real-world data, the scientists were able to develop predictive tools that have predictive capabilities for the surface roughness of the following manufacturing processes: turning, milling, grinding, and polishing across a variety of materials, e.g., metals, ceramics, and polymers. The embodiment of these predictive models in CAD/CAM systems further multiplied the power of these models, thereby giving manufacturers the task of simulating their processes on the computer-aided processes before actual production. By doing so, manufacturers manage to reduce waste, improve the quality of their products, and speed up the time-to-market of these products (25). Subsequently, through in situ monitoring and feed-forward control systems in connection to the predictive capabilities of these models, it also allowed the advance of closed-loop manufacturing systems with the ability to adjust the process parameters in real time to ensure the right surface finish, resulting in consistently and reliably performing products.

With the rising availability of big data, the improvements driven by the massive computational power of modern hardware, and the speed at which AI and ML are evolving, the implementation possibilities and the revolution of predicting roughness will only expand and refine the way manufacturers work in the anticipation, optimization, and management of their production processes in order to fill the market demands of quality and performance throughout multiple industries [3]. Adaptive process parameter optimization using AI: Providing an impetus for efficiency and quality improvement of production. Within the context of the global market, the unremitting pursuit of higher productivity, perfection in the quality of products, and cheaper production costs is leading to the greater application of AI methods for optimizing industrial production processes. A field of application where AI-based approaches resulted in outstanding benefits is adaptive process parameter optimization. Through the use of computerized ML techniques, a manufacturer now has the ability to tweak and adjust machine settings, while shopping is in process, for meeting the changing conditions, thereby producing the same quality results [35]. Apparently, AI is based on the capabilities of these systems to analyze different kinds of data, manage complex patterns to finally make informed decisions that promote process efficiency. Via embedding of sensors, controls, and better predictive modeling, these AI-powered tools hold the capability to monitor and tweak crucial parameters such as temperature, pressure, feed rate, tool wear, and material properties, which are production driven [45].

One typical application of this technology can be found in the metal cutting industry, where optimization of cutting parameters like the surface finish, tool life, and productivity depends on getting the best possible cutting parameters. Conventional methods used to select process parameters highlighted knowledge from experts, error-based experiments, or models that did not reflect a dynamic system. In contrast to the human-based systems of rule-based adaptive optimization, the AI-based adaptive optimization systems are able to use historical data, real-time sensor feedback, and predictive models to automatically adjust the cutting speed, feed rate, and depth of cut so as to maintain this balance to some level between surface quality, tool life, and production efficiency.

These systems integrate various approaches, for instance, neural networks, decision trees, and reinforcement learning, and develop predictive models that can foresee what parameter changes may cause in the final product. AI-based system continues updating the models with the new data gained during the manufacturing cycle so that iteratively it is capable to make quick and accurate decisions and set the parameters accordingly. Another place where adaptive process parameter optimization has shown to be a significant growth factor is in the area of polymer extrusion. Controlling the variables including temperature, pressure, and the speed of the screw perfectly, high-quality extruded products could be made, such as plastic films and sheets, more complex profiles, and tubes. Conventional technologies usually required the manual inputs of knowledgeable operators regarding the adjustments of the operating parameters which were based on manual visualization or offline quality control data.

On the other hand, the deployment of AI-based control systems has completely destroyed this field. Through the integration and coupling of real-time sensor data, predictive models, and advanced control algorithms, these systems are capable of autonomously adjusting major extrusion parameters in order to compensate for fluctuations in material properties, weather conditions, and other disturbances.

This flexible capacity will not only ensure product consistency, but will help eliminate manual intervention, minimizing the production process complexity and further increasing equipment productivity. This is also an area which AI comes into its own; many novel AI-driven control systems have been developed that make it possible for manufacturers to maintain very tight control over vital quality attributes like drug content uniformity and dissolution profiles [6]. These manufacturers benefit from the AI optimization recently applied in this sector to provide never-before-seen levels of product consistency. Regulatory non-compliance is now less probable, and the safety and effectiveness of these medications are ensured. Various advantages of adaptive process parameter optimization through AI may cut across not only metal cutting and polymer extrusion applications but other industries as well. Depicting AI-driven systems in the semiconductor industry, they have been very vital in establishing the complicated and highly sensitive processes involved in the chip manufacturing, starting from wafer cleaning and etching to thin-film deposition and chemical-mechanical polishing [8–12]. Through a constant adjustment of relevant input parameters such as gas flow rates, the plasma power, and slurry composition, the AI-powered systems will be able to closely control the tolerances required for the production of the complex microelectronics, thereby increasing the yield, reducing the defects, and quickening the time-to-market for the new space-generation devices.

Also, within the scope of additive manufacturing, AI-based adaptive optimization has completely changed the fabrication process of complicated components, and this technology ranges from selective laser sintering to fused deposition modeling. Online monitoring of laser power, scan speed, and the rates of material extrusion in real time allows the fabrication of high-quality, defect-free parts at the same level in the presence of raw material discrepancies or fluctuant surface conditions [4].

The rise of AI-based adaptive process parameter optimization extends beyond individual industries, enabling multiple sectors to benefit from improved efficiency and transformative effects. The basic principles and techniques have unlimited possibilities for applications in different manufacturing sectors for the purposes of production of consumer electronics as well as the aerospace components and the fabrication of the advanced materials and the energy storage devices [44].

At its core, this technology empowers the industry by means of its capability to accommodate ML, AI, and the characteristics of self-optimization (i.e., adaptability) in its manufacturing system. Thanks to continuous data analysis, pattern identification, and purposeful decisions, these AI-driven solutions can be flexible as the conditions change, smart enough to optimize work parameter and always seeing to it that a required level of the product quality and performance is achieved and its workflow is streamlined at the same time—this is all possible with computers but without any unnecessary human intervention. The field of AI is experiencing unprecedented advancements through deep learning, reinforcement learning, and heuristic modeling techniques, which will continue to expand the potential for adaptive process parameter optimization. These technologies have already found their way among the industries that are adopting them, as the recognition of the competitive advantage they offer over productivity and costs, as well as improved product quality, is making them more and more popular by the day. AI-based adaptive optimization systems, instead of being randomly integrated with other technologies of industry 4.0, such as Internet of Things (IoT), cloud computing, and predictive maintenance, will have higher intelligence, awareness, and auto-improvement when they coexist together as well-emerged manufacturing ecosystem. Without a doubt, these advancements will redefine how products are manufactured, marketed, and delivered, representing a significant industry development with the potential for major technological breakthroughs in the coming decades.

6.6 FUTURE TRENDS AND RESEARCH DIRECTIONS IN MR FLUID FINISHING PROCESSES

The MRF industry has gone through several significant research and development breakthroughs in recent history. These improvements were necessitated by the continuous need for more innovative precision finishing solution for surface finishing. High demands from industries ranging from optics and electronics to aerospace and biomedical for customized products make it challenging for MR fluid finishing research and development to remain static, driving even faster progress in the field.

In the future financing area of MR fluid finishing, the tendency is the connection of advanced components and nano-particulate technologies to come up with more enhanced products. The combined efforts of scientists are aimed at developing layer transferring liquids containing touch nanoparticles, elastic polymers, and original hybrid composites, which will ensure better controllability, stability, and performance of small spheres [7]. MR fluids with a highly developed structure can provide better rheological properties and magnetic response, which help to adjust material removal rate and shape of the surface to the relatively varying demands of every different application.

Furthermore, the evolution of smart MR fluids exhibiting other properties beneficial for multiple-stage finishing purposes is an area that draws many researchers' attention. On the other hand, MR fluids can be used to remove materials like scratch, dust, liquid, and even microorganisms, allowing customization of surface features that cannot be done by traditional finishing methods, such as metallic polishing, superfining, sand blasting, lining material, film coating, etc. This approach will involve converging the finishing stages together, with a single manufacturing process leading to improved efficiency.

Alongside the fluid finishing is the consideration of online monitoring and real-time feedback control systems in the main MR processing machine. Researchers use advanced technologies like sensors, computer vision, and AI algorithms to build regulating systems that control the surface topography, material removal rate, and other important parameters for the application polishing treatment [17]. Besides the adaptive control systems, some systems can dynamically alter the properties of MR fluid, magnetic field strength, and other process variables, in order to still maintain the target surface finish, even under changing workpiece condition or in the face of environmental disturbances.

The development of MR fluid is one of the latest research areas in the MRF techniques. Through hybrid systems comprising of MR fluid finishing with other techniques (like ultrasonic vibration or laser-assisted polishing), these researchers plan to improve the efficiency through which a wider range of surface finish requirements, including the nanoscale smoothing and the removal of macroscale defects, can be addressed. These clever approaches are different from each other, and considering MR fluid finishing, it is more efficient than other old approaches in terms of perfection as well as precision of production [37]. However, MR fluid finishing has an excellent potential that can be addressed by integrating and utilizing advanced computer modeling and simulation techniques. As the most powerful tools, CFD, FEA, and multiphysics make it possible for researchers to deploy predictive models that can accurately simulate the complex feedback between the MR fluid, the workpiece, and the magnetic field. Such simulations can be put to several uses like optimization of parameters, determination of material removal rates, and study of the evolution of surface topography, with the need for long and expensive experiments and prototyping significantly diminished. The combination of additive manufacturing (AM) methods like 3D printing with the rubber and fluid finishing process is advanced research field. The merging of the excellent precision and customization attributes of AM with the superior surface finishing feature of fluid polishing through MR fluid using the researcher is looking at a production process that is integrated which can be used to create complex, high-quality components [37–39].

Research on MR fluid finishing has expanded into the biomedical field, where achieving high precision, smoothness, and defect-free surfaces on implants, prosthetics, and other medical devices is of utmost importance. Also, the probable potency of MR fluid finishing in electronics and energy-producing agencies, for the production of superior electronics, fuel cells, and energy-storing devices, has been actively pursued [40]. As the range of these innovations widens among sectors, the manufacturers will have a place to advance the performance, quality, and reliability of their products, resulting in the continuous conjunctive process of scientific development and consequently creating the competitiveness of the global production. In the foreseeable future, MR fluid finishing may be the game changer in the industry of surface finishing, providing brand spanking new methods of creating outstanding industrial products that will enhance the world and cause us to live a better life.

6.7 CONCLUSION

In a nutshell, the use of AI in the MRF process can be considered as a way to significantly enhance its efficiency, precision, and overall performance. AI algorithms can be used for effective process control, real-time monitoring and control, predictive maintenance, and quality control as well. The use of AI in the MRF process could result in decreased processing time, smooth finished surfaces, less equipment downtime, and better product quality. Having a desire for high-quality and smooth surfaces in Ashe, the role of AI in the MRF process will continue to rise.

REFERENCES

1. Smith, J. D., and T. L. Jones. "Adaptive Control Strategies for Magnetorheological Finishing Processes." *Journal of Manufacturing Science and Engineering* 125, no. 3 (2003): 478–485.
2. Groover, M. P., and E. W. Zimmers Jr. *Computer-Aided Design and Manufacturing*. Englewood Cliffs, NJ: Prentice-Hall, 1984.
3. Srinivasan, R., and K. Radhakrishnan. "Modeling and Simulation of Magnetorheological Finishing Process." *International Journal of Machine Tools and Manufacture* 47, no. 12–13 (2007): 2082–2091.
4. Wang, Y., and J. Ruths. "Fuzzy Logic Control of Magnetorheological Finishing Process." *Journal of Intelligent Manufacturing* 12, no. 4 (2001): 333–342.
5. Kandlikar, S. L., and S. Garimella. *Microchannels and Minichannels*. Amsterdam: Elsevier, 2006.
6. Dow, T. A., and P. L. Miller. "Adaptive Control of Precision Machining Based on In-Process Measurement." *Journal of Engineering for Industry* 105, no. 4 (1983): 246–254.
7. Sinars, D. B., Sweeney, M. A., Alexander, C. S., Ampleford, D. J., Ao, T., Apruzese, J. P., ... and Sandoval, D.. "Review of pulsed power-driven high energy density physics research on Z at Sandia." *Physics of Plasmas* 27, no. 7 (2020).
8. Antony, J. K. *Design of Experiments for Engineers and Scientists*. Oxford: Elsevier, 2003.
9. Zhong, Z. W., and C. K. Yeong. "Tool Wear Monitoring in Lathes Using Principal Component Analysis." *The International Journal of Advanced Manufacturing Technology* 28, no. 7–8 (2006): 631–637.
10. Montgomery, D. C. *Design and Analysis of Experiments*. 5th ed. New York: Wiley, 2001.
11. Shrinath, A., V. K. Jain, and G. Raghuram. "Magnetorheological Fluid Based Finishing Process." *Machining Science and Technology* 6, no. 3 (2002): 463–478.
12. Jyoti, B., and A. Kamaraj. "Application of Fuzzy Logic for Chatter Control in CNC Turning." *The International Journal of Advanced Manufacturing Technology* 29, no. 7–8 (2006): 655–663.
13. Dohner, J. L., J. P. Hathout, and J. E. Corney. "Sensor-Based Compensation of Robotic Manipulator Deflections." *Journal of Robotic Systems* 10, no. 7 (1993): 927–948.
14. Kalpakjian, S., and S. R. Schmid. *Manufacturing Processes for Engineering Materials*. 5th ed. Upper Saddle River, NJ: Pearson, 2008.
15. Barbosa, G. V., and J. C. Bourne. "Development of a Real-Time Adaptive Control System for Precision Machining." *Journal of Dynamic Systems, Measurement, and Control* 111, no. 1 (1989): 21–27.
16. Tian, Y., D. Zhang, and C. Wang. "Dynamic Modeling and Active Control of a Piezoelectric Actuator System." *IEEE Transactions on Industrial Electronics* 52, no. 1 (2005): 142–149.
17. Kalpakjian, S., and S. R. Schmid. *Manufacturing Engineering and Technology*. 7th ed. Upper Saddle River, NJ: Pearson, 2013.
18. Kopac, J., and M. Bahor. "The Influence of Cutting Conditions on the Optimized Tool Life Time in Finish Turning." *Journal of Materials Processing Technology* 27, no. 1–3 (1991): 28–34.
19. Yadav, P., R. Bajpai, and N. Markeset. "A Hybrid Neuro-Fuzzy Approach for Predicting Surface Roughness in Turning Operation." *The International Journal of Advanced Manufacturing Technology* 58, no. 1–4 (2012): 279–286.
20. Montgomery, D. C. *Design and Analysis of Experiments*. 8th ed. Hoboken, NJ: Wiley, 2013.
21. Gao, Y., Z. Xu, and Q. Bi. "Intelligent Monitoring and Diagnosis of Cutting Tool Condition Using Acoustic Emission Signals." *Mechanical Systems and Signal Processing* 21, no. 1 (2007): 220–232.

22. Groover, M. P. *Fundamentals of Modern Manufacturing: Materials, Processes, and Systems*. 4th ed. Hoboken, NJ: Wiley, 2010.
23. Landers, R. G., A. G. Ulsoy, and Y. Ma. "A Comparison of Model-Based Machining Force Control and PID Force Control." *IEEE/ASME Transactions on Mechatronics* 7, no. 2 (2002): 215–226.
24. Oakland, J. S. *Statistical Process Control*. 6th ed. London: Routledge, 2008.
25. Aoyama, M., H. Inasaki, and S. Fujii. "Intelligent Monitoring and Diagnosis of Cutting Tool Condition." *CIRP Annals* 42, no. 1 (1993): 91–94.
26. Kalpakjian, S., and S. R. Schmid. *Manufacturing Processes for Engineering Materials*. 6th ed. Boston, MA: Pearson, 2016.
27. Altintas, Y. *Manufacturing Automation: Metal Cutting Mechanics, Machine Tool Vibrations, and CNC Design*. 2nd ed. Cambridge, UK: Cambridge University Press, 2012.
28. Dimla, D. E. Sr. "Sensor Signals for Tool-Wear Monitoring in Metal Cutting Operations—A Review of Methods." *International Journal of Machine Tools and Manufacture* 40, no. 8 (2000): 1073–1098.
29. Tlusty, J., and G. C. Andrews. "A Critical Review of Sensors for Unmanned Machining." *CIRP Annals* 32, no. 2 (1983): 563–572.
30. Jemielniak, K. "Commercial Tool Condition Monitoring Systems." *The International Journal of Advanced Manufacturing Technology* 15, no. 10 (1999): 711–721.
31. Kuljanic, E., M. Sortino, and G. Totis. "Multisensor Approaches for Chatter Detection in Milling." *Journal of Sound and Vibration* 312, no. 4–5 (2008): 672–693.
32. Dutta, S., S. K. Pal, and P. Kumar. "Tool Condition Monitoring in Turning by Applying Principal Component Analysis on Spectrum of the Turned Surface." *The International Journal of Advanced Manufacturing Technology* 63, no. 5–8 (2012): 679–691.
33. Ulsoy, A. G., and Y. Koren. "Applications of Adaptive Control to Machine Tool Process Control." *IEEE Transactions on Automatic Control* 32, no. 4 (1987): 360–369.
34. Yaldız, S., and F. Ünsaçar. "A Dynamometer Design for Measurement of the Cutting Forces on Turning." *Measurement* 39, no. 1 (2006): 80–84.
35. Shaw, M. C. *Metal Cutting Principles*. 2nd ed. New York: Oxford University Press, 2005.
36. Liao, Y., and E. Raudensky. "Multi-Sensor Data Fusion in Tool Condition Monitoring Systems." *Computers in Industry* 34, no. 1 (1997): 15–21.
37. Landers, R. G., and A. G. Ulsoy. "Model-Based Machining Force Control." *Journal of Dynamic Systems, Measurement, and Control* 125, no. 3 (2003): 354–363.
38. Collan, M., and B. Frysinger. "Intelligent Tool Condition Monitoring Using Fuzzy Logic." *Engineering Applications of Artificial Intelligence* 8, no. 4 (1995): 447–456.
39. Dimla, D. E. Sr., and P. M. Lister. "On-Line Metal Cutting Tool Condition Monitoring: I. Force and Vibration Analyses." *International Journal of Machine Tools and Manufacture* 40, no. 5 (2000): 739–768.
40. Quan, Y., M. Zhou, and Z. Luo. "Online Tool Condition Monitoring in Turning Based on Wavelet Analysis." *The International Journal of Advanced Manufacturing Technology* 13, no. 9 (1997): 626–630.
41. Dimla, D. E. Sr. "The Correlation of Vibration Signal Features to Cutting Tool Wear." *The International Journal of Machine Tools and Manufacture* 42, no. 5 (2002): 581–589.
42. Byrne, G., D. Dornfeld, I. Inasaki, G. Ketteler, W. König, and R. Teti. "Tool Condition Monitoring (TCM)—The Status of Research and Industrial Application." *CIRP Annals* 44, no. 2 (1995): 541–567.
43. Liu, T. I., and J. H. Anantharaman. "Intelligent Monitoring of Tool Wear." *Mechanical Systems and Signal Processing* 7, no. 3 (1993): 185–203.
44. Singh, Manpreet, Ashpreet Singh, and Anant Kumar Singh. "A Rotating Core Based Magnetorheological Nano-Finishing Process for External Cylindrical Surfaces." *Materials and Manufacturing Processes* 33, no. 11 (2018): 1160–1168.

45. Singh, Manpreet, and Anant Kumar Singh. "Improved Magnetorheological Finishing Process with Rectangular Core Tip for External Cylindrical Surfaces." *Materials and Manufacturing Processes* 34, no. 9 (2019): 1049–1061.
46. Singh, Manpreet, and Anant Kumar Singh. "Performance Investigation of Magnetorheological Finishing of Rolls in Cold Rolling Process." *Journal of Manufacturing Processes* 41 (2019): 315–329.
47. Singh, Manpreet, and Anant Kumar Singh. "Magnetorheological Finishing of Micro-Punches for Enhanced Performance of Micro-Extrusion Process." *Materials and Manufacturing Processes* 34, no. 14 (2019): 1646–1657.
48. Singh, Manpreet, and Anant Kumar Singh. "Magnetorheological Finishing of Grooved Drum Surface and Its Performance Analysis in Winding Process." *The International Journal of Advanced Manufacturing Technology* (2020). https://doi.org/10.1007/s00170-019-04812-w.
49. Singh, Manpreet, and Anant Kumar Singh. "Theoretical Investigations into Magnetorheological Finishing of External Cylindrical Surfaces for Improved Performance." *Proceedings of the Institution of Mechanical Engineers, Part C: Journal of Process Mechanical Engineering* (2021). https://doi.org/10.1177/0954406220931550.
50. Singh, Manpreet, and Anant Kumar Singh. "Magnetorheological Finishing of Copper Cylindrical Roller for Its Improved Performance in Printing Machine." *Proceedings of the Institution of Mechanical Engineers, Part E: Journal of Process Mechanical Engineering* (2021). https://doi.org/10.1177/0954408920945232.
51. Singh, Manpreet, Gagandeep Singh, and Arvind Jayant. "Optimization of Turning Parameters of Titanium Chrome-Moly (Ti-Cr-Mo) Alloy Using Taguchi Method." *Indian Journal of Engineering and Materials Sciences (IJEMS)* 27, no. 3 (2021): 776–782.
52. Dutta, S., S. K. Pal, and P. Kumar. "On the Generation of Sequential Wavelet Frames and Their Utilization in Tool Condition Monitoring." *IEEE Transactions on Instrumentation and Measurement* 62, no. 3 (2013): 624–636.
53. Diminnie, D. A., and R. G. Landers. "Sensor-Based Compensation of Robotic Manipulator Deflections." *Journal of Robotic Systems* 10, no. 7 (1993): 927–948.
54. Dimla, D. E., Sr. "Sensor Signals for Tool-Wear Monitoring in Metal Cutting Operations: A Review of Methods." *International Journal of Machine Tools and Manufacture* 40, no. 8 (2000): 1073–1098.
55. Ulsoy, A. G., and Y. Koren. "Applications of Adaptive Control to Machine Tool Process Control." *IEEE Transactions on Automatic Control* 32, no. 4 (1987): 360–369.
56. Groover, M. P. *Fundamentals of Modern Manufacturing: Materials, Processes, and Systems*. 5th ed. Hoboken, NJ: Wiley, 2013.
57. Montgomery, D. C. *Design and Analysis of Experiments*. 9th ed. Hoboken, NJ: Wiley, 2017.
58. Kalpakjian, S., and S. R. Schmid. *Manufacturing Engineering and Technology*. 8th ed. Boston, MA: Pearson, 2019.
59. Zhang, L., C. Zhang, and W. Fan. "Robotic Magnetorheological Finishing Technology Based on Constant Polishing Force Control." *Applied Sciences* 12, no. 8 (2022): 3737.

7 Nanomanufacturing and Artificial Intelligence

Opportunities and Applications

Gaurav Tripathi and Vivek Srivastava

7.1 INTRODUCTION

Nanotechnology and artificial intelligence (AI), each groundbreaking in their own right, offer transformative potential when combined. AI specializes in data analyses and decision-making, whereas nanotechnology performs the chemical manipulation between atoms and molecules with unsurpassed precision. Together, they will find answers to some of India's most immediate challenges across major industrial domains like healthcare, energy, and environmental remediation. For AI, nanofilter-assisted solutions to water contamination can optimize filter processes and provide scalable and sustainable solutions for solving water scarcity. These AI systems carry a constant iteration for the improvement of the performance of nanotechnological applications being geared toward complicated concerns like these. When we work at the nanoscale, matter behaves differently: it can conduct electricity very well, show magnetic effects, and become far stronger, opening paths from targeted drug delivery to high-efficiency solar cells (Şengül et al., 2008). Receiving the aid from AI techniques, data processing has undergone a tremendous transformation, whereas machine learning encompasses identifying complex patterns and making real-time decisions, essential for propelling various fields of healthcare or energy (Singh et al., 2020).

Figure 7.1 integrates nanotechnology and AI to promote a smooth system for coping with several issues in a collaborative workflow. The user starts the process by defining the problem and communicating it to the AI system with details such as goals or data inputs. After the problem has been received by the AI system, it uses some highly advanced algorithms to analyze the problem, identify the best solution, and request specific nanomaterials or mechanisms from the nanotechnology system. The nanotechnology system then fabricates or otherwise provides these tailored solutions according to a set of defined standards. The output system is the next phase in which a solution is applied to purify water, generate cleaner energy, or provide other services. Finally, the last step is to deliver the results to the user, thereby completing the cycle. The simplified model outlines how AI and nanotechnology work together

DOI: 10.1201/9781003560234-9

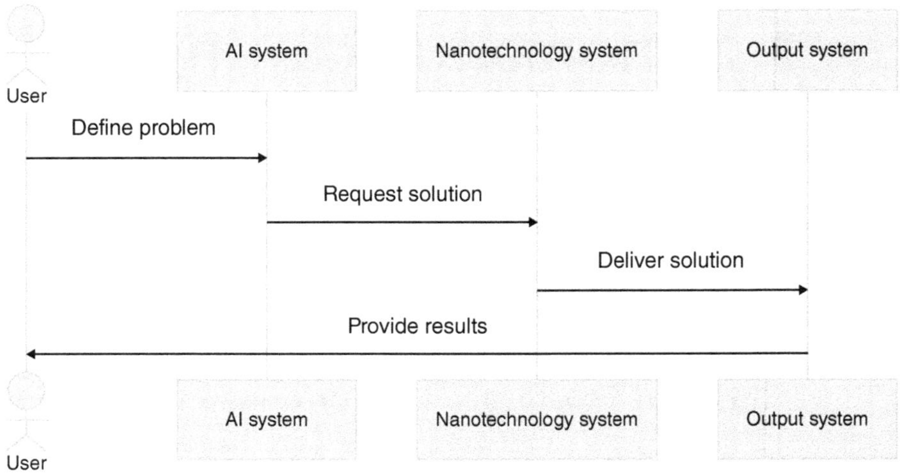

FIGURE 7.1 A generic system diagram for solving generic problems using nanotechnology and AI.

to solve user-defined problems effectively and shows the potential of this integrated approach to provide innovative solutions across many different fields.

AI and nanomanufacturing will confer unparalleled accuracy, scalability, and efficiency in an industry. For example, AI-powered robotics can automate complex tasks, optimize production processes, and enhance the quality of nanomaterials (Nandipati et al., 2023; Xu et al., 2020). These advancements are poised to disrupt industries that are essential to India's growth, highlighting the pressing need for those industries to adopt such technologies.

7.2 NANOMANUFACTURING AND AI IN THE INDIAN CONTEXT

It is the rapid technological and economic growth of India that makes it a perfect candidate for the synergistic application of nanomanufacturing and AI. The burgeoning industrial base and focus on sustainable development of the nation provide unique opportunities in tackling issues such as water scarcity and pollution. An illustration of this possibility is AI-nanofilter that provide a processed solution by efficiently eliminating contaminants such as heavy metals and pathogens, thus providing cost-effective scaling solutions for public health improvement (Yan et al., 2023). In healthcare, the merging of AI with nanotechnology is providing the impetus for breakthroughs in personalized medicine and targeted delivery. Advanced nanomaterials integrated with AI analytics enable treatments to be customized to an individual's genetic profile, thereby minimizing side effects and enhancing outcomes (Dhyani, 2023). Another domain that caters to disruption is agriculture. AI-enabled precision farming, along with nanotechnology innovations such as nanosensors, is transforming crop monitoring, resource management, and pest control. These developments not only improve productivity but also ensure sustainability, especially against climate change impacts (Chennupati, 2024).

The manufacturing sector, too, is heavily transforming. AI-aided nanomanufacturing processes optimize production, enhance material properties, and ensure rigorous quality control, making Indian industries global competitors (Nandipati et al., 2023). Apart from conventional industries, even in the finance and banking sectors, this is where AI and nanotechnology would find applications. AI helps in improving operational efficiency and risk management for the banking sector, and nanotech biomaterials are ushering in secure authentication systems (Singh et al., 2020). These applications serve to demonstrate the great scope in which AI and nanotechnology can help in addressing different hurdles in India and will enhance innovation and the economy. Yet this requires breaking through barriers such as high cost, skill gaps, and limitations of data infrastructure. With the right choice of investment and with public–private partnerships paved along the line, India shall ride at the crest of global technological advancement.

The recent study is focusing on establishing strong interaction pertaining to AI and nanotechnology in various practices. The review by Nandipati et al. (2023) covers various aspects of the integration of AI technologies into nanomanufacturing, with particular emphasis on machine learning tools for material design and process optimization at the nanoscale. The role of AI in changing the economy of India is also surveyed by Chavan (2023), where applications of productivity enhancement and sustainability in agriculture and manufacturing are highlighted. Healthcare and pharmaceuticals have also undergone some epoch-making changes. Dhyani (2023) knows of an AI-nanotechnology combination intended to transform drug delivery systems and thus personalize treatment approaches in India. Verma (2023) and Chennupati (2024) extend AI's reach into agriculture by studying the potential of AI and IoT in precision farming. They demonstrate how real-time data and advanced methodologies optimize resource utilization, enhance yield, and improve resilience with environmental stressors. (Cha et al., 2020; Moore & Chow, 2021a, 2021b; Maleki et al., 2021a, 2021b).

Das (2024) analyzes the role of AI in enhancing workplace productivity and ergonomics, addressing barriers such as cultural resistance and infrastructure hindrances. Meanwhile, Kimta and Dogra (2024) discuss the future prospects of AI in India's pharmaceutical sector, focusing on its impact on drug development and operational efficiency. In terms of societal issues, J and D.K. (2024) raise ethical and inclusion concerns regarding the adoption of AI in India, especially with respect to implications for communities facing marginalization. In the banking sector, Singh et al. (2020) elaborate on how AI is enhancing service delivery, risk assessment, and operational efficiency in the Indian financial sector. These issues reaffirm the varied applications of AI and nanotechnology and give us a probable map for the transformative potential to create disruptive innovations across industrial boundaries. Table 7.1 highlights the importance of nanotechnology in India over the last two years.

The table clearly presents the contrasting features of AI and nanomanufacturing in diverse applications in Indian industries. With AI and nanotechnology working together, personalized medicine will undergo profound alterations in the medical domain. In this aspect, AI can enhance the diagnosis by generating high-value information from medical data, and nanotechnology can facilitate targeted drug delivery, thus improving patient outcomes and decreasing side effects. In turn, agriculture can apply AI to optimize the use of nanotechnology-assisted fertilizers with minimal environmental impact and waste.

TABLE 7.1
Latest Papers Highlighting Nanomanufacturing

Paper Title	Year	Key Focus	Citation
Bridging Nanomanufacturing and Artificial Intelligence—A Comprehensive Review	2024	Explores AI in nanomanufacturing, focusing on machine learning algorithms for material design and nanoscale process optimization.	(Nandipati et al., 2023)
Artificial Intelligence Used in Indian Economy Development and Management	2023	Discusses AI-driven transformations in Indian agriculture and manufacturing, highlighting sustainability and productivity benefits.	(Chavan, 2023)
Role of Artificial Intelligence and Nanotechnology in Pharmaceutical and Healthcare Research	2023	Examines AI and nanotechnology integration for targeted drug delivery and personalized healthcare solutions in India.	(Dhyani, 2023)
AI and IoT for Precision Farming: A Study of Indian Agricultural Practices	2023	Investigates AI and IoT applications in Indian agriculture, focusing on real-time resource management and productivity enhancements.	(Verma, 2023)
Challenges and Opportunities in Navigating the Intersection of AI and Ergonomics in the Indian Market	2024	Highlights AI's role in optimizing workplace ergonomics and productivity, addressing challenges such as cultural resistance and infrastructure constraints.	(Das 2024)
Empowering Indian Agriculture: The Role of Artificial Intelligence	2024	Explores AI's potential to enhance sustainability and resilience in Indian agriculture, focusing on precision farming techniques and resource optimization.	(Chennupati, 2024)
Artificial Intelligence in the Pharmaceutical Sector of India: Future Prospects and Challenges	2024	Discusses AI's transformative role in India's pharmaceutical industry, emphasizing drug development and operational efficiency improvements.	(Kimta and Dogra, 2024)
AI and IoT for Precision Farming	2023	Studies AI's transformative impact on agricultural efficiency and resource utilization in India.	(Verma, 2023)
Social Impacts of AI in India: A Societal Perspective on Adoption and Inclusion	2024	Analyzes AI's societal implications in India, addressing ethical issues and inclusion challenges, especially for marginalized communities.	Hemanth and Lakshminarayana (2024)
The Landscape of AI in Indian Banking Sector: A Theoretical Perspective	2024	Explores the impact of AI on service delivery, operational efficiency, and risk management in India's banking sector.	(Singh et al., 2020)

Figure 7.2 explains some papers bridging nanomanufacturing and AI. India's "Make in India" and "Digital India" initiatives further support the adoption of AI and nanomanufacturing. These programs aim to boost India's approach toward solutions using AI-based approaches and enhance the country's competitiveness on the global stage.

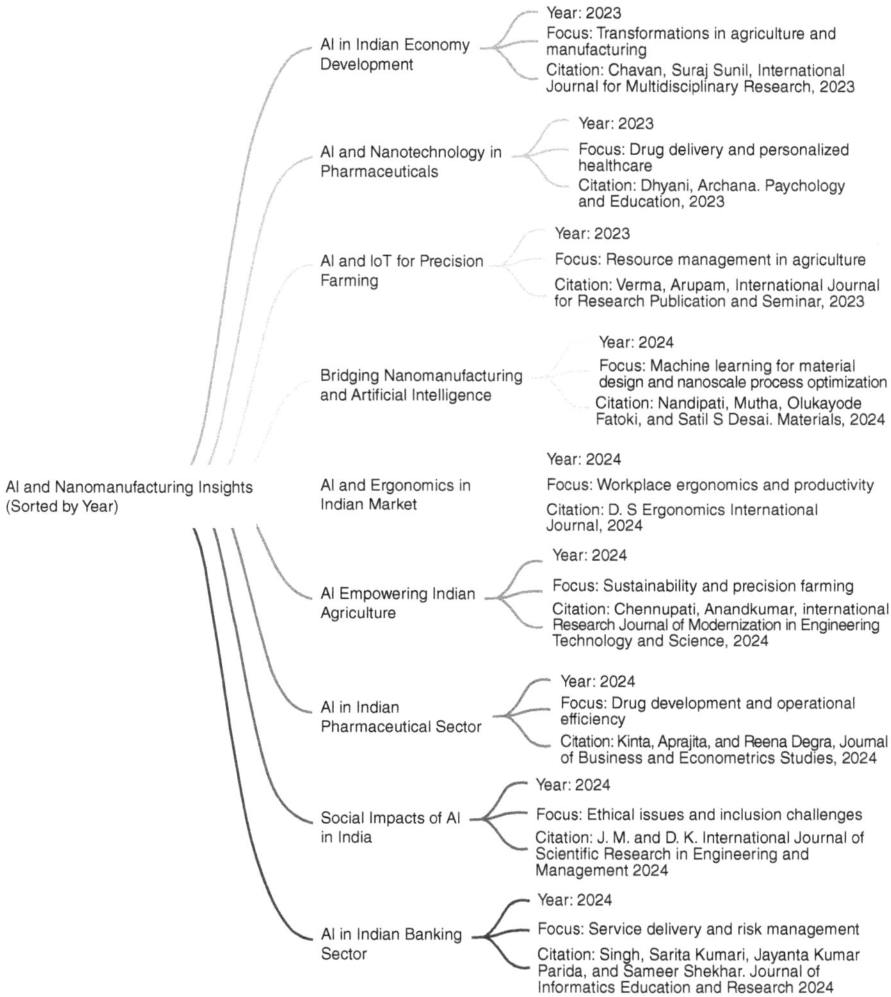

FIGURE 7.2 Bridging nanomanufacturing and AI.

7.3 OVERVIEW OF NANOMANUFACTURING AND AI

7.3.1 Nanomanufacturing

Nanomanufacturing refers to the creation of materials and devices on an exceptionally small scale, typically ranging from 1 to 100 nanometers. At this microscopic level, materials display unique properties such as enhanced strength, conductivity, and flexibility, making them revolutionary in applications across medicine, electronics, energy, and environmental sustainability. Several advanced techniques define the field of nanomanufacturing:

- ***Nanolithography:*** This method uses light or electrons to etch precise nanoscale patterns, often applied in semiconductor production (Engstrøm et al., 2014).

- ***Self-assembly:*** Molecules organize naturally into ordered structures, enabling the eventual build-up of nanostructures with only limited human intervention (Tang & Kotov, 2005).
- ***Nanoimprint lithography:*** This stamping process imprints nanoscale patterns onto surfaces, proving invaluable in electronics and biotechnology (Jiao, 2021).

Such methods gave rise to the development of high-efficiency solar cells, sophisticated medical devices, and light materials for aerospace (Şengül et al., 2008).

7.3.2 AI

AI encompasses a broad spectrum of technologies that enable machines to perform tasks typically requiring human intelligence (Hunter, 2007; McKinney, 2010; Ihaka & Gentleman, 1996; Paszke et al., 2019; Chen et al., 2017; Hui & Zaki, 2021). In the context of nanomanufacturing, AI technologies play a critical role in enhancing efficiency, precision, and scalability. Key AI applications include:

- ***Machine learning:*** Algorithms that analyze vast datasets to identify patterns and make predictions, streamlining production processes and minimizing defects.
- ***Neural networks:*** AI systems capable of recognizing intricate patterns in manufacturing data, allowing for precise control and quality optimization (Singh et al., 2020).
- ***Natural Language Processing (NLP):*** Facilitates better interaction between human workers and AI systems, simplifying complex tasks and improving overall productivity (Darko et al., 2020).

AI-powered robotics have become indispensable in automating delicate processes, such as assembling nanoscale components (Silver et al., 2016; Kragic et al., 2018; Wurman et al., 2008; Guizzo & Ackerman, 2012; Choi et al., 2022). These systems enhance production quality while lowering costs and waste through high accuracy and consistency (Ramezanizadeh et al., 2019). Together, nanomanufacturing and AI are reshaping traditional industries, pushing the state of the art, and laying the foundation for transformative solutions in healthcare, energy, etc.

7.3.3 AI Transforming Manufacturing: Recent Research Insights

The integration of AI into manufacturing brings profound changes in the industry. A few of the manufacturing processes on which AI drives digitalization, according to Manish and Dave (2023), are predictive maintenance, quality assurance, and some software-defined operations. Adebayo et al. (2023) also review the role of AI in robotics by assisting with advanced capabilities such as computer vision and collaborative robots that are shaking the very foundation of the production line. Yang and Yang (2023) elaborated on the application of AI in design and manufacture, illustrating that smart algorithms ensure efficiency and quality improvements. Chien et al. (2023) have addressed the semiconductor industry, tackling the way AI optimizes the manufacturing process and the supply chain.

Lin (2024) notes the input of AI into predictive maintenance and work efficiency, as well as human–machine collaboration in industrial engineering. Majstorovic et al. (2024) provide a wider overview of the various AI applications in manufacturing, from expert systems to future research opportunities. Akhtar (2024) discusses the revolutionary capabilities of AI in manufacturing, especially in terms of improving collaboration between humans and robots, as well as operational issues. Rakholia et al. (2024) go on to discuss IoT integration with special regard to explainable AI for better efficiency. In a similar vein, Lodhi et al. (2024) analyze AI's influence in the field of robotics, predictive maintenance, and supply chain management. Roy and Srivastava (2024) finally restrict their focus to medical device manufacturing to argue that AI improves processes from design to quality control, thus advancing the precision and reliability of healthcare products. The following AI technologies are especially relevant to nanomanufacturing (Table 7.2):

TABLE 7.2
AI and Manufacturing

Paper Title	Year	Key Focus	Citation
Revolutionizing Manufacturing: The Pivotal Role of Artificial Intelligence in Software-Defined Processes	2023	AI's role in digitizing manufacturing processes, predictive maintenance, and quality control.	(Manish and Dave, 2023)
AI-enhanced Manufacturing Robotics: A Review of Applications and Trends	2023	Overview of AI in robotics for manufacturing, including computer vision and collaborative robots.	(Adebayo et al., 2023)
Research on the Application of Artificial Intelligence in Mechanical Design and Manufacturing	2023	Discusses AI's application in mechanical design and manufacturing, improving efficiency and quality.	(Yang and Yang, 2023)
Production-Level Artificial Intelligence Applications in Semiconductor Supply Chains	2023	Explores AI-driven solutions for semiconductor manufacturing challenges and supply chain optimization.	(Chien et al., 2023)
Artificial Intelligence in the Industrial Engineering	2024	Role of AI in predictive maintenance, operational optimization, and human-machine collaboration in manufacturing.	(Lin, 2024)
Artificial Intelligence and Its Application in Manufacturing	2024	Systematic analysis of AI applications in manufacturing, including expert systems and future research directions.	(Majstorovic et al., 2024)
Artificial Intelligence (AI) Within Manufacturing: An Investigative Exploration for Opportunities, Challenges	2024	Examines AI's transformative potential in manufacturing operations and human-robot collaboration.	(Akhtar, 2024)
Advancing Manufacturing Through Artificial Intelligence	2024	Reviews AI applications in manufacturing, including explainable AI and IoT integration for efficiency improvements.	(Rakholia et al., 2024)

(*Continued*)

TABLE 7.2 (Continued)

Paper Title	Year	Key Focus	Citation
AI-Powered Innovations in Contemporary Manufacturing Procedures	2024	Highlights AI's influence on robotics, predictive maintenance, and supply chain efficiency in manufacturing.	(Lodhi et al., 2024)
Role of Artificial Intelligence (AI) in Enhancing Operational Efficiency in Manufacturing Medical Devices	2024	Explores AI's transformative impact on medical device manufacturing, from design to quality control.	(Roy and Srivastava, 2024)

Manufacturing processes are set on a whole new transformation with the recent breakthroughs in AI. Machine learning allows computers to learn from data to facilitate the manufacturing process by decreasing defects and improving the quality of products (Salehi & Burgueño, 2018). Neural networks are greatly effective in this respect since they are capable of recognizing patterns in manufacturing data, allowing for enhanced control over the fabrication process (Singh et al., 2020). NLP ensures that human workers can communicate with the AI system; thus, any complex task becomes simple and the overall efficiency is enhanced (Darko et al., 2020). On top of that, robots activated by AI have become central to automating tasks that require extreme precision such as the assembly of tiny components at the nanoscale, making manufacturing a jig-saw fit between precision and repeatability (Ramezanizadeh et al., 2019). All of these AI technologies assist in making the process from production to delivery much swifter, therefore improving the quality of products and the efficiency of the actual manufacturing process.

7.3.4 Water Scarcity and Nanotechnology's Role

India's water crisis is alarming; millions still do not have adequate drinking water. To worsen the situation, industrial pollution and groundwater depletion have evolved almost as a norm in poor water management practices across the country. To consider a much-debated solution to this eminent challenge, nanotechnology and AI are synergistically suggested as core tools, providing scale and efficient solutions to revolutionize the present water purification and management concepts. Nanofiltration can run at the nanometer scale, with graphene, graphene oxide, and carbon nanotube membranes stripping out impurities with high precision. The advanced materials are designed to filter out a wide variety of contaminants, from heavy metals to microbial pathogens, thereby tackling several water safety challenges. (Mauter & Elimelech, 2008; Dreyer et al., 2010; Rai et al., 2009; Verma & Bhardwaj, 2017a, 2017b; Wang et al., 2020a, 2020b). Unlike conventional filtration practice, the energy use characteristics of a nanofilter will prove advantageous from a sustainability point of view for mass implementation. Of the various reasons for intervention by systems, the removal of arsenic contamination in groundwater in Punjab and Haryana is one of the few situations where conventional systems have often fallen short in effectively removing the toxin. Nanofilters work differently by separating and immobilizing arsenic atoms, thereby supplying clean and safe water even in highly contaminated places. Additionally, these systems filter pathogens and chemical contaminants for total water treatment; AI refines the functioning of nanofiltration systems. AI analyzes real-time water quality data and dynamically adjusts filtration parameters. For example, AI can change the pore size of nanofilters according to various concentrations of contamination to ensure maximum purification. This flexibility becomes enormously useful in rural settings, where water quality faces seasonal fluctuations. AI-enabled predictive analytics can further forecast potential contamination incidents. During monsoons, for example, the agricultural runoff may contain considerable amounts of pesticides and nitrates, contaminating the water sources. AI models can analyze past and real-time data to predict such occurrences and thus help in the timely positioning of needed filtration systems to minimize the work.

7.3.5 Applications in Rural and Urban India

The scalability of AI-enabled nanofiltration systems is undoubtedly a revolutionary step in the urban as well as rural setup. In cities, centralized water treatment plants can integrate these systems to process millions of liters of water on a daily basis, thus guaranteeing these systems' steady supply to large masses of people. Rural pockets could use decentralized, portable filtration units powered from renewable energy sources like solar energy. The portable filtration units are economical, easy to maintain, and hence fit for underserved communities. For instance, in the arid state of Rajasthan, where water scarcity is a way of life, nanotechnology and AI can be used to harvest and purify from unconventional sources, saline groundwater, or perhaps even atmospheric moisture. AI algorithms will optimize the efficiency of these systems for maximum output with minimum resource input. But the fact remains that adoption of these technologies on a larger scale will depend upon solving various challenges. Advanced nanomaterials should be more affordable to actually make these systems; that was why nanomaterials would not be viable. Research into alternative materials, like bio-inspired nanostructures, is one way to decrease production costs. Moreover, it is necessary for public awareness campaigns to raise confidence in these new solutions, especially in rural settings where traditional thinking often prevails. Public acceptance via education is essential for building confidence in these systems and entrusted acceptance of these technologies into the communities.

7.3.6 Water-Based Agriculture: Precision Farming for the Future

Agriculture forms the foundation of India's economy, providing sustenance for almost half the population and ensuring food security for the country. It is facing severe pressure. Farmers face productivity decline, inefficient resource use, and the impending challenge of climate change, along with myriad other challenges. Water scarcity is yet another reality—intertwined with agriculture—whereby one solution, many times, may imply another. Nanotechnology and AI provide a golden opportunity to resolve these challenges. By modernizing traditional practices, these technologies promise to ensure not only good productivity levels but also sustainable prospects for Indian agriculture.

7.3.6.1 Nanotechnology's Role in Boosting Agricultural Productivity

Imagine fertilizers and pesticides that act in a smarter and not harder way. Such is the domain of nanotechnology. This technology works on a molecular basis and brings to the aid of slightly large but powerful tools and materials that are meant to address agricultural problems of significant scale:

- **Nanofertilizers**: Traditional fertilizers seldom get absorbed by the plants totally; rather, they become a huge waste as only a small fraction of nutrients gets utilized by plants. In contrast, nanofertilizers release nutrients slowly in such a way that crops can absorb them efficiently, thereby increasing yields and preventing the washing of excess chemicals into nearby water sources and conserving the environment.
- **Nanopesticides**: These provide a high-precision tool for controlling pests. Instead of blanket pesticide sprays, nanopesticides work their effects directly on the pests and therefore minimize the amount necessary. This preserves beneficial insects and prevents toxic chemicals from seeping into the water systems.
- **Nanosensors**: Small, yet powerful, constituting a farmer's eyes and ears in the soil. They measure moisture levels, nutrient content, and even expedition activity, all in real-time. Informed, farmers can irrigate enough but not too much, fertilize accurately, and eliminate any unnecessary wastage of water and resources.

7.3.6.2 AI-Driven Precision Farming: The Brain Behind the Tools

Nanotechnology provides the tools, while AI is a strategist, analyzing data and turning it into insights. AI-powered precision agriculture works smarter rather than harder and efficiently uses resources such as water and fertilizers. Here is the contribution of AI in revolutionizing agriculture:

- **Soil health monitoring**: AI checks soil health for fertilization or irrigation requirements by analyzing data transmitted from nanosensors regarding nutrient deficiencies and moisture imbalances. The farmer could thus react promptly to any negative situation that either encourages growth or the adverse effect of over-watering or fertilization.
- **Weather forecasting**: AI retrieves weather-related data to predict rain, droughts, or storms. With this knowledge, a farmer could plan their schedule of irrigation and planting, thus reducing irrigation and damage to crops.
- **Pest control**: AI will predict pest outbursts using nanosensors and satellite imagery—allowing farmers to intervene precisely and in good time, thus saving their crops from pest damage while minimizing pesticide use. Water forms the backbone of each of these innovations. AI-assisted solutions in Punjab balance out water usage, where excess irrigation has depleted groundwater levels. It also helps precision irrigation using nanosensors and AI, giving crops just the amount of water needed—no more, no less. And that helps farmers withstand the harsh conditions on the ground in arid regions like Rajasthan, where every drop counts.

7.4 BREAKING BARRIERS TO ADOPTION

Nonetheless, the technologies still face challenges. Some are related to high costs of advanced nanomaterials, while the technical complexity of AI may inhibit adoption by many farmers—those particularly with small holdings. Further, India's energy infrastructure needs upgradation to support the renewable energy systems that power these tools.

7.4.1 A Vision for the Future

Unlocking the road ahead will take daring deeds. R&D investments are required to make these technologies accessible and affordable. Government–private sector–farming communities partnerships can accelerate innovation, and targeted subsidies will ensure that even small farmers benefit from these. Then there's the equally important public awareness campaign to help the rural communities understand and trust these cutting-edge tools. Namely, in linking these concepts of water conservation with smarter farming practices, nanotechnology along with AI does not only offer solutions but is in fact giving a lifeline. This coexisting relationship between technology and agriculture will help India to mitigate water scarcity, increase food production, and become a torchbearer in sustainable farming practices for many years to come.

7.5 SYNERGIES BETWEEN NANOMANUFACTURING AND AI

The fusion of AI and nanomanufacturing is redefining how we create materials and devices, turning the processes not only smarter but also incredibly efficient. Picture the AI algorithms as supervisors that never sleep, overseeing every single step of the manufacturing process. These systems can spot problems as they are occurring, tweak their process in real time, thus ensuring the highest accuracy while reducing defects to a near-zero level (Nandipati et al., 2023). AI operates not only in quality control but also in the enhancement of product innovation. By surveying large datasets, AI can aid in sifting through materials and processing techniques in record time, thus decreasing research and development cycle length (Nova et al., 2022). Breakdowns that would have taken several months or even years may now be completed in weeks. Cost reduction is another benefit that AI provides, especially in high-tech industries such as nanomanufacturing. AI reduces waste while minimizing production costs by optimizing the processes (Dong et al., 2019). In addition, scaling is easy. AI robots can transfer operations from a small lab setup to full-scale industrial production, all without a hitch (Xu et al., 2020). These advances powered by AI are making nanomanufacturing faster, more accurate, and very cheap. As the technologies mature, they promise to unlock new opportunities across industries, from healthcare and electronics to energy and environmental sustainability. The AI-nanomanufacturing synergy is not only for efficiencies but also for setting the stage for a completely new innovation.

7.6 AI-DRIVEN APPLICATIONS OF NANOMANUFACTURING IN INDIA

The integration of AI into the layout of nanomanufacturing environments has changed the ways in which advanced materials and devices are created so that they are now possible with utmost precision, efficiency, and quality. This process is transformational and is manifested in several key steps, each one an integral part of assuring seamless and optimized production. The first step involves data preparation, which is captivatingly referred to as the foundation of any AI-driven process. From sensors embedded in production lines, high-quality, real-time data are gathered. Still, data alone are not enough; careful treatment to cleanse and standardize makes certain that AI models can use it efficiently. Think of this as tidying up a cluttered desk such that the toolbox works better.

Next comes the **model development**, the time when AI really starts to become marvelous. Machine learning algorithms with advanced neural networks are built using historical production data. The trained models will now detect trends, recognize inefficiencies, and foresee optimal settings for the manufacturing process. They act toward fine-tuning all aspects of manufacturing practice accepted with the utmost insight and least-to-know holding of wastes, somewhat like an expert advisor. Then the last step of real-time monitoring takes production into another arena. Monitoring is done by AI models in real time in the manufacturing process and varying it when needed. For instance, in the case of nanolithography—for creating intricate nanoscale patterns—AI continuously adjusts ray parameters to ensure the final product meets specifications in real time. Continued scrutiny by AI quite literally assures the quality and almost entirely negates errors. It is this symbiosis of human ingenuity mixed with so-called machine intelligence. AI not just improves the speed and efficiency of nanomanufacturing; it takes the precision and consistency to levels that redefine the art. From next-generation medical devices to advanced materials for renewable energy, AI-driven nanomanufacturing has applications that revolutionize entire industries and position India at the forefront of technological innovation.

7.6.1 Addressing Water Purification Challenges

Access to clean drinking water continues to be one of India's pressing concerns. Heavy metal contamination, fortifying pathogens, and industrial pollutants pose serious health risks to millions. Fortunately, the combination of nanotechnology and AI presents a monumental opportunity to solve this crisis. Leading this innovation are nanofilters made from newer materials such as graphene oxide and carbon nanotubes. These filters, working at the level of molecules, are designed to trap and remove the minute contaminants with great efficacy. Further, the ingenious use of AI to augment their performance makes them groundbreaking. AI optimizes filtration processes so that maximum purification is achieved with minimal energy expenditures and costs, thus facilitating easy access to safe drinking water for communities in India. The following describes the mechanism of AI-optimized nanofilters.

1. ***Initializing the AI model:*** To train the AI model, an extensive database was built, incorporating water quality parameters as well as performance indicators for different nanofilter designs. This corpus furnishes the AI with elementary knowledge that it uses in making its predictions with great accuracy.

2. ***Loading nanofilter design parameters:*** The nanofilter structural-functional characteristics such as pore size and material composition will be fed into the AI model—just enough to give a general idea for tuning the filters to specific water samples.
3. **Processing water samples**: For each water sample, the following steps are performed:
   - Water quality measuring: Sensors make assessments of critical parameters like pH, total heavy metals, total dissolved solids, or microbial content.
   - Data analysis and prediction: AI analyzes water quality data and predicts the compromised nanofilter design that will remove those specific contaminants present.
   - Fine-tuning nanofilter design: The nanofilter design is adjusted based on AI insights so that it fits absolutely with the requirements of the water sample.
   - Water purification: The sample is thereby filtered through the optimized nanofilter for maximum contaminant removal.
   - Quality testing: The purified water samples are tested rigorously to check whether they fall within the limits of safety.
   - Learning and improvement: The process data are stored and used to retrain the AI model so that over time the model performs even better.

7.6.2 Optimizing Water Filtration with AI: A Step-by-Step Framework

Systematic-dynamical AI advancement of water filtration relates to constant learning and divergence in developing smarter and more effective filtration systems. The process proceeds as follows:

7.6.2.1 Initialization: Laying the Foundation

The journey begins by setting up the AI model:

- First, the model is trained on water quality and previous filtration results: Detail that allows the AI to determine the best nanofilter configurations for certain kinds of contaminants.
- Next, the core design parameters of the nanofilters, such as pore size and material properties, are loaded into the system for further fine-tuning.

7.6.2.2 Processing Water Samples: From Data to Action

Once the system has been set, it will begin the processing of water samples through the following stages:

- ***Measurement of water quality*:** Sensors are measuring the key parameters of each water sample, such as its pH, heavy metal contamination, and microbial contamination.
- ***Prediction of the optimal configuration*:** After measuring, the AI analyzes the data and finds the nanofilter configuration that is best suited for dealing with contaminants present in the sample.
- ***Adjustment of nanofilter design*:** The design of the filter is then altered according to the AI prediction, to achieve the highest purification level possible.

- ***Filtering the water*:** The adjusted nanofilter is applied to purify the sample.
- ***Quality testing*:** The purified water is tested to confirm that it meets safety and quality standards.

7.6.2.3 Data Storage: Building a Knowledge Base

Every step of the process generates valuable data. The system records details such as:

- The initial quality of the water sample.
- The specific nanofilter configuration applied.
- The quality of the purified water.

These performance data are stored, forming a growing knowledge base that will inform future operations.

7.6.2.4 Model Retraining: Getting Smarter over Time

The final and most critical step involves using the stored data to retrain the AI model. With each new batch of samples, the model becomes more adept at making predictions, continuously refining its ability to optimize nanofilters for a wide range of water quality scenarios.

7.6.2.5 An Iterative Process for Continuous Improvement

This framework uses an iterative design. The filtration system learns and becomes smarter with every iteration using feedback loops between machine learning and

```
// Initialize AI model with training data on water quality and filtration performance
DECLARE AIModel
AIModel <-- InitializeModel(TrainingData)

// Load nanofilter design parameters
DECLARE NanofilterDesignParameters
NanofilterDesignParameters <-- LoadDesignParameters()

// For each water sample
DECLARE WaterSample, WaterQualityData, OptimalNanofilterConfiguration, PurifiedWaterQuality

FOR EACH WaterSample IN WaterSamples
    // Measure water quality parameters (e.g., pH, contaminant levels)
    WaterQualityData <-- MeasureWaterQuality(WaterSample)

    // Input water quality data into AI model
    // AI model predicts optimal nanofilter configuration
    OptimalNanofilterConfiguration <-- AIModel.Predict(WaterQualityData)

    // Adjust nanofilter design based on AI predictions
    AdjustNanofilterDesign(NanofilterDesignParameters, OptimalNanofilterConfiguration)

    // Filter water sample through optimized nanofilter
    PurifiedWaterQuality <-- FilterWater(WaterSample, OptimalNanofilterConfiguration)

    // Measure purified water quality
    PurifiedWaterQuality <-- MeasureWaterQuality(PurifiedWaterQuality)

    // Store performance data for model retraining
    StorePerformanceData(WaterQualityData, OptimalNanofilterConfiguration,
PurifiedWaterQuality)
NEXT WaterSample
```

FIGURE 7.3 Pseudocode for water problem.

real life. Resulting in cleaner water, optimized processes, and tackling India's water challenges sustainably. The pseudocode for the water problem has been illustrated in Figure 7.3.

7.7 CHALLENGES AND FUTURE DIRECTIONS

While the potential of nanotechnology and AI to change industries and touch human life is enormous, turning those possibilities into realities is not without challenges. India is faced with a plethora of challenges—technical, educational, and regulatory—that need to be addressed in tandem to harness the entire potential of those technologies. Opportunities to overcome the aforementioned hurdles will be provided through strategic interventions and synergy among various sectors, thereby paving a path toward sustainable development and innovation. Nanotechnology and AI sit at the confluence of extremely diverse disciplines, requiring specialists in engineering, data science, biology, chemistry, and so forth. The glaring problem faced by India in this modern-day scenario is the very inadequate supply of personnel trained for operating at this interdisciplinary front. The gap can be closed only by transforming current educational institutions and their curricula. Interdisciplinary programs that combine courses in nanotechnology, AI, and applied sciences can generate individuals with the essential skills and knowledge to make an impact in this ever-changing field. Skill India and similar initiatives could play important roles in providing training in emerging technologies. Moreover, industry association is critical to ensuring the currency of programs while providing students with the hands-on experience that complements real-life applications.

7.7.1 Creating a Robust Data Infrastructure

AI thrives on data; one of the major reasons that makes an AI virtually ineffective is the existence of fragmented and inconsistent data systems. For nanotechnology and AI-related applications to function in a near-perfect environment in India, the country shall have to take some steps in creating a standardized ecosystem of data which is cohesive and, therefore, well-functioning. The Digital India initiative appears to have a strong reason to initiate this change. In this context, the government can enable seamless integration of AI solutions by establishing common processes for data collection, storage, and sharing across sectors. Good-quality, real-time data shall enable even faster and better decision-making as well as R&D for nanotechnology.

7.7.2 Establishing Regulatory Frameworks

- Like any transformative technology, rapid growth in the area of nanotechnology and AI comes along with many pressing ethical and safety concerns. It becomes essential to set up a thorough regulatory framework that will help manage the risks involved without stifling innovation. Regulation must therefore prioritize the following aspects:
- *Ensuring safety standards for nanomaterials*: Especially in sensitive applications like healthcare and agriculture, rigorous safety testing and approval processes are necessary to protect public health.

- *Promoting ethical AI use*: Transparent guidelines should govern AI decision-making, ensuring accountability and fairness.
- *Addressing environmental sustainability*: The emergence of nanotechnology has made regulation of the disposal of nano waste paramount in limiting ecological damage.

Appropriate concealment is believed to establish trust within the stakeholders while promoting responsible use of these advanced technologies.

7.8 ENCOURAGING PUBLIC–PRIVATE COLLABORATION

No one sector can single-handedly push for this transformation. Due to the intricacies involved in nanotechnology and AI applications, strong public–private partnerships are needed. Such partnerships can expedite the innovation process and bring advanced solutions to the marketplace by combining their resources and expertise. For example, research organizations can focus on developing advanced nanomaterials, leaving their commercialization and large-scale deployment in the hands of private industry. When these partnerships have government backing through grants, tax incentives, or policy frameworks, they flourish into a vibrant ecosystem of innovation. One of the oft-ignored challenges in taking up emerging technologies is the perspective that the public holds about them. Ill-conceived or distrusting notions in the public arena can serve to slow down the process of acceptability of transformative solutions such as nanotechnology and AI. Targeted awareness campaigns are crucial to filling this gap. Citizens need to be made aware of the goods and safe adoption of these technologies in order to build trust and their widespread use, particularly among rural and deprived communities, where skepticism is bound to be higher. Through research, innovation comes to the fore, and there is no negotiation about the need for a robust funding regime to support R&D. India's very next level of prioritization should be investment in frontier research by creating centers and innovation hubs devoted to furthering the exploration of nanotechnology and AI applications. This kind of environment will give imagination and experimentation platforms to solve the country's own problems and stay abreast with global players.

7.9 A VISION FOR THE FUTURE

Getting past these barriers, in contrast, means ensuring the innovation that positions India as a world leader in nanotechnology and AI. Some of the pressing barriers lie in data quality and availability. In the context of AI-driven nanomanufacturing, only data of high quality and structured formats can be called the lifeblood, whereas a lousy data infrastructure can risk defeating even the most sophisticated systems developed for nanotechnological applications (Singh et al., 2020). Interdisciplinary collaboration is significant here, too. By working across fields, you ensure solutions that are holistic, leveraging the best of AI and nanotechnology (Nandipati et al., 2023). Regulatory and ethical challenges are further aspects of the whole scheme to be handled ahead, by providing governments with a toolbox for adequate and safe uses (Jakupović et al., 2018). With that, India would empower sustainable development, quality living, and long-term economic growth by directing all these efforts to break the bottlenecks on its way. Once overcome, these challenges will ensure that India is well placed in the global market, producing a brighter future for everyone. Realizing the full potential of AI and nanomanufacturing depends on future research focusing on three critical aspects:

7.9.1 Advanced AI Algorithms are in the Process of Being Developed

Certain specific algorithms are needed, which are intricately related to the sometimes bizarre demands involved in nanomanufacturing. These should be systems capable of handling complex data inputs that give precise control over nanoscale manufacturing processes (Wang et al., 2021).

7.9.2 Multisectoral Collaboration

These good partnerships foster innovation, with researchers and AI developers on the one hand and industry stakeholders on the other. Knowledge and resource sharing, and the expertise brought to bear on unsolved interdisciplinary problems facilitate moving breakthroughs from lab-scale to practical applications (Jiao, 2021).

7.9.3 Answering Ethics and Regulations

Especially, transparent regulatory guidelines are necessary for public safety, ethical obligations, and the management of possible risks that advanced technologies pose (Mellit & Kalogirou, 2008).

7.10 CONCLUSION

The AI–nanomanufacturing integration offers an opportunity for India to respond to key issues such as water purification, work efficiency, and healthcare innovation. This high degree of precision combined with intelligence enables extraordinary developments in various sectors toward a better future for millions. However, the road to realizing this potential requires collaboration, foresight, and commitment. By investing in research, modernizing education, and building robust regulatory frameworks, India can lead the global charge in innovation. Embracing these transformative technologies not only positions the nation as a leader in nanotechnology and AI but also ensures sustainable development, elevating the quality of life for all its citizens.

REFERENCES

Adebayo, Riliwan Adekola, Obiuto, Nwankwo Constance, Olajiga, Oladiran Kayode, and Festus-Ikhuoria, Igberaese Clinton. (2023). "AI-enhanced manufacturing robotics: A review of applications and trends." *World Journal of Advanced Research and Reviews*, 21(3), 2060–2072.

Akhtar, Zarif Bin. (2024). "Artificial intelligence (AI) within manufacturing: An investigative exploration for opportunities, challenges, future directions." *Metaverse*, 5(2), 2731.

Cha, J. H., Yang, S., Oh, J., Choi, S., Park, S., Jang, B. C., Ahn, W., & Choi, S. (2020). Conductive-bridging random-access memories for emerging neuromorphic computing. *Nanoscale*, 12(27), 14339–14368.

Chavan, S. S. (2023). Artificial intelligence used in Indian economy development and management. *International Journal for Multidisciplinary Research*, *5*(5), 364–370. https://doi.org/10.36948/ijfmr.2023.v05i05.6836

Chen, A., Chatterjee, S., & Li, S. (2017). Nanomaterial-based electrochemical sensors for biomedical applications. *Chemical Society Reviews*, 46(8), 2025–2120.

Chennupati, A. (2024). Empowering Indian agriculture: The role of artificial intelligence. *International Research Journal of Modernization in Engineering, Technology and Science*, 6(4). 10.56726/IRJMETS53243

Chien, Chen-Fu, Ehm, H., Fowler, John W., Kempf, Karl G., Mönch, Lars, and Wu, Cheng-Hung. (2023). "Production-level artificial intelligence applications in semiconductor supply chains." *IEEE Transactions on Semiconductor Manufacturing*, 36(4), 560–569.

Choi, J. B., Nguyen, P. C. H., Sen, O., Udaykumar, H., & Baek, S. S. (2022). Artificial intelligence approaches for energetic materials by design: State of the art, challenges, and future directions. ArXiv, abs/2211.08179.

Darko, A., Chan, A., Adabre, M. A., Edwards, D., Hosseini, M., & Ameyaw, E. (2020). Artificial intelligence in the AEC industry: Scientometric analysis and visualization of research activities. *Automation in Construction*, 112, 103081.

Das, S. (2024). Challenges and opportunities in navigating the intersection of AI and ergonomics in the Indian market. *Ergonomics International Journal*, *8*(3), 000331. https://doi.org/10.23880/EOIJ-16000331

Dhyani, A. (2023). Role of artificial intelligence and nanotechnology in pharmaceutical and healthcare research. *Psychology and Education Journal*, 55(1).

Dong, K., Peng, X., & Wang, Z. L. (2019). Fiber/fabric-based piezoelectric and triboelectric nanogenerators for flexible/stretchable and wearable electronics and artificial intelligence. *Advanced Materials*, 32, 1902549.

Dreyer, D. R., Park, S., Bielawski, C. W., & Ruoff, R. S. (2010). The chemistry of graphene oxide. *Chemical Society Reviews*, 39(1), 228–240.

Engstrøm, D. S., Porter, B., Pacios, M., & Bhaskaran, H. (2014). Additive nanomanufacturing—A review. *Journal of Materials Research*, 29, 1792–1816.

Guizzo, E., & Ackerman, E. (2012). How rethink robotics built its new baxter robot worker. *IEEE Spectrum*, 18, 2012.

Hemanth, J., & Lakshminarayana, K. (2024). Social impacts of AI in India: A societal perspective on adoption and inclusion. *International Journal of Scientific Research in Engineering and Management*, 8(8), 1–6. https://doi.org/10.55041/IJSREM37144

Hui, K. S., & Zaki, M. I. (2021). *Synthesis, Properties, and Applications of Zeolites*. Springer.

Hunter, J. D. (2007). Matplotlib: A 2D graphics environment. *Computing in Science & Engineering*, 9(3), 90–95.

Ihaka, R., & Gentleman, R. (1996). R: A language for data analysis and graphics. *Journal of Computational and Graphical Statistics*, 5(3), 299–314.

Jakupović, A., Kovacevic, Z., Gurbeta, L., & Badnjević, A. (2018). Review of artificial neural network application in nanotechnology. *2018 7th Mediterranean Conference on Embedded Computing (MECO)*, 1–4.

Jiao, P. (2021). Emerging artificial intelligence in piezoelectric and triboelectric nanogenerators. *Nano Energy*, 88, 106227.

Kragic, D., Khansari, M., & Kragic, D. (2018). Collaborative robots and the future of work. *IEEE Robotics and Automation Magazine*, 25(2), 79–84.

Lin, Xuze. (2024). "Artificial intelligence in the industrial engineering." *Advances in Operation Research and Production Management*, 1, 1–16.

Lodhi, Shahrukh Khan, Gill, Ahmad Yousaf, and Hussain, Ibrar. (2024). "AI-powered innovations in contemporary manufacturing procedures: An extensive analysis." *International Journal of Multidisciplinary Sciences and Arts*, 3(4), 15–25.

Majstorovic, V., Dimitrijević, Sonja, Simeunović, Vladimir, and Stosic, Dragan. "Artificial Intelligence and its Application in Manufacturing." In *Artificial Intelligence in Industry 4.0: The Future That Comes True*, 2024.

Maleki, A., Haghighi, A., & Mahariq, I. (2021a). Machine learning-based approaches for modeling thermophysical properties of hybrid nanofluids: A comprehensive review. *Journal of Molecular Liquids*, 322, 114843.

Maleki, A., Haghighi, A., & Mahariq, I. (2021b). Machine learning-based approaches for modeling thermophysical properties of hybrid nanofluids: A comprehensive review. *Journal of Molecular Liquids*, 322, 114843.

Manish, D., & Dave, K. (2023). Revolutionizing manufacturing: The pivotal role of artificial intelligence in software-defined processes. *International Research Journal of Modernization in Engineering Technology and Science*, 5(12), 430–438. https://doi.org/10.56726/IRJMETS47054

Mauter, M. S., & Elimelech, M. (2008). Environmental applications of carbon-based nanomaterials. *Environmental Science & Technology*, 42(16), 5843–5859.

McKinney, W. (2010). Data structures for statistical computing in python. In *Proceedings of the 9th Python in Science Conference*, Vol. 445, pp. 51–56.

Mellit, A., & Kalogirou, S. (2008). Artificial intelligence techniques for photovoltaic applications: A review. *Progress in Energy and Combustion Science*, 34, 574–632.

Moore, J. A., & Chow, J. (2021a). Recent progress and applications of gold nanotechnology in medical biophysics using artificial intelligence and mathematical modeling. *Nano Express*, 2, 022001.

Moore, J. A., & Chow, J. (2021b). Recent progress and applications of gold nanotechnology in medical biophysics using artificial intelligence and mathematical modeling. *Nano Express*, 2, 022001.

Nandipati, M., Fatoki, O., & Desai, S. (2023). Bridging nanomanufacturing and artificial intelligence—A comprehensive review. *Materials*, 17(7), 1621.

Nova, M. V., Lin, T. P., Shanehsazzadeh, S., Jain, K., Ng, S. C. Y., Wacker, R., Chichakly, K. J., & Wacker, M. (2022). Nanomedicine Ex Machina: Between model-informed development and artificial intelligence. *Frontiers in Digital Health*, 4, 799341.

Paszke, A., Gross, S., Massa, F., Lerer, A., Bradbury, J., Chanan, G., … & Chintala, S. (2019). PyTorch: An imperative style, high-performance deep learning library. *Advances in Neural Information Processing Systems*, 32, 8026–8037.

Rai, M., Yadav, A., & Gade, A. (2009). Silver nanoparticles as a new generation of antimicrobials. *Biotechnology Advances*, 27(1), 76–83.

Rakholia, R., Suárez-Cetrulo, A. L., Singh, M., & Carbajo, R. S. (2024). Advancing manufacturing through artificial intelligence: Current landscape, perspectives, best practices, challenges, and future direction. *IEEE Access*, 12, 131621–131637. https://doi.org/10.1109/ACCESS.2024.3458830

Ramezanizadeh, M., Alhuyi Nazari, M., Ahmadi, M., Lorenzini, G., & Pop, I. (2019). A review on the applications of intelligence methods in predicting thermal conductivity of nanofluids. *Journal of Thermal Analysis and Calorimetry*, 138, 827–843.

Roy, Rishabh, and Srivastava, Alpana (2024). "Role of artificial intelligence (AI) in enhancing operational efficiency in manufacturing medical devices." *The Journal of Multidisciplinary Research*, 35–40.

Salehi, H., & Burgueño, R. (2018). Emerging artificial intelligence methods in structural engineering. *Engineering Structures*, 171, 170–189.

Şengül, H., Theis, T., & Ghosh, S. (2008). Toward sustainable nanoproducts. *Journal of Industrial Ecology*, 12, 329–359.

Silver, D., Huang, A., Maddison, C. J., Guez, A., Sifre, L., Van Den Driessche, G., … & Hassabis, D. (2016). Mastering the game of go with deep neural networks and tree search. *Nature*, 529(7587), 484–489.

Singh, A. V., Rosenkranz, D., Ansari, M., Singh, R., Kanase, A., Singh, S. P., Johnston, B., Tentschert, J., Laux, P., & Luch, A. (2020). Artificial intelligence and machine learning empower advanced biomedical material design to toxicity prediction. *Advanced Intelligent Systems*, 2, 2000084.

Tang, Z., & Kotov, N. (2005). One-dimensional assemblies of nanoparticles: Preparation, properties, and promise. *Advanced Materials*, 17, 951–962.

Verma, N., & Bhardwaj, A. (2017a). Biosensors for heavy metals. *Biometals*, 30(4), 461–491.

Verma, N., & Bhardwaj, A. (2017b). Biosensors for heavy metals. *Biometals*, 30(4), 461–491.

Verma, M. (2023). Artificial-intelligence role in modern science: Aims, merits, risks and its applications. *International Journal of Trend in Scientific Research and Development*, *7*(5), 335–342.

Wang, J., Lu, P., Huang, Z., Su, X., Zhang, L., & Gu, W. (2020a). Electrochemical sensors for environmental analysis: From designed nanomaterials to practical applications. *TrAC Trends in Analytical Chemistry*, 124, 115804.

Wang, J., Lu, P., Huang, Z., Su, X., Zhang, L., & Gu, W. (2020b). Electrochemical sensors for environmental analysis: From designed nanomaterials to practical applications. *TrAC Trends in Analytical Chemistry*, 124, 115804.

Wang, N., Yan, W., Qu, Y., Ma, S., Li, S. Z., & Qiu, M. (2021). Intelligent designs in nanophotonics: From optimization towards inverse creation. *PhotoniX*, 2, 1–35.

Wurman, P. R., D'Andrea, R., & Mountz, M. (2008). Coordinating hundreds of cooperative, autonomous vehicles in warehouses. *AI Magazine*, 29(1), 9–9.

Xu, J., Wang, X., Wang, C., Yuan, L., Chen, W., Bao, J., Su, Q., Xu, Z., Wang, C., Wang, Z., Shan, D., & Guo, B. (2020). A review on micro/nanoforming to fabricate 3D metallic structures. *Advanced Materials*, 33, 2000893.

Yan, X., Yue, T., Winkler, D., Yin, Y., Zhu, H., Jiang, G., & Yan, B. (2023). Converting nanotoxicity data to information using artificial intelligence and simulation. *Chemical Reviews*, 123(13), 8575–8637.

Yang, Xianyu, and Yang, Yujing. (2023). "Research on the Application of Artificial Intelligence in Mechanical Design and Manufacturing." *2023 13th International Conference on Information Technology in Medicine and Education (ITME).*

8 Nanomaterials
Synthesis Strategies and Emerging Applications Across Science and Industry

Ankit Agarwal, Navneet Goyal, Amandeep and Suman Dhingra

8.1 INTRODUCTION

In recent decades, nanotechnology has developed as a transformative and rapidly advancing field with far-reaching implications across science and engineering. Materials designed with at least one dimension in the nanoscale range (1–100 nm) are known as nanomaterials, and they are at the centre of this technological revolution (Malik, Muhammad, and Waheed 2023). Due to their unique size-dependent properties such as quantum effects, high surface area, and enhanced mechanical, optical, and electrical characteristics, nanomaterials have opened new frontiers in research and applications. Unlike their bulk counterparts, nanomaterials often exhibit entirely different behaviours, making them attractive for innovative solutions in various domains including energy, electronics, healthcare, and environmental science (Gohar et al. 2024).

Nanotechnology is inherently interdisciplinary, drawing knowledge from chemistry, physics, materials science, biology, and engineering (Kumar et al. 2024a). This convergence allows for the design, fabrication, and manipulation of nanomaterials for highly specific and often multifunctional uses. As such, understanding the methods used to synthesize nanomaterials is crucial, as the chosen synthesis route strongly influences their properties, morphology, stability, and performance in applications. The two main categories of synthesis techniques are top-down, which reduces bigger materials to nanoscale structures, and bottom-up, which builds things up molecule by molecule or atom by atom (Zain Ul Abidin et al. 2024). Every approach has pros and downsides, and the selection is frequently influenced by the intended use, financial concerns, and scalability.

Recent literature has witnessed an explosion of studies detailing various synthesis techniques, characterization tools, and novel applications of nanomaterials (Prasad et al. 2024; Kumar et al. 2024b; Shahzad, Mardare, and Hassel 2024; Tiwari, Bystrzejewski, and Zhu 2024). However, for young researchers and those new to the field, this vast and fragmented information can be overwhelming. While we acknowledge the continuous emergence of new techniques and highly specialized materials,

DOI: 10.1201/9781003560234-10

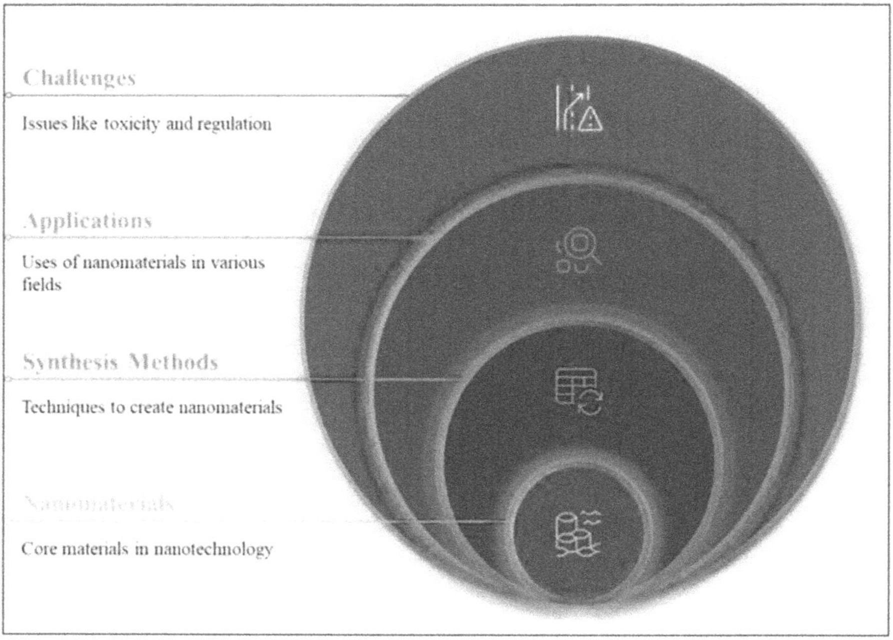

FIGURE 8.1 Synthesis and applications of nanomaterials.

this chapter limits its scope to broadly adopted methods such as mechanical milling, lithography, sol-gel processing, chemical vapour deposition (CVD), hydrothermal synthesis, and biosynthesis.

In addition to synthesis, the growing list of nanomaterial applications is of great significance. In electronics, nanomaterials are key to device miniaturization and performance enhancement, supporting the development of faster, more efficient semiconductors, sensors, and display technologies (Habeeb Dolapo Salaudeen et al. 2024). In biomedicine, nanoparticles are revolutionizing drug delivery systems, enabling targeted therapy and reducing side effects, while also contributing to advanced imaging and diagnostic tools (Al-Thani et al. 2024). In the energy sector, nanomaterials improve the efficiency of energy storage devices like lithium-ion batteries and enable breakthroughs in fuel cells and photovoltaic systems (Mahmud 2023).

Despite their promise, nanomaterials pose challenges related to toxicity, environmental impact, reproducibility, and regulatory oversight (Singh et al. 2019). This review aims to bridge the knowledge gap by presenting a coherent picture of the synthesis and application landscape of nanomaterials. Figure 8.1 depicts the synthesis and applications of nanomaterials.

8.2 SYNTHESIS OF NANOMATERIALS

8.2.1 Top-Down Approaches

In top-down approaches to nanomaterial production, bulk materials are broken down into nanoscale structures by mechanical, chemical, or physical means. While top-down approaches start with macroscopic materials and gradually shrink their size, bottom-up approaches build nanostructures from atomic or molecular antecedents (Lin et al. 2022). This method offers the advantage of utilizing well-established industrial processes and is particularly useful for producing nanostructured thin films, nanocomposites, and particles with controlled size and shape. In the context of current trends and future directions, top-down approaches are gaining relevance for their scalability, integration with microfabrication technologies, and application in high-throughput production systems (George, Jose, and Anju 2025; Gaspar et al. 2020). Figure 8.2 depicts the various types of Top-Down approaches.

8.2.1.1 Mechanical Milling

Mechanical milling, particularly high-energy ball milling, is a widely adopted top-down technique for synthesizing nanomaterials such as nanocrystalline metals, alloys, and ceramics. The process involves repeated fracturing, welding, and re-welding of powder particles under intense mechanical forces. It is favoured for its simplicity, scalability, and cost-effectiveness. However, limitations include potential contamination from milling media and inconsistent particle size distribution. Recent advancements, including improved milling equipment and controlled atmospheres like inert gas environments, have helped mitigate these challenges, enabling better control over material purity, morphology, and structural refinement during the synthesis of high-performance nanomaterials (Etafa et al. 2024).

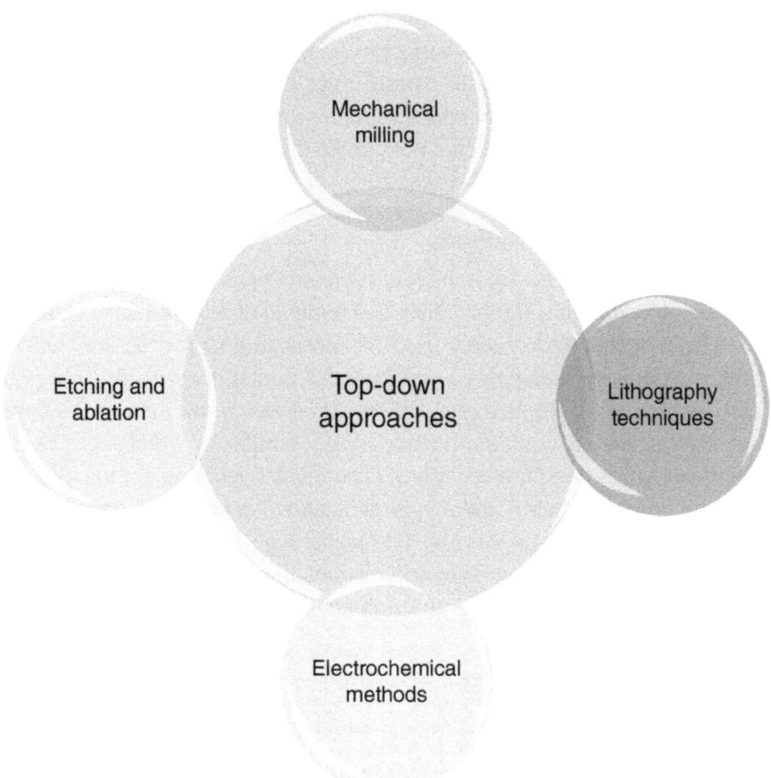

FIGURE 8.2 Various types of top-down approaches.

8.2.1.2 Lithography Techniques

In making nanoscale features on surfaces, photolithography, electron-beam lithography (EBL), and nanoimprint lithography play important roles. The semiconductor industry generally applies these methods to manufacture integrated circuits and microelectromechanical systems (MEMS). Lithography offers high precision and pattern fidelity, but it often requires cleanroom environments and sophisticated equipment, limiting its accessibility for bulk nanomaterial production (Basu et al. 2025). Nanoimprint lithography, a mechanical method, replicates nanoscale features by pressing a patterned mould onto a substrate. Although these techniques offer exceptional precision and pattern fidelity, they often require cleanroom conditions, costly instrumentation, and complex processing steps, which can limit their scalability and affordability for bulk nanomaterial production. Nonetheless, they remain indispensable for high-end nanodevice fabrication.

8.2.1.3 Etching and Ablation

Chemical and physical etching techniques, such as reactive ion etching (RIE) and plasma etching, are widely utilized in nanofabrication to selectively remove material from surfaces, enabling the creation of intricate nanoscale features with high precision. These methods use chemically reactive plasma or ions to etch specific regions, making them ideal for pattern transfer in microelectronics and MEMS devices. Laser ablation, another top-down technique, employs high-energy laser pulses to vaporize material from a solid target, often in vacuum or liquid environments, producing nanoparticles or creating patterned surfaces (Asadizanjani, Reddy Kottur, and Dalir 2025). These approaches offer fine control over depth, shape, and feature resolution, supporting the fabrication of nanostructures for optics, sensors, and biomedical devices. Their versatility and effectiveness make them essential in both research and industrial nanomanufacturing processes.

8.2.1.4 Electrochemical Methods

Top-down electrochemical methods, including electrochemical machining and anodization, are effective techniques for the controlled removal of material via electrochemical reactions. These methods are widely used for fabricating nanostructures with precise geometries, particularly nanoporous surfaces and thin films with tunable pore size, depth, and distribution (Campbell et al. 2025). Anodization, for instance, is commonly employed to create anodic aluminium oxide (AAO) templates, which serve as scaffolds for growing nanowires or nanotubes. Electrochemical machining, on the other hand, is used to pattern complex micro- and nanoscale features in conductive materials. These techniques offer advantages such as low thermal damage, high precision, and reproducibility. They are especially valuable in applications like sensors, filtration membranes, batteries, and templates for further bottom-up nanomaterial synthesis.

8.2.2 Bottom-Up Approaches

The bottom-up approach is a fundamental and widely adopted strategy in the synthesis of nanomaterials, where atoms, molecules, or molecular clusters self-assemble into well-defined nanostructures (Gaspar et al. 2020). Unlike top-down methods that reduce bulk material into nanoscale dimensions, bottom-up techniques build nanomaterials from the atomic or molecular level upward (Bora et al. 2025). This method allows for precise control over composition, morphology, crystallinity, and functional properties, making it particularly suitable for applications requiring high uniformity and tailor-made nanostructures.

Modern nanotechnology and materials science regard the bottom-up synthesis as its core as it allows the production of nanomaterials that exhibit quantum-size effects, large surface-to-volume ratio, and new physicochemical properties. As the need to develop environmentally sustainable, economically viable, and scalable synthesis methods continues to grow, bottom-up methodologies are quickly changing by means of advances in green chemistry, automation, and combination with artificial intelligence (AI) (Hosny et al. 2025; Mehta 2025; Han et al. 2025). Figure 8.3 depicts the various types of bottom-up approaches.

8.2.2.1 Sol-Gel Method

The sol-gel process is a highly versatile and widely adopted bottom-up technique for synthesizing metal oxide and ceramic nanomaterials. It involves the transformation of metal alkoxides or inorganic salts through hydrolysis and polycondensation reactions, leading to the formation of a colloidal solution or "sol." As the reaction progresses, the sol gradually transitions into a semi-solid, porous "gel" network

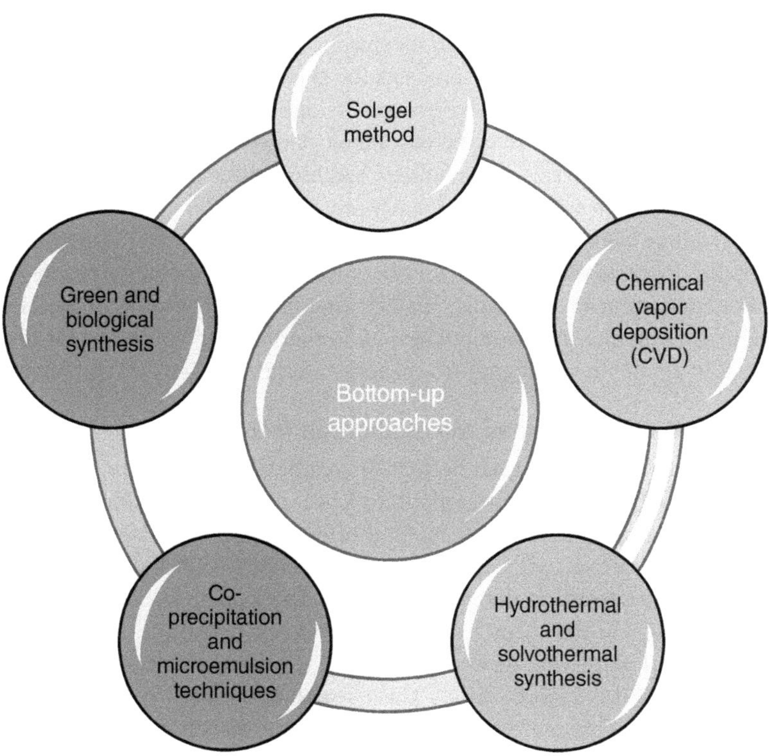

FIGURE 8.3 Different types of bottom-up approaches.

comprising both liquid and solid phases. This method allows precise control over the chemical composition, homogeneity, and microstructure of the resulting materials. The gel can be further dried and heat-treated to obtain nanoscale powders, coatings, or monoliths. The sol-gel process is valued for its low processing temperatures, compatibility with complex shapes, and applicability in fields such as optics, catalysis, sensors, and biomedical coatings.

8.2.2.2 CVD

CVD is a bottom-up synthesis technique widely used to produce high-purity, well-adhered nanomaterials in the form of thin films or nanostructures. In this process, volatile precursor gases are introduced into a reaction chamber, where they thermally decompose or react on a heated substrate surface, forming a solid material layer. CVD offers excellent control over film thickness, composition, and uniformity, making it suitable for fabricating nanoscale coatings, semiconductors, carbon nanotubes, and graphene. The process can be tailored through variations such as plasma-enhanced CVD (PECVD) and metal-organic CVD (MOCVD), which enable deposition at lower temperatures or with complex chemistries. Due to its scalability, precision, and material versatility, CVD is essential in microelectronics, photovoltaics, and protective coatings for advanced technologies.

8.2.2.3 Hydrothermal and Solvothermal Synthesis

Hydrothermal and solvothermal synthesis are solution-based bottom-up techniques carried out in sealed autoclaves under elevated temperature and pressure conditions. In hydrothermal synthesis, water serves as the solvent, whereas solvothermal methods use organic solvents to dissolve precursors and facilitate reactions. These techniques are particularly effective for producing nanomaterials with well-controlled crystal structure, particle size, morphology, and high phase purity. The high-pressure environment enhances solubility and reactivity, enabling the growth of uniform and highly crystalline nanostructures such as nanorods, nanowires, and quantum dots. Additionally, these methods allow for the synthesis of materials that are difficult to obtain under ambient conditions. Widely used in the fabrication of metal oxides, sulphides, and complex nanocomposites, hydrothermal and solvothermal processes are critical in applications ranging from energy storage to catalysis and photonics.

8.2.2.4 Co-Precipitation and Microemulsion Techniques

Co-precipitation is a widely used bottom-up synthesis technique in which multiple components precipitate simultaneously from a homogeneous solution, forming oxide or mixed-metal nanoparticles. This method offers a simple, cost-effective route for producing nanomaterials with controlled stoichiometry and uniform composition. It is commonly employed in the synthesis of magnetic nanoparticles, catalysts, and ceramic powders. On the other hand, microemulsion techniques utilize nanoscale droplets—formed by mixing immiscible liquids with surfactants—as confined reaction environments or "nanoreactors." These droplets limit the space for nucleation and growth, allowing precise control over particle size, morphology, and distribution. Microemulsions are especially useful for creating monodisperse nanoparticles and core-shell structures. Both methods are valued for their scalability, low-temperature operation, and suitability for producing nanomaterials for biomedical, environmental, and energy-related applications.

8.2.2.5 Green and Biological Synthesis

Recent advancements in green chemistry have enabled the development of environmentally friendly bottom-up synthesis methods for nanomaterials, utilizing biological entities such as plant extracts, bacteria, fungi, and algae. These biogenic approaches eliminate the need for toxic reducing agents and hazardous solvents typically used in conventional chemical synthesis, thereby minimizing environmental impact and health risks. The biological components act as both reducing and stabilizing agents, enabling the formation of nanoparticles under mild conditions often at room temperature and atmospheric pressure significantly reducing energy consumption (Joy et al. 2025). These eco-friendly methods align closely with sustainable development goals (SDGs) by promoting safer, cleaner, and renewable routes for nanomaterial production. Biogenic nanomaterials are being explored for various applications, including drug delivery, catalysis, and environmental remediation, highlighting their potential in both scientific innovation and environmental protection.

8.3 APPLICATIONS OF NANOMATERIALS

Almost every important scientific and industrial field has made use of nanomaterials due to their exceptional physicochemical characteristics, which include a high surface area-to-volume ratio, quantum effects, enhanced reactivity, tunable optical properties, and exceptional mechanical strength. As the synthesis methods of nanomaterials continue to evolve, their applications are expanding in both scope and complexity (Saravanan et al. 2022). The current era of nanotechnology is characterized by a shift from laboratory-scale demonstration to real-world implementation, supported by advances in manufacturing, scalability, and regulatory approval (Kulkarni, Ayachit, and Aminabhavi 2022; Lakhani, Bhanderi, and Modi 2024). This section explores the current and emerging applications of nanomaterials, structured by major industry domains. Figure 8.4 depicts the various applications of Nanomaterials

8.3.1 Electronics and Semiconductors

Nanomaterials have revolutionized the electronics industry by enabling unprecedented levels of miniaturization, enhanced processing speeds, and reduced power consumption. Their unique electrical, mechanical, and thermal properties have made materials like carbon nanotubes, graphene, silver, and copper nanoparticles integral to next-generation electronic devices. Carbon-based nanostructures are being used to develop nanoscale transistors and flexible interconnects, pushing the limits of Moore's law (Gupta 2025). Conductive nanoparticles, such as silver and copper, serve as essential components in printed electronics for creating low-cost, flexible circuits on wearable and disposable substrates.

Furthermore, nanomaterials like silver nanowires, graphene, and carbon nanotube networks are replacing brittle indium tin oxide (ITO) in transparent conductive films for touchscreens, OLED displays, and solar panels. Their high conductivity, flexibility,

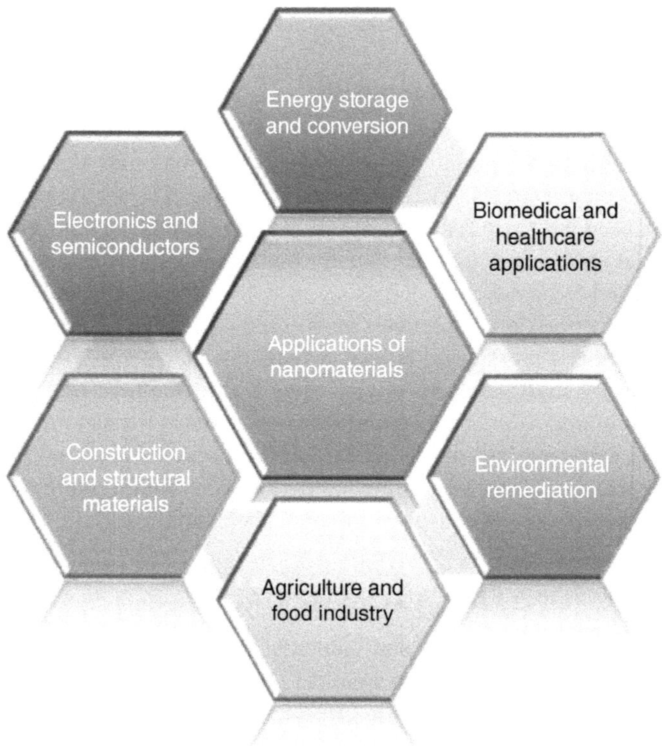

FIGURE 8.4 Importance of nanomaterials in engineering application.

and optical transparency allow for the fabrication of foldable, stretchable, and rollable electronic devices. Nanomaterials also play a vital role in energy-efficient transistors, sensors, and emerging quantum devices, contributing to advancements in AI, Internet of Things (IoT), and quantum computing. Their integration is driving the development of smarter, lighter, and more energy-efficient electronic systems, heralding a new era in nanoelectronics.

8.3.2 Energy Storage and Conversion

Nanomaterials have significantly advanced the performance and efficiency of energy conversion and storage technologies, particularly in batteries, fuel cells, and solar cells. Their unique nanoscale properties such as high surface area, tunable chemical composition, and controlled porosity enable more efficient ion and electron transport, which is critical for enhancing energy density, faster charge/discharge rates, and long-term durability. In lithium-ion batteries, nanostructured materials like silicon nanowires, graphene, and metal oxide nanoparticles are used as anode and cathode components to accommodate volume changes and improve conductivity, thereby extending cycle life and increasing storage capacity (Kiani 2025).

Similarly, in fuel cells, nano catalysts such as platinum nanoparticles dispersed on carbon supports offer higher catalytic activity and greater surface exposure, reducing the amount of precious metals required while enhancing reaction kinetics. In the realm of solar energy, nanomaterials like quantum dots, perovskite nanocrystals, and TiO_2 nanostructures are utilized to improve light absorption, charge separation, and overall power conversion efficiency. Their ability to absorb a broader spectrum of sunlight and facilitate rapid charge movement has made them central to next-generation photovoltaic devices. Moreover, their compatibility with flexible substrates enables the development of lightweight, portable, and wearable energy systems. Overall, nanomaterials are playing a transformative role in the evolution of energy systems by making them more compact, efficient, and adaptable to a wide range of applications.

8.3.3 Biomedical and Healthcare Applications

Nanomaterials are at the forefront of innovation in biomedical applications, particularly in targeted drug delivery, diagnostics, biosensing, and regenerative medicine, due to their exceptionally small size, high surface area, and ability to be functionalized with specific biological ligands. In targeted drug delivery, nanoparticles such as liposomes, polymeric carriers, gold nanoparticles, and dendrimers can be engineered to carry therapeutic agents directly to diseased cells or tissues such as tumours, thereby minimizing side effects and enhancing treatment efficacy. Their surfaces can be modified with antibodies, peptides, or aptamers to recognize and bind selectively to molecular markers expressed on the target site. In diagnostics, nanomaterials like quantum dots and superparamagnetic nanoparticles offer enhanced imaging capabilities for early disease detection, providing high-resolution contrast in MRI, CT, and fluorescence imaging. In biosensing, nanostructures such as carbon nanotubes, graphene, and metallic nanostructures enable ultra-sensitive and rapid detection of biomolecules, pathogens, or genetic material due to their high conductivity and unique optical or electrochemical properties. Furthermore, in regenerative medicine, nanomaterials are used to develop scaffolds that mimic the extracellular matrix, promoting cell adhesion, growth, and tissue regeneration. Materials like hydroxyapatite nanoparticles and nanofibrous polymers are being explored for bone and nerve tissue engineering. The ability of nanomaterials to interact at the molecular and cellular levels, combined with their multifunctionality, is transforming the landscape of modern medicine by enabling more personalized, precise, and effective healthcare solutions.

8.3.4 Environmental Remediation

Nanomaterials provide highly efficient, cost-effective, and versatile solutions for environmental remediation, particularly in the purification of air, water, and soil. Their exceptionally high surface area, tunable porosity, and enhanced chemical reactivity allow for superior interaction with a wide range of pollutants. In water purification, nanomaterials such as carbon nanotubes, graphene oxide, and metal oxide nanoparticles (e.g., TiO_2, ZnO, Fe_3O_4) are used for the adsorption and removal of heavy metals, dyes, organic contaminants, and microbial pathogens. Photocatalytic nanoparticles like TiO_2, when activated by UV or visible light, can degrade persistent organic pollutants through advanced oxidation processes, rendering them harmless. In air purification, nanostructured catalysts and adsorbents are incorporated into filters or coatings to capture particulate matter, volatile organic compounds (VOCs), and toxic gases such as NOx and SOx. Materials like activated carbon nanofibers and metal-organic frameworks (MOFs) are particularly effective due to their tailored pore structures and functional groups. For soil decontamination, nanoscale zero-valent iron (nZVI) is widely applied to degrade chlorinated compounds and immobilize heavy metals through redox reactions.

Moreover, silver and copper nanoparticles exhibit strong antimicrobial properties, enabling the disinfection of water and surfaces by disrupting microbial cell membranes. These nanomaterials can be integrated into membranes, composites, or coatings, offering scalable and sustainable alternatives to conventional treatment methods. As a result, nanotechnology plays a crucial role in advancing environmental sustainability by providing innovative tools to combat pollution and restore ecosystem health in a more targeted, efficient, and environmentally friendly manner.

8.3.5 Agriculture and Food Industry

Nanotechnology is playing a transformative role in agriculture by addressing key challenges such as excessive pesticide use, inefficient nutrient delivery, and significant post-harvest losses. Nanomaterials enable the development of smart agrochemical formulations that allow for the controlled and targeted release of fertilizers and pesticides, thereby reducing environmental contamination and minimizing input waste. Nanoencapsulation techniques protect active ingredients from premature degradation and release them in response to specific environmental triggers such as pH, temperature, or moisture levels, ensuring their availability precisely when and where needed.

Additionally, nanoparticles can improve the bioavailability and uptake of nutrients by plants due to their small size and high reactivity, which facilitates easier penetration through plant tissues and enhances absorption efficiency. This targeted delivery system not only boosts crop productivity but also reduces the frequency and quantity of chemical applications, leading to more sustainable and eco-friendly farming practices. Furthermore, nanotechnology is being explored for post-harvest applications, including nano sensors for spoilage detection and antimicrobial coatings to extend the shelf-life of produce. Collectively, these innovations contribute to improved agricultural sustainability, resource efficiency, and food security in the face of growing global demands.

8.3.6 Construction and Structural Materials

Nanomaterials are significantly advancing the construction industry by enhancing the mechanical strength, durability, and functional performance of building materials. When incorporated into cement, concrete, or coatings, nanomaterials such as nano silica, carbon nanotubes, and graphene oxide refine the microstructure, reduce porosity, and improve load-bearing capacity, resulting in stronger and more resilient structures. Some nanomaterials impart self-healing properties, enabling materials to autonomously repair microcracks through mechanisms like hydration of unreacted cement particles or release of encapsulated healing agents, thereby extending the service life of infrastructure.

Additionally, nanocoating with hydrophobic, anti-corrosive, or antimicrobial properties protects surfaces from environmental degradation, moisture, and biological growth. Nanomaterials also contribute to energy-efficient buildings by enhancing thermal insulation (e.g., aerogels), light reflection, and solar energy absorption in smart glazing systems. These improvements reduce long-term maintenance costs, minimize energy consumption, and contribute to the development of sustainable and smart infrastructure systems suitable for modern urban environments.

8.4 CHALLENGES AND FUTURE PROSPECTS

The field of nanomaterials has made tremendous progress over the past few decades, driven by advancements in synthesis techniques, characterization tools, and application demands across multiple disciplines. However, as nanomaterials transition from laboratory research to widespread industrial and societal use, several critical challenges must be addressed to fully realize their potential. Simultaneously, the future prospects of nanomaterials are promising, with innovations on the horizon poised to reshape manufacturing, healthcare, energy, and environmental sustainability. Figures 8.5 and 8.6 show the various challenges and future scope of nanomaterials

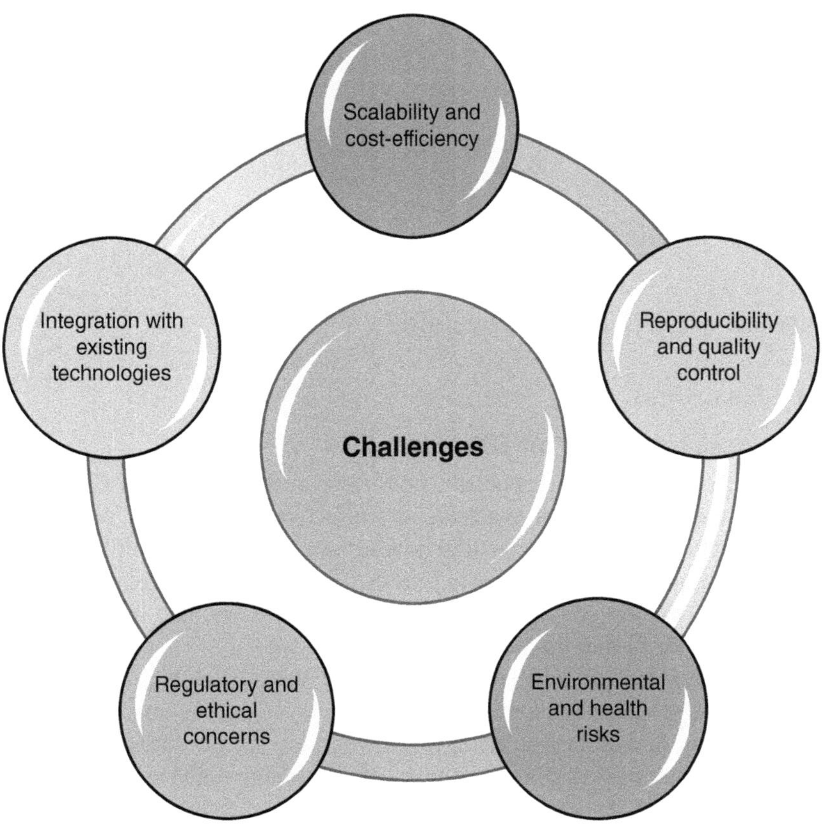

FIGURE 8.5 Various challenges for the synthesis of nanomaterials.

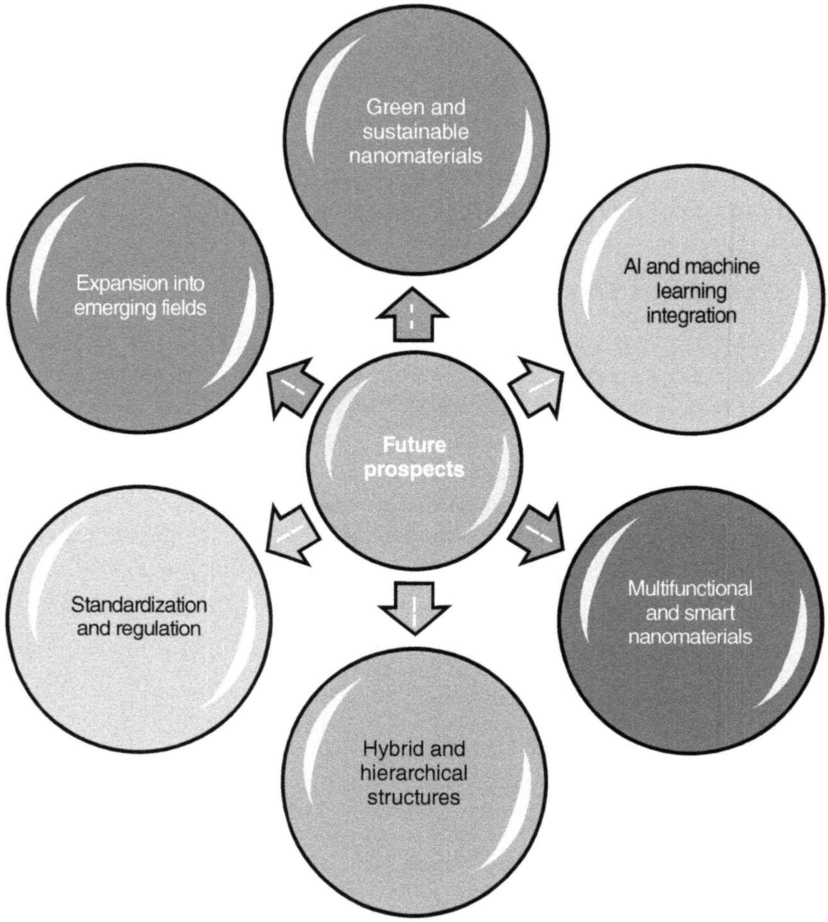

FIGURE 8.6 Future prospect of nanomaterials.

8.4.1 Challenges

8.4.1.1 Scalability and Cost-Efficiency

Many synthesis techniques, especially bottom-up methods such as CVD, sol-gel processing, and hydrothermal synthesis, are effective at laboratory scale but face obstacles when scaled up for industrial production. Maintaining uniformity, purity, and yield at large scales while keeping production costs low remains a significant challenge. High energy consumption, long processing times, and the need for specialized equipment further limit commercial scalability.

Additionally, many of these techniques require specialized, high-cost equipment and controlled atmospheres, making large-scale deployment economically and technically demanding. Addressing these limitations is critical for the commercialization of nanomaterials, and ongoing research focuses on developing scalable, energy-efficient, and cost-effective alternatives without compromising material performance or structural integrity.

8.4.1.2 Reproducibility and Quality Control

Achieving reproducibility in nanomaterial synthesis—ensuring consistent size, shape, surface chemistry, and structural integrity—is essential for their reliable performance in real-world applications such as electronics, biomedicine, and energy systems. However, even minor fluctuations in synthesis parameters such as temperature, pH, precursor concentration, or reaction time can result in significant variations across different batches. This inconsistency poses challenges for large-scale production and commercial adoption. The absence of standardized synthesis protocols, uniform characterization techniques, and robust quality control frameworks further complicates efforts to integrate nanomaterials into industrial manufacturing processes. Without reproducibility, scalability and regulatory approval become difficult, especially in fields where material performance must meet stringent specifications. Addressing this issue requires international collaboration to establish benchmarking standards, automated synthesis systems, and reproducible, scalable fabrication techniques.

8.4.1.3 Environmental and Health Risks

Nanomaterials may react unpredictably with biological systems and ecosystems because they have a large surface area and reactivity. Issues such as bioaccumulation, cytotoxicity, and long-term environmental persistence are not yet fully understood. Regulatory agencies are still developing comprehensive risk assessment protocols, and the lack of standardized toxicity testing methods slows regulatory approvals and public acceptance.

Moreover, the absence of standardized toxicity testing protocols and comprehensive risk assessment frameworks presents a major barrier to regulatory approval and public trust. Regulatory bodies are actively working to develop guidelines, but the lack of harmonized global standards hampers progress. Ensuring safe, responsible development of nanotechnologies requires coordinated efforts in toxicology, policy-making, and environmental monitoring.

8.4.1.4 Regulatory and Ethical Concerns

The regulatory landscape for nanomaterials remains fragmented and underdeveloped in many countries. There is a lack of universally accepted guidelines for labelling, handling, and disposing of nanomaterials. Furthermore, ethical concerns surrounding privacy (e.g., in nano sensors), human enhancement, and environmental justice present additional complexities that must be addressed transparently. These issues demand transparent governance and inclusive dialogue among scientists, policymakers, industry, and the public. Developing a comprehensive and adaptive regulatory framework, informed by scientific evidence and ethical considerations, is essential to ensure the responsible and equitable advancement of nanotechnology across sectors.

8.4.1.5 Integration with Existing Technologies

Nanomaterials often need to be integrated with traditional materials, systems, and manufacturing processes. This requires compatibility with legacy infrastructure, which is not always straightforward. Issues such as interfacial bonding, stability, and mechanical mismatch must be resolved for successful integration into devices, coatings, and composites. Furthermore, adapting large-scale fabrication processes to accommodate nanoscale components without altering equipment or production lines adds complexity. Achieving seamless integration demands interdisciplinary approaches involving materials science, surface engineering, and process optimization. Advancing this integration is crucial for realizing the full potential of nanotechnology in industries such as aerospace, electronics, automotive, and construction, where hybrid material systems are increasingly in demand.

8.4.2 Future Prospects

8.4.2.1 Green and Sustainable Nanomaterials

Future research is expected to focus heavily on environmentally friendly and sustainable synthesis methods, such as biological synthesis using plant extracts, microorganisms, and waste-derived precursors. Green chemistry principles are being applied to develop nanomaterials with minimal ecological impact, recyclable properties, and low carbon footprints. Additionally, researchers are developing biodegradable and recyclable nanomaterials to support circular economy models and reduce long-term environmental persistence. Emphasis is also being placed on lowering the carbon footprint of production processes through low-temperature, solvent-free, and renewable-resource-based techniques. Advancing these eco-conscious approaches is essential for scaling nanotechnology in a sustainable, socially responsible manner while maintaining material performance and functionality.

8.4.2.2 AI and Machine Learning (ML) Integration

AI and ML are emerging as powerful tools to accelerate nanomaterial discovery, optimize synthesis conditions, and predict material properties. These tools can analyse large datasets to uncover structure–property relationships, simulate complex reactions, and design new materials with desired functionalities faster than traditional trial-and-error methods. Unlike traditional trial-and-error approaches, AI and ML enable data-driven material design, allowing researchers to simulate reaction pathways, screen potential precursors, and tailor properties for specific applications. These tools can also accelerate the discovery of novel nanomaterials by generating predictive models that guide experimental work, reducing time and resource consumption. Integration of AI/ML with high-throughput experimentation and advanced characterization techniques holds great promise for advancing the design, synthesis, and application of nanomaterials across a wide range of industries.

8.4.2.3 Multifunctional and Smart Nanomaterials

The development of multifunctional nanomaterials capable of performing multiple tasks simultaneously (e.g., sensing, self-healing, and actuation) is a key trend. Such materials will play vital roles in next-generation technologies including wearable electronics, autonomous systems, and responsive biomedical implants. Their integration is particularly promising in wearable electronics, where flexibility, sensitivity, and durability are critical, and in autonomous systems that require real-time feedback and adaptability. In the biomedical field, multifunctional nanomaterials are being developed for responsive implants that can monitor healing, deliver drugs, and repair themselves. The convergence of material science with electronics, biotechnology, and soft robotics is driving innovation in this area, paving the way for next-generation devices that are intelligent, efficient, and highly integrated.

8.4.2.4 Hybrid and Hierarchical Structures

Combining top-down and bottom-up approaches, researchers are designing hierarchical nanostructures that incorporate multiple types of materials (e.g., organic-inorganic hybrids) for enhanced performance. These hybrid materials offer tunable properties and open new applications in energy storage, catalysis, and nanoelectronics. Such hierarchical architectures are particularly valuable in energy storage devices, where they improve charge transport and mechanical stability, and in catalysis, where they increase active surface area and selectivity. In nanoelectronics, these materials enable the fabrication of components with greater functionality and miniaturization potential. The synergy between structural complexity and material diversity offers vast opportunities for innovation, paving the way for next-generation technologies that demand high efficiency, multifunctionality, and material adaptability.

8.4.2.5 Standardization and Regulation

With growing commercial interest, there is a strong push toward developing standardized protocols, certification frameworks, and international regulatory guidelines for nanomaterials. This will facilitate safer production, transport, application, and disposal while building trust among consumers and industries. Moreover, standardized testing methods and material characterization protocols will enhance transparency, reproducibility, and quality assurance, which are critical for both industrial adoption and public acceptance. By fostering greater trust among stakeholders—including manufacturers, consumers, researchers, and regulators—such frameworks will play a key role in supporting the responsible and sustainable development of nanotechnology on a global scale.

8.4.2.6 Expansion into Emerging Fields

Nanomaterials are expected to play a transformative role in emerging areas such as quantum computing, space exploration, personalized medicine, and synthetic biology. Their integration into advanced systems will require interdisciplinary collaboration, combining knowledge from physics, biology, chemistry, and engineering. Additionally, in synthetic biology, they serve as scaffolds and functional components for engineered biological systems. The successful integration of nanomaterials into these advanced applications demands interdisciplinary collaboration, blending insights from physics, chemistry, biology, and engineering to design materials and systems that are efficient, scalable, and responsive to complex functional demands.

8.5 CONCLUSION

This review has comprehensively examined the synthesis techniques and diverse applications of nanomaterials, offering insight into current practices as well as the future trajectory of the field. Two major synthesis strategies—top-down and bottom-up approaches—form the foundational methodologies for producing nanostructures.

Etching, lithography, and mechanical milling are examples of top-down techniques. Precision and scalability are advantages of this, especially in semiconductor production and microfabrication. On the other hand, bottom-up methods like sol-gel synthesis, CVD, hydrothermal processes, and green/biological synthesis provide greater control over the size, shape, and crystallinity of the particles, making them highly suitable for energy, biomedical, and catalytic applications. The choice of method is often dictated by the desired application, scalability, cost, environmental considerations, and structural requirements.

The applications of nanomaterials have grown exponentially, shifting from laboratory-scale innovation to real-world deployment. They have enabled more efficient energy storage systems, advanced drug delivery platforms, ultra-sensitive sensors, and environmentally sustainable solutions. Their role in developing high-strength lightweight materials, self-healing composites, and smart coatings is particularly noteworthy in emerging technologies. Additionally, the integration of nanomaterials with AI, ML, and the IoT is beginning to define a new era of smart and adaptive material systems.

Despite the tremendous progress, the field of nanomaterials still faces several critical challenges. These include the need for environmentally friendly and economically viable synthesis methods, long-term stability of nanostructures in operational environments, potential toxicity and health risks, and regulatory hurdles related to commercialization and large-scale deployment. Addressing these issues requires interdisciplinary collaboration, robust safety evaluation frameworks, and innovation in material design and processing.

Looking ahead, the future direction of nanomaterials research is expected to emphasize sustainability, functionality, and scalability. There will be increased focus on green synthesis techniques, bio-compatible materials, and circular economy practices. Moreover, the application of AI and data-driven modelling is anticipated to accelerate material discovery and process optimization, leading to faster development cycles and smarter design of nano systems.

In conclusion, nanomaterials hold immense promise for solving some of the most pressing global challenges from energy shortages and climate change to healthcare and sustainable manufacturing. With continuous advancements in synthesis techniques and a deeper understanding of nanoscale phenomena, nanomaterials are poised to remain at the forefront of scientific and technological innovation for decades to come.

REFERENCES

Al-Thani, Alshayma N., Asma Ghafoor Jan, Mohamed Abbas, Mithra Geetha, and Kishor Kumar Sadasivuni. 2024. "Nanoparticles in Cancer Theragnostic and Drug Delivery: A Comprehensive Review." *Life Sciences* 352 (September): 122899. https://doi.org/10.1016/j.lfs.2024.122899

Asadizanjani, Navid, Himanandhan Reddy Kottur, and Hamed Dalir. 2025. "Etching Techniques and Applications in Advanced IC Packaging." In, 61–79. https://doi.org/10.1007/978-3-031-86102-4_4

Basu, Prithvi, Jyoti Verma, Vishnuram Abhinav, Ratneshwar Kumar Ratnesh, Yogesh Kumar Singla, and Vibhor Kumar. 2025. "Advancements in Lithography Techniques and Emerging Molecular Strategies for Nanostructure Fabrication." *International Journal of Molecular Sciences* 26 (7): 3027. https://doi.org/10.3390/ijms26073027

Bora, Tonmoy J., Subir Biswas, Arpita Devi, Kusum K. Bania, and Magdi E. A. Zaki. 2025. "Top-Down Approaches of Nanomaterial Synthesis." In, 425–52. https://doi.org/10.1007/978-3-031-84643-4_14

Campbell, Eric, Alex Brown, Huynh Tam Minh Nguyen, Kelin He, Munkhbayar Batmunkh, and Yu Lin Zhong. 2025. "Recent Advances in Selective Chemical Etching of Nanomaterials for High-Performance Electrodes in Electrocatalysis and Energy Storage." *Small* 21 (8). https://doi.org/10.1002/smll.202409552

Etafa, Yonas, Sampandam Elangovan, Kebede Legesse, and Misganu Chewaka. 2024. "Synthesis and Fabrication Techniques for Nanomaterials." In, 241–66. https://doi.org/10.4018/979-8-3693-6326-3.ch011

Gaspar, Vítor M., Pedro Lavrador, João Borges, Mariana B. Oliveira, and João F. Mano. 2020. "Advanced Bottom-Up Engineering of Living Architectures." *Advanced Materials* 32 (6). https://doi.org/10.1002/adma.201903975

George, Mariat, Rosemary Jose, and T. R. Anju. 2025. "Bottom Up Approaches in Nanomaterial Synthesis." In, 453–81. https://doi.org/10.1007/978-3-031-84643-4_15

Gohar, Osama, Muhammad Zubair Khan, Iram Bibi, Nadia Bashir, Urooj Tariq, Manahil Bakhtiar, Muhammad Ramzan Abdul Karim, Farman Ali, Muhammad Bilal Hanif, and Martin Motola. 2024. "Nanomaterials for Advanced Energy Applications: Recent Advancements and Future Trends." *Materials & Design* 241 (May): 112930. https://doi.org/10.1016/j.matdes.2024.112930

Gupta, Shilpa. 2025. "Review—Challenges and Innovations in Nano-Computing and Reliable Interconnect Technologies." *ECS Journal of Solid State Science and Technology* 14 (1): 011003. https://doi.org/10.1149/2162-8777/ada825

Habeeb Dolapo Salaudeen, Rebecca Dupe Akinniranye, Michael Ibukun Kolawole, and Suleiman Yahaya. 2024. "Nanomaterials in Electronics: Advancements and Challenges in High-Performance Devices." *World Journal of Advanced Research and Reviews* 24 (1): 830–45. https://doi.org/10.30574/wjarr.2024.24.1.3116

Han, Zongsu, Yihao Yang, Joshua Rushlow, Jiatong Huo, Zhaoyi Liu, Yu-Chuan Hsu, Rujie Yin, et al. 2025. "Development of the Design and Synthesis of Metal–Organic Frameworks (MOFs) – from Large Scale Attempts, Functional Oriented Modifications, to Artificial Intelligence (AI) Predictions." *Chemical Society Reviews* 54 (1): 367–95. https://doi.org/10.1039/D4CS00432A

Hosny, Shimaa, Lamiaa Z. Mohamed, Mona S. Ragab, Qusi K Alomoush, Ehab M. Abdalla, and Samar A. Aly. 2025. "Nanomaterials in Biomedical Applications: Opportunities and Challenges—A Review." *Chemical Papers* 79 (5): 2657–78. https://doi.org/10.1007/s11696-025-03937-5

Joy, Sindhu Rachel, T. R. Anju, Jameel M. Al-Khayri, and Shri Mohan Jain. 2025. "Nanotechnology Era: Nanomaterial Green Synthesis." In, 1–30. https://doi.org/10.1007/978-3-031-84643-4_1

Kiani, Maryam. 2025. "Nanomaterials Unveiled: Catalyzing the Future of Technological Innovation." https://doi.org/10.2139/ssrn.5086632

Kulkarni, Madhusudan B., Narasimha H. Ayachit, and Tejraj M. Aminabhavi. 2022. "Recent Advancements in Nanobiosensors: Current Trends, Challenges, Applications, and Future Scope." *Biosensors* 12 (10): 892. https://doi.org/10.3390/bios12100892

Kumar, R. Deepak Suresh, L. Hrithick Kumar, S. K. Jeeva Roshini, Jobin Varghese, and Lokesh Singh. 2024a. "Nanotechnology in Mechanical Engineering - A Review." *Materials Science Forum* 1143 (December): 47–70. https://doi.org/10.4028/p-abdB7G

Kumar, Rinku, Priyanshu Kumar, Saurabh Gupta, Gourav Paliwal, Manoj Kumar, and Milan Singh. 2024b. "A Review on Advancements in Nanomaterial Synthesis, Characterization, and Applications: Towards Sustainable and Innovative Solutions." *Brazilian Journal of Physics* 54 (4): 137. https://doi.org/10.1007/s13538-024-01528-5

Lakhani, Pratikkumar, Dhavalkumar Bhanderi, and Chetan K. Modi. 2024. "Nanocatalysis: Recent Progress, Mechanistic Insights, and Diverse Applications." *Journal of Nanoparticle Research* 26 (7): 148. https://doi.org/10.1007/s11051-024-06053-9

Lin, Hengyu Kengsley, Tian-Hao Yan, Sajid Bashir, and Jingbo Louise Liu. 2022. "Synthesis of Nanomaterials Using Bottom-up Methods." In *Advanced Nanomaterials and Their Applications in Renewable Energy*, 61–110. Elsevier. https://doi.org/10.1016/B978-0-323-99877-2.00003-5

Mahmud, Md Zobair Al. 2023. "A Concise Review of Nanoparticles Utilized Energy Storage and Conservation." *Journal of Nanomaterials* 2023 (November): 1–14. https://doi.org/10.1155/2023/5432099

Malik, Shiza, Khalid Muhammad, and Yasir Waheed. 2023. "Nanotechnology: A Revolution in Modern Industry." *Molecules* 28 (2): 661. https://doi.org/10.3390/molecules28020661

Mehta, Sourabh. 2025. "Introduction to Nanoparticles, Synthesis Characterization and Clinical Status." In *Nanoparticles in the Management of Atherosclerosis*, 83–214. Cham: Springer Nature Switzerland. https://doi.org/10.1007/978-3-031-46752-3_2

Prasad, Rai Dhirendra, Rai Surendra Prasad, Rai Bishvendra Prasad, Saurabh R Prasad, Shashi Bhushan Singha, Rai Jitendra Prasad, Pramod Sinha, et al. 2024. "A Review on Modern Characterization Techniques for Analysis of Nanomaterials and Biomaterials." *ES Energy & Environment*. https://doi.org/10.30919/esee1087

Saravanan, A., P. Senthil Kumar, R.V. Hemavathy, S. Jeevanantham, Marie Jyotsna Jawahar, J.P. Neshaanthini, and R. Saravanan. 2022. "A Review on Synthesis Methods and Recent Applications of Nanomaterial in Wastewater Treatment: Challenges and Future Perspectives." *Chemosphere* 307 (November): 135713. https://doi.org/10.1016/j.chemosphere.2022.135713

Shahzad, Khurram, Andrei Ionut Mardare, and Achim Walter Hassel. 2024. "Accelerating Materials Discovery: Combinatorial Synthesis, High-Throughput Characterization, and Computational Advances." *Science and Technology of Advanced Materials: Methods* 4 (1). https://doi.org/10.1080/27660400.2023.2292486

Singh, Ajay Vikram, Peter Laux, Andreas Luch, Chaitanya Sudrik, Stefan Wiehr, Anna-Maria Wild, Giulia Santomauro, Joachim Bill, and Metin Sitti. 2019. "Review of Emerging Concepts in Nanotoxicology: Opportunities and Challenges for Safer Nanomaterial Design." *Toxicology Mechanisms and Methods* 29 (5): 378–87. https://doi.org/10.1080/15376516.2019.1566425

Tiwari, Santosh K., Michał Bystrzejewski, and Yanqiu Zhu. 2024. "Synthesis, Properties, and Applications of Carbon-Encapsulated Metal Nanoparticles." *Coordination Chemistry Reviews* 520 (December): 216125. https://doi.org/10.1016/j.ccr.2024.216125

Zain Ul Abidin, Muhammad, Muhammad Ikram, Sawaira Moeen, Ghazanfar Nazir, Mohammed Benali Kanoun, and Souraya Goumri-Said. 2024. "A Comprehensive Review on the Synthesis of Ferrite Nanomaterials via Bottom-up and Top-down Approaches Advantages, Disadvantages, Characterizations and Computational Insights." *Coordination Chemistry Reviews* 520 (December): 216158. https://doi.org/10.1016/j.ccr.2024.216158

Part 3

Applications Across Industries

9 Nanomaterials in Contemporary Science
Applications and Cognate Risks

Maitree Singh and Gurpreet Kaur

9.1 INTRODUCTION

For thousands of years, *Homo sapiens* have built civilizations by taking inspiration from Mother Nature, evident in the form of innovations, discoveries, and inventions. For every other question, the answer lies in the lap of nature. For instance, a simple natural phenomenon like the falling of an apple inspired Newton to fabricate the three laws of motion, which connected so many dots in the world of physics. Like this, scientists have always been enticed with the notion of minimization of sizes on the atomic scale and the related changes in properties as the aftermath, thus marking a start of the uprising of "Nanotechnology."

Nanotechnology is studied as a branch of science and engineering which revolves around the fabrication of structures by playing with atoms and molecules at the nanoscale. These studied structures are named as nanomaterials. The dimensions of these materials expand between 1 and 100 nanometers, which is the very reason for their unique properties (Sajid 2021, 25). There is no single conclusive definition for explaining nanomaterials, as few scientists consider this term when the size is few nanometers while others differ in using this term when anything is less than a micrometer (Baig et al. 2021, 6). At least half of the total number of particles must fit into one of the mentioned standards:

i. Retaining one or more exterior dimensions between 1 and 100 nanometers (nm).
ii. Showcasing an elongated shape, such as a rod or fiber.
iii. Possessing a platelet-like shape with one outer dimension lesser than 1nm and the other two more than 100nm (Gao and Lowry 2018, 9).

However, Borm et al. (Borm et al. 2006, 3) focus on some of the important definitions in light of contemporary advancements, like that of the Europische Akademie in recent times, which explains nanotechnology as a consortium of functional systems through sub-groups having certain size-dependent properties ruling individually or as a collective arrangement.

DOI: 10.1201/9781003560234-12

The Royal Society draws attention to nanoscience as studying of phenomena and modulating matter at atomic or molecular scale where the claims vary with that of the macroscopic materials (Hobson and Guy 2014, 434–36).

Nanotechnology refers to identification, production, and consumption of structures, devices, and systems in which shape and size are amended at the nanometer scale (Meyer and Kuusi 2006, 181–99).

Nanomaterials can prevail in a surrounding dimension of lower than 1 nm looking like 1×10^{-9} m in contrast to a human hair which has a thickness of approximately 80,000 nm, while DNA double helix shows a diameter of 2 nm (Gajanan and Tijare 2018, 5). Nanomaterials can be seen in the macrocosm, or could also be isolated as obtained byproducts from variety of processes such as ignition reactions. They can even be engineered for a specific concern keeping the ulterior motive in mind. These can exist as organic and inorganic materials (e.g., ashes from volcanic eruption, finer particles of dust and sand, sea spray), biomolecules (e.g., protein structures in viruses), minerals (e.g., in clay and shells), colloids (e.g., in milk, blood, and fog) as well as many others that can be summed up in a central theme (Nsw 2023) (Sharma et al. 2015, 44). Nanomaterials can be easily produced by many novel processes like sol-gel or hydrothermal techniques, giving them a range of structural forms such as carbon nanotubes, graphene, quantum dots, and nanomaterials of metal oxides like gold (Khan et al. 2017, 12). Fascinatingly, the origin of nanomaterials can be traced back to 2000 years, i.e., in the period of the Roman Empire, where evidently a mass of Au nanoparticles was employed to create brilliant colors (Pitkethly 2004, 7). These evolved further owing to advancements in modern computing power along with several new methodologies combined with characterizations such as atomic force microscopy (AFM) (Pitkethly 2004, 7).

Nanoparticles are usually susceptible to being crystalline or amorphous in nature. The two are worlds apart regarding the physio-chemical, biological, and magnetic parameters. Significant research on zinc oxide (ZnO) nanocrystals prove them to be more bio-compatible than amorphous forms because of the larger disorderliness of native atomic bonds, causing them to be more reactive (Sapno et al. 2001, 16) Reis et al. 2015, 84). Nanocrystals are, on the other hand, showcases atomic periodicity with better crystalline arrangements, ensuring fewer defects (Hoang and Ganguli 2012, 25) (Moreau et al. 2013, 25) (Jiang et al. 2008, 2).

It is indeed remarkable that the nanomaterials can be tailored by changing the physical, chemical, optical, and biological parameters of their bulk counterparts. These changes come about with diverse configurations giving rise to new cutting-edge technologies (Razavi et al. 2018, 39; Janfaza et al. 2019, 186; Hashemzadeh et al. 2020, 10; Garimella and Eltorai 2016, 14; Hashemzadeh et al. 2019, 11). Researchers in Japan observed that a metal such as copper or silver, having good capacity for conducting electricity and heat, will lose such behavior when its dimension shrinks to the nanoscale, becoming non-conducting and non-thermal at this level (Shi et al. 2015, 1–23). Although many, health is the most promising sector among others that nanotechnology will affect, such as cosmetics, aerospace, food inspection, and others. In the application of healthcare, nanomaterials take the lead in drug delivery, many biosensors, bone tissue surgery, and so on. In addition, good SPF sunscreens with quality nanomaterials are equally essential, considering how much they protect from the cancerous UV-radiation. The carbon nanotubes have always been favored for morphing of aircraft wings in aerospace industry.

9.2 A COMPARATIVE STUDY BETWEEN NANOMATERIALS AND BULK MATERIALS

Table 9.1 shows how nanomaterials differ from bulk materials in many aspects like their size, features, and applications.

9.3 CLASSIFICATION OF NANOMATERIALS

Table 9.2 shows how nanomaterials can be classified into various categories depending on the mentioned parameters.

9.4 SYNTHESIS OF NANOMATERIALS

Rapid progress in science has opened the door for different ways in which nanomaterials can create a comparatively wider range of shapes and show distinctive properties. These processes can mainly be divided into two categories, viz., top-down and bottom-up. The top-down and bottom-up methods depend on the primary phase of the desired material. If this phase of the material is solid, top-down methods will be adopted. Alternatively, if the primary phase of the material is liquid or gaseous, bottom-up methods will be followed.

TABLE 9.1
Comparison Between Nanomaterials and Bulk Materials

Nanomaterials	Bulk Materials
Material lying within the range between 1nm and 100nm on a nanoscale.	Materials possessing size above 100nm, be it any dimension.
These can only be observed and analyzed through specialized instruments.	One can see it clearly with naked eye or a microscope.
Substantial surface-to-volume ratio aids performance, such as in catalysis and biosensors.	Lower surface-to-volume ratio facilitates a greater performance in phenomenon such as catalysis.
Larger aggregation of atoms or molecules bestows unique properties to the materials.	Lower percentages of atoms or molecules are noted on the surface.
Able to showcase unique, distinctive scattering properties.	Bulk materials possess ordinary scattering properties.
Physical and chemical properties can be manipulated with a change in size and shape.	Physical and chemical properties of these materials cannot be altered.
These can show confined energy states in the band structure in case of semiconductors.	They may not show confined energy states in the band structure in case of semiconductors.
The rate of adsorption and absorption of molecules is much more in these materials.	These materials tend to acquire a lower rate of adsorption and absorption.
Examples of nanomaterials are: carbon quantum dots, nanorods, nanosheets.	Examples of bulk materials are: cement, ore, salts.

TABLE 9.2
Classification of Nanomaterials

Classification Based On	Types	Size Distribution	Examples
Structural configuration	I. Organic nanomaterials	1nm–100nm	Dendrimers, liposomes
	II. Inorganic nanomaterials		Gold nanoparticles, silver nanoparticles
	III. Carbon-based nanomaterials		Fullerenes, graphene
	IV. Composite nanomaterials		Carbon nanotube polymer, graphene metal-oxide composites
Origin	I. Natural nanomaterials	1nm–100nm	Volcanic ash, smoke
	II. Artificial nanomaterials		Carbon nanotubes, Quantum dots
Number of dimensions	I. 0D	<100nm	Clusters, nanospheres
	II. 1D	<100nm	Nanowires, nanorods
	III. 2D	<100nm	Thin films, nanoplates
	IV. 3D	>100nm	Polycrystals, foams
Porosity	I. Microporous nanomaterials	0.5nm–2nm	Zeolites, silica nanoparticles
	II. Mesoporous nanomaterials	2nm–50nm	Mesoporous silica nanoparticles
	III. Macroporous nanomaterials	50nm to several micrometers	Macroporous silica nanomaterials, macroporous polymer nanomaterials
Amount of toxicity	I. Fiber-like nanomaterials	Tens to hundreds of nanometers	Electrospun nanofibers, cellulose nanofibers
	II. Biopersistent granular nanomaterials	1nm–100nm	Nano zero-valent iron (nZVI), titanium oxide nanoparticles (TiO_2)

9.4.1 Top-Down Synthesis of Nanomaterials

This approach is subject to using physical forces or chemical treatments responsible for reducing the larger to smaller particles in the nanoscale dimension range, as indicated in Figure 9.1. Some of the techniques employed under top-down approaches include laser ablation method, mechanical milling method, etching, thermal decomposition method, sputtering method, and nanolithography. These are briefly explained in the following subsections.

9.4.1.1 Laser Ablation Method

Laser ablation method is favored as a sustainable mechanism for the production of nanomaterials. A high-power pulsed laser beam is focused on the target material, producing an extensive electric field that also heats the target area. As a consequence,

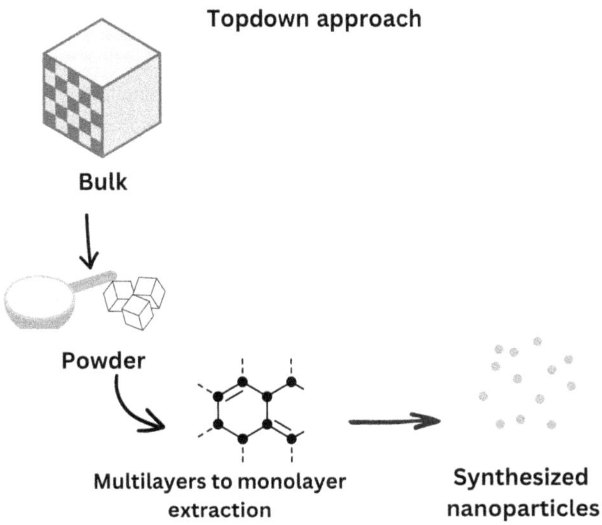

FIGURE 9.1 Schematic diagram showing top-down synthesis of nanomaterials.

the material evaporates, converting it into a plasma state, further rapidly expanding and cooling, leading to the formation of nanoparticles. This method ensures the purity of metal to a higher degree and facilitates the ease in controlling the nanoparticle sizes (Zhang and Guan 2014, 125–69).

9.4.1.2 Mechanical Milling Method

This technique of mechanical milling is also called ball milling. Such a simple procedure is more favored in terms of the large-scale industrial production of nanoparticles. The outside force is one of the most important ruling components for generating a heavy impact due to the dropping of the balls from the highest point of the chamber. The desired material is pounded into utterly finer particles in the powdered form.

9.4.1.3 Etching Method

Etching method comes under the top-down approach techniques for the synthesis of nanomaterials. This is a physical process in which selective removal of material from the desired sample is performed in order to create nanomaterials. Usually, the surface areas that show the greatest lattice distortion are etched out.

9.4.1.4 Thermal Decomposition Method

Thermal decomposition method, which is an endothermic process, can be employed in the synthesis of the nanoparticles. In this technique, the desired sample is subjected to very intense heating, which leads to the disintegration of chemical bonds into smaller ones when it reaches the threshold temperature for that specific compound (Mekuye and Abera 2023, 4). This method of chemical reaction is mainly used for the production of metal-oxide nanomaterials.

9.4.1.5 Sputtering Method

Sputtering is a physical phenomenon used for the production of nanoparticles. By utilizing the plasma energy, atoms from the required material can be removed with this process (Mekuye and Abera 2023, 4). The particles which are ejected due to this collision get deposited on the substrate. Sputtering method is widely popular for producing nanomaterials with a wide range of shapes like rods, circular, etc. (Baig et al. 2021, 6; Bu and Hu 2016, 28).

9.4.1.6 Nanolithography

Lithography falls under the top-down technique for the synthesis of nanomaterials. Using this technique, the light-sensitive material is imprinted with the desired shape, then painstakingly a portion of the material is eliminated to have the intended structural configuration (Mekuye and Abera 2023, 4). Various sources of radiation, like X-ray, are employed after reviewing the motive behind the necessity of that material. One of the complexities in this method is the demand for a more precise and costly instrument.

9.4.2 Bottom-Up Synthesis of Nanomaterials

The bottom-up approach involves the synthesis of nanoparticles from clusters which are combination of the atoms as shown in Figure 9.2. This method is also referred to as the constructive method. Few of the methods which come under the bottom-up technique include spinning method, pyrolysis method, sol-gel method, chemical vapor deposition (CVD) method, and biosynthesis method. These methods are briefly explained in the following subsections.

9.4.2.1 Spinning Method

The spinning method, which comes under the bottom-up approach, can be utilized in the synthesis of nanoparticles. It is also known for its cost-effectiveness. In this process, nanoparticles like the magnetic ones are made with the help of a spinning

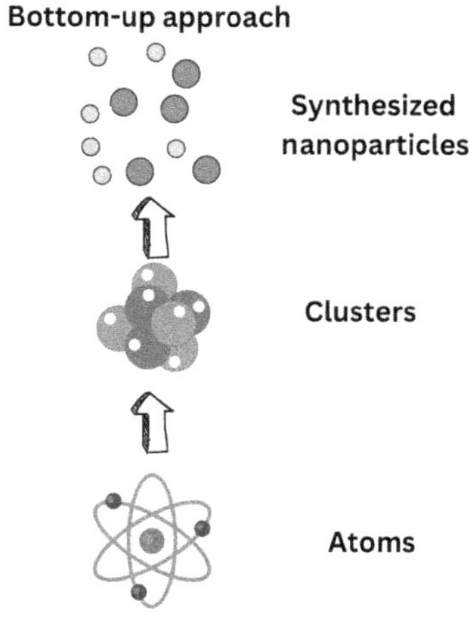

FIGURE 9.2 Schematic diagram showing bottom-up synthesis of nanomaterials.

disk reactor. The rotating disk is situated inside a reactor where important parameters like temperature can be managed (Mekuye and Abera 2023, 4).

9.4.2.2 Pyrolysis Method

For the extensive production of nanomaterials across the industries, the pyrolysis method is the most favored one. It is counted among one of the most efficient processes with a larger production capacity. When a precursor existing in a liquid or gaseous state is allowed to pass through an opening with a larger pressure, the resulting product, which is solid, is subjected to air to extract the nanoparticles (Mekuye and Abera 2023, 4).

9.4.2.3 Sol-Gel Method

Sol-gel method is a wet-chemical and economic method which is opted widely for the manufacturing of nanoparticles like zinc oxide (ZnO) with a diverse range of shapes and sizes. In this process, a colloidal solution called sol is created which later serves as a precursor for the gel. The product thus formed is a solid created from small molecules. It is easier to control the temperature in this method (Bokov et al. 2021, 1).

9.4.2.4 CVD

CVD is a vacuum deposition technique that has produced nanomaterials of the highest purity. It involves the following two important stages: transportation and deposition (Bube 2001, 873–79). The gaseous substances are transported into a chamber inside which the reaction occurs (Mekuye and Abera 2023, 4). The heated substrate is brought into contact with this chemical reaction, whereupon the thin film of the nanomaterial is deposited onto the surface. When the heated substrate comes in contact with this chemical reaction, a thin film of the nanomaterial is deposited on the surface. However, the complexity and high cost of instruments required render this technique disadvantageous.

9.4.2.5 Biosynthesis Method

Biosynthesis is known as an eco-friendly and green approach for the production of the non-harmful nanoparticles possessing biological properties (Mekuye and Abera 2023, 4). This includes natural products derived from plants, fungi, algae, and bacterial species of the flora and fauna. This approach purely depends on the biocompatibility property; hence, it is mostly favored for biomedical applications (Kokima et al. 2024, 1–24).

9.5 PROPERTIES OF NANOMATERIALS

Nanomaterials, because of their small size, possess some unique properties. These properties make them possible to implement in many kinds of applications while taking up less space. Few of those properties can be enlisted as:

- Higher surface area
- Good catalytic activity
- Melting temperatures changes according to side variation
- Optical properties particularly observed in semiconductors
- Showcasing greater electrical conductivity

- Magnetic properties
- Quantum confinement

These materials have remarkable optical features in terms of larger light absorption and scattering capacity. They have gained a significant amount of attention for showing quantum confinement, plasmonic behavior, and phenomena like supermagnetism. Because of their minimal size, there is more transparency in coatings while maintaining high material strength along with being lightweight. As compared to the bulk materials, these due to having larger surface area give greater electrical conductivity, along with increasing the catalytic activity.

9.6 APPLICATIONS OF NANOMATERIALS

Few applications of nanomaterials as shown in Figure 9.3 are highlighted in the following subsections.

9.6.1 Nanomaterials in Food Safety

The happiness index and well-being of people residing in a nation are a reflection of the ongoing progress of a nation. Among many such factors, health acts as the backbone for ensuring people's welfare, as a healthy mass is able to perform better.

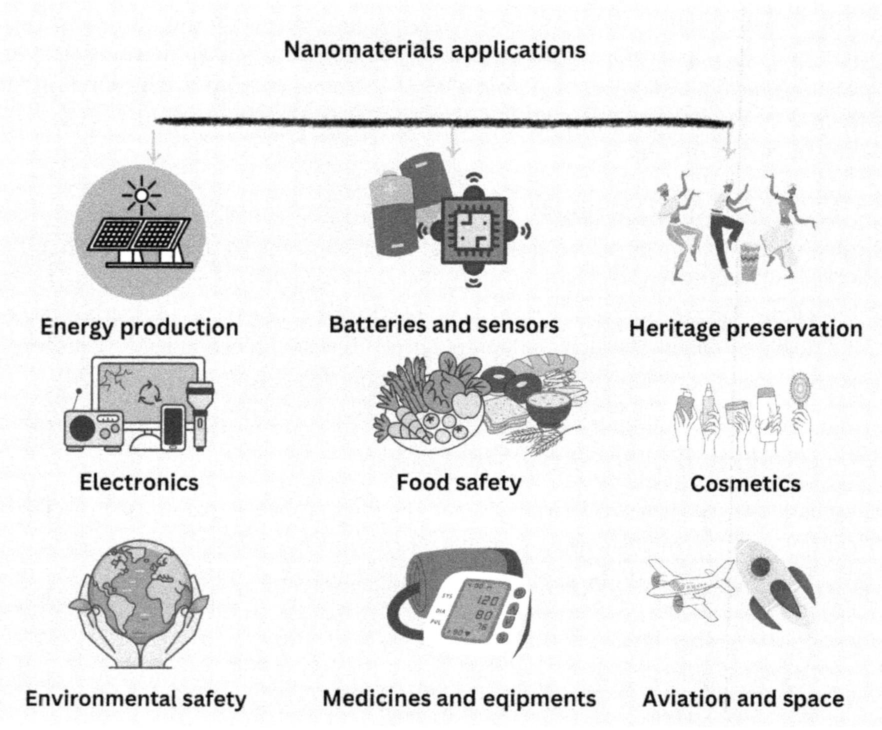

FIGURE 9.3 Applications of nanomaterials in various industries.

Exercises, vitamins, yoga, eating wholesome, organic and nutritional food, etc., are the ruling factors in an individual's holistic health. In this scenario, the food safety comes into the bigger picture. Food safety is a crucial parameter for ensuring that the food the public consumes is effectively free from contamination and is safe to consume. There are many prominent health conditions such as diarrhea, cancer, diabetes, etc., which are just a few of the diseases linked to agents such as bacteria, viruses, parasites, or even chemicals that may prove to be fatal. It is essential for industries to maintain very high standards regarding safety checks. The incorporation of nanomaterials in the food industry has been a real blessing lately. These nanomaterials are employed to enhance food safety, enrich food quality, support more extended shelf life with proper packaging, etc. These nanomaterials help in creating a smart packaging system which helps in detecting any type of toxic substances at the earliest, monitoring the condition of food, or even releasing antimicrobial agents like silver to halt the growth of harmful bacteria in order to prevent the food from spoiling (Bowman and Fitzharris 2007, 31). Moreover, nanomaterials can also be utilized as food additives, which boost the texture, flavor, and nutritional value of food. Nanoencapsulation is a technique being used for confining vitamins, minerals, or proteins within nanoparticles, which overall helps in fulfilling the nutritional requirements of the human body without compromising the taste (Singh et al. 2017, 8). Likewise, nanoemulsion is formulated by mixing two or more liquids like oil and water into a stable and uniform mixture. These nanoemulsions are used for the delivery of nutraceutical, coloring, flavoring agents and antimicrobial compounds as reported by Mohammad et.al. (2022, 2). The nanolaminates can be blended as edible coatings and films which act as a barrier for moisture, lipids, gas, etc., thus keeping the food products like fruits, chocolates, vegetables, meats, confectionery products, or dairy products fresh for consumption (Liakos et al. 2018, 143–163). The use of nanosensors in making an electronic tongue or nose aids in inspecting the condition of food by signaling through the gases released by the food items (Garcia et al. 2005, 113). There are countless instances where nanomaterials have become an important part in the food sector.

9.6.2 Nanomaterials in Environmental Protection

Environmental protection has been termed probably the most talked about and yet the most critical subject in the most recent years. Mainly, human activities jeopardize countless ecosystems with their delicate biodiversity, which supports life on this planet. There are so many problems, such as pollution, deforestation, the extinction of many great floras and faunas, global warming, and soil erosion, among others. Air and water pollution are two of the greatest threats that have to be constantly monitored and resolved. Air pollution is mostly caused by the exhaust of poisonous gases from traffic and the burning of fossil fuels, industrial exhaust, and other conditions which favor respiratory diseases. Other bodies of water are contaminated and turned into centers for breeding harmful pathogens, which cause severe forms of gastrointestinal diseases. The nanomaterials show a very promiscuous future in resolving these issues and would be in favor of promoting sustainable practices and improving environmental conditions. Strictly limited on earth are the sources of fresh water;

therefore, maintaining the purity of drinking water amidst pollution is too significant for us all. Desalination can be done more efficiently with the help of nanoporous materials which ease the removal of salts and other toxins from seawater while also ensuring cost-effectiveness. Zerovalent iron (ZVI) nanoparticles are popular for in situ reductive immobilization of heavy metals in soil as reported by Ibrahim et al. (Ibrahim et al. 2016, 23). Metal-based materials such as zinc, silver, or magnetite nanoparticles and polymeric nanoabsorbents have been reported to be more efficient for liquid waste management (Dermatas et al. 2018, 36). For tackling the problem of waste generation from extensive usage of fossil fuels, nanocatalysts support the storage of oxygen, hence ensuring complete combustion of fuel without any trace of waste. This greatly reduces the greenhouse emission (Chausali et al. 2023, 12).

9.6.3 Nanomaterials in Sensors, Fuel Cells, and Batteries

Nanomaterials possess the advantageous characteristics of larger surface area, tunable electronic properties, and catalytic activities. In the past few decades, fuel cells, batteries, and sensors have advanced in a noteworthy manner in material sciences and energy harvesting technologies. Nanomaterials have even enhanced the integration of these technologies by increasing their efficiency, durability, and performance. Platinum nanoparticles serve as electrocatalysts because of their higher surface areas, facilitating the conversion of hydrogen and oxygen into water and electricity, respectively (Lamsal et al. 2020, 1139). The higher surface area improves the catalytic productivity, thus facilitating smoother energy conversion. Carbon nanotubes (CNTs) and graphene are very suitable for improving electron transportation rate and durability in fuel cells when fused with electrodes (Kumar et al. 1966, 20). Metal-oxide nanoparticles such as zinc oxide or titanium oxide when employed in sensors improve sensitivity and selectivity which detect low concentration of analytes. Gold nanoparticles are preferred in biosensors as they can easily target biomolecules due to their biocompatibility property (Kumar et al. 1966, 20). In comparison to traditional graphite anodes, silicon nanoparticles are proven to increase the capacity and longevity of the batteries when fused with anodes and cathodes (Daulay et al. 2022, 6).

9.6.4 Nanomaterials in Energy Harvesting

Energy harvesting is accentuated on the shift toward sustainability as far as natural resources are concerned. The increasing pressure on fossil fuels and rising demand for energy-efficient and eco-friendly technologies has made this the most prominent sector, catering from consumer electronics to industrial automation. Several commonly consumed forms of energy, such as solar energy, hydro energy, thermal energy, vibrational energy, radio frequency energy, and kinetic energy, fall under this category in the present times. The nanomaterials have emerged as game-changing technology owing to the unique tailor-made properties they can possess. Nanomaterials offer innovative solutions for the capture and conversion of energies in an efficient manner. Quantum dots and perovskite nanoparticles help in maximizing the absorption of sunlight in the solar cells, hence producing electricity. Nanoporous and nanostructured materials are used in thermoelectric generators across the industries where electricity is produced from the heat energy. Nanofluids can absorb more solar radiation through the absorber in a solar collector, for greater heat transfer (Hamzat et al. 2022, 266). Nanomaterials are utilized in piezoelectric energy harvesters for serving many purposes like monitoring sensors, tire pressure sensors, and pacemakers (Phillips 2021, 3).

9.6.5 Nanomaterials in Medical Sector

Nanomaterials are now an integral part of medical science. It has countless biomedical applications. The minimization in their size makes them easier to get inside tissues, aiming specific spots where treatment is required. Some of the boons of the usage of nanoparticles in biomedical field are:

- Stabilization of drugs
- Efficacy
- Less side effects
- Reduce overall costs for treatments
- Greater biocompatibility
- Good solubility and absorption rate
- Sensitivity and accuracy of medical instruments
- Cost-effective instrumentation
- Quick response

The nanoparticles are utilized in the bio-imaging techniques like X-ray, positron emission tomography (PET), ultrasound, magnetic resonance imaging (MRI), computed tomography (CT) scans, etc. (Lai et al. 2012). Biosensors are presently ruling the medical industry as they help in detecting biomolecules like enzymes, tissues, microorganisms, acids, cells, etc. The infusion of nanoparticles like zinc oxide in these physio-chemical detectors has made the things way easier for providing the patients with early detection and correct treatment. Drug is delivered inside a human or animal body to provide a therapeutic effect with the help of mechanisms like diffusion, osmosis, and ion exchange. Nanoparticles play a pivotal role in this by taking charge of the agents and targeting the delivery site while also minimizing the toxicity. They act as a vehicle on which the drugs are encapsulated or chemical bonding takes place (McNamara and Tofail 2016, 2). Franco-Molina et.al. investigated the implementation of colloidal silver in the prospect of treating breast cancer (Franco-Molina et al. 2010, 29). Zhao et.al. reported about several kinds of bioactive nanomaterials suitable for wound healing, neurodegenerative disease therapy, cancer treatment, and as biocatalyst (Zhao et al. 2021, 1).

9.6.6 Nanomaterials in Cosmetic Industry

Recent trends indicate a continuing and even increasing growth of the cosmetics industry, which almost approaches a global figure of around \$500 billion. This industry is divided into several different product formats like skincare, makeup, hair care, fragrances, and personal hygiene. Cosmetics have had a long tradition as many discovered their usage from thousands of years back, during which ancient civilizations like the Romans, Egyptians, and Greeks were involved. Because of the effective technical and cultural advances, the cosmetic industry has undergone several transformations. Nanotechnology, for instance, allows for modifying and creating products with smoother texture, better absorption and efficacy as demonstrated by the usage of nanoparticles in sunscreens and anti-aging creams. Silica nanoparticles give the powder and foundation a silky texture which helps in reducing the fine lines and pores. Nanopigments are useful in developing dispersion of the pigments in the eyeshadows which brings more vibrant colors. The higher surface area of nanomaterials allows more transportation of desired ingredients through the skin barrier (Ahmad et al. 2018, 68). Skincare products like serum and essence have quite an amount of liposome nanoparticles and nanoemulsions. Fusion of nanoparticles in the makeup products also reduces the heavy makeup feel. Some of the most used nanomaterials in the cosmetic industry are ZnO, TiO_2, Silica, Gold, and silver nanoparticles (Gupta et al. 2022, 8). Nano-hydroxypatite is favored for incorporating it into oral hygiene products like toothpaste and mouthwash.

9.6.7 Nanomaterials in Aviation and Space Technology

The aviation and space industries have experienced stark growth in the last few years. Global passenger traffic, technological progress, expanding commercial opportunities, and lower fuel costs are some of the key factors in the rapid growth of the aviation industry. The miniaturization technology has accelerated the deployment of small satellites, which are now smaller and quicker to build. Such satellites are put to several uses—global broadcast Internet, agricultural information, disaster management and monitoring, climate oversight, and even national security. In this, nanomaterials have also made some greater contributions due to their special properties of high strength-to-weight ratios, greater thermal stability, and electrical conductivity. The combination of incredible strength while being lightweight is shown by nanomaterials like carbon nanotubes (Abbasi et al. 2020, 795–825). This feature improves the payload capacity. Nanomaterial composites increase the durability of the structures used in aircraft and spacecraft. Nanomaterials like graphene are used for their excellent thermal conductivity in cosmo flight missions as it is required to keep a check on extremely high temperatures during phenomena like entry and exit through the earth's atmospheric layer (Hu 2024, 60). Some of the nanomaterials possess the self-healing property which is indeed useful for repairing minor inconveniences which may lead to accidents. The use of nanomaterials in wing surface coating is known to reduce drag which improves the aerodynamics with good fuel efficiency. In a space mission where every gram counts, the miniaturization of nanomaterials comes into play. These are also used to create more accurate sensors for environmental monitoring which is crucial in space missions. For electrostatic discharge, EMI shielding and low-friction nanotubes and nanoparticles having nanographite and nanoaluminum are being used. When nanocarbon quantum dots are fused with protective coatings, they are shown to improve performance (Weerasinghe et al. 2023, 13).

9.6.8 Nanomaterials in Cultural Heritage Protection

Cultural heritage is just like a hieroglyph honoring all the ancient civilizations which have existed since the beginning of human evolution. Therefore, keeping them protected is important, since they act as a mirror to peek into the past happenings. Cultural heritage protection involves safeguarding sites, artifacts, and practices that hold historical, artistic, social, or spiritual significance for a community or a nation. This covers ancient monuments, historic buildings, archaeological sites, traditional music, languages, and rituals. Preservation of cultural heritage is vital to conserving the identity and history of peoples and fostering respect and cultural exchanges. Nanomaterials such as nano-silica and nano-calcium hydroxide are used to merge fragile materials like stone, wood, and plaster. These penetrate into the depth of the microstructure of the material, hence enhancing its mechanical features with strong binding. Nano-lime has always been favored for the restoration of age-old frescoes and wall paintings without leaving any damage. Photocatalytic nanoparticles like titanium dioxide (TiO_2) are implemented as photocatalytic coatings which break down organic contaminants on surfaces when exposed to light, making them easier to clean (Chobba et al. 2023). As acidic degradation is a very common problem in paper and textile artifacts, nano-magnesium hydroxide can neutralize acids by stopping further degradation, extending the shelf life. Nanoparticles of cerium oxide acts as corrosion inhibitors which prevents oxidation. Anti-graffiti nanocoatings are applied to prevent vandalism in places of cultural importance. Alkoxysilane-based formulations, specifically tetraethoxysilane (TEOS) and methyltrimethoxysilane (MTMOS), are two of the most universally used stone consolidants, largely due to their ability to penetrate easily inside the porous matrix (Sierra-Fernandez et al. 2017, 67). Thymol-loaded SiO_2 nanoparticles possess hydrophobic properties along with biocidal capability to fight against bacteria. These nanoparticles also protect the ancient stones, ensuring a significant reduction in material loss (Kanth and Soni 2022, 5).

9.6.9 Nanomaterials for Optoelectronic Devices

Optoelectronic devices are those which can generate, detect, and control light. Optoelectronic devices include LEDs, solar cells, photodetectors, lasers, etc., which are easier for interaction with light. The important properties of nanomaterials that make them useful in optoelectronic devices are

- Enhanced optical emission due to quantum confinement
- Larger quantum efficiencies
- Controlling wavelengths with great precision
- Ability for good light absorption
- Directional interaction with light

Nanomaterials serve a wider range of applications. Quantum dots due to its good electrical and optical properties are used in displays like QLEDs, solar cells, photodetectors, etc. Transition metal dichalcogenides (TMDs) have direct bandgaps in monolayer form, which makes them apt for thin-film transistors and light-emitting devices. Nanowires made from materials like silicon and gallium nitride (GaN) due to their higher surface area have excellent charge transport qualities and strong light matter interaction. Halide perovskites are good nanomaterials for their high absorption coefficients, tunable bandgaps, and high charge carrier mobilities. Plasmonic nanoparticles have the ability to confine and increase electromagnetic fields at their surfaces, which can be used in solar cells, sensors, etc. TiO_2 nanomaterials are used in UV photodetectors, memory resistors, etc. (Ge et al. 2023, 13). Cds and CdSe nanoparticles are a good option for creating photovoltaic devices (Loo et al. 2005, 5).

9.7 HAZARDS OF NANOMATERIALS

Nanomaterials certainly possess huge applications in our daily lives today. These come with their own share of threats, though. Some nanoparticles metamorphose into a toxic form when inhaled deep in the lungs. This can cause many severe respiratory concerns like inflammation, fibrosis, or lung cancer. Although skin acts as a natural barrier for the human body, nanomaterials when penetrated in excessive amounts can trigger potential allergies. If ingested, it may also lead to gastrointestinal toxicity because of its flow through the bloodstream. Some nanomaterials are capable of causing cell damage, leading to cytotoxicity or genotoxicity which causes mutations. When the concentration of nanomaterials increases in the bodies of the aquatic animals, it might put a stop to their growth, reproduction, and survival. These materials also possess a risk of contaminating groundwater or igniting sudden combustion due to sudden reaction. For the risk mitigation of airborne diseases due to nanomaterials utilization of contaminant systems, proper ventilation and filtration can be of help. Regular monitoring and detection of contamination due to the nanoparticles in the water bodies should be done. Along with this, proper guidance and awareness about the safe usage and disposal of nanoparticles should be encouraged, specifically in workplaces.

9.8 CONCLUSION

Richard Feynman's discussion about the manipulation of materials at the atomic or molecular level opened the doors for nanotechnology. Today, nanomaterials with many of their exceptional properties are ruling the industry worth billions like the semiconductor industry. As the demand for various ranges of nanomaterials grows in the market, many new cost-effective methods for their production are being devised. This chapter highlights few of those methods along with the extensive applications of the nanomaterials in various sectors. However, there is still not much research done about the toxicity and safety assessment of nanomaterials. It might be challenging, but it is still very essential in order to generate systems and products which are safe to use and have good proficiency.

REFERENCES

Abbasi, S., Peerzada, M. H., Nizamuddin, S., & Mubarak, N. M. (2020). Functionalized Nanomaterials for the Aerospace, Vehicle, and Sports Industries. In *Elsevier Handbook of Functionalized Nanomaterials for Industrial Applications.*

Ahmad, Usama, Ahmad, Zeeshan, Khan, Ahmed, Akhtar, Juber, Singh, Satya, & Ahmad, Farhan. (2018). Strategies in Development and Delivery of Nanotechnology Based Cosmetic Products. *Drug Research*, 68(10), 545–552.

Baig, N., Irshad K., & Falath, W. (2021). Nanomaterials: A Review of Synthesis Methods, Properties Recent Progress, and Challenges. *Materials Advances*, 2(6), 1821–1871.

Bokov, D., Jalil, A. T., Chupradit, S., Suksatan, W., Ansari, M. J., Shewael, I. H., Valiev, G. H., & Kianfar, E. (2021). Nanomaterial by Sol-Gel Method: Synthesis and Application. *Advances in Materials Science and Engineering*, 2021(1), 5102014.

Borm, P. J. A.., Robbins, D., Haubold, S., Kuhlbusch, T., Fissan, H., Donaldson, K., Schins, R., Stone, V., Kreyling, W., Lademann, J., Krutmann J., Warheit, D., & Oberdorster, E. (2006). The Potential Risks of Nanomaterials: A Review Carried Out for ECETOC. *Particle and Fibre Toxicology*, 3, 11.

Bowman, D. M., & Fitzharris, M. (2007). Too Small for Concern? Public Health and Nanotechnology. *Australian and New Zealand Journal of Public Health*, 31(4), 382–384.

Bu, Ian Y. Y., & Hu, T. H. (2016). P-type ZnO Films Deposited by DC Reactive Magnetron Sputtering Using Codoping Process. *Journal of Materials Science Materials in Electronics*, 28(3), 2852–2858.

Bube, R. H. (2001). Cadmium Sulfide and Telluride. In *Elsevier Encyclopedia of Materials: Science and Technology* (2nd Ed.).

Chausali, Neha, Saxena, Jyoti, & Prasad, Ram. (2023). Nanotechnology as a Sustainable Approach for Combating the Environmental Effects of Climate Change. *Journal of Agriculture and Food Research*, 12, 100541.

Chobba, M. B., Weththimuni, M. L., Messaoud, M., Urzi, C., Maalej, R. & Licchelli. M. (2023). Silver Nanoparticles in the Cultural Heritage Conservation. In *IntechOpen Self-Assembly of Materials and Their Applications.*

Daulay, A., Andriayani, N., Marpongahtun, N., & Gea, S. (2022). Synthesis and Application of Silicon Nanoparticles Prepared From Rice Husk for Lithium-ion Batteries. *Case Studies in Chemical and Environmental Engineering*, 6, 100256.

de Melo Reis, É., de Rezende, A. A. A., Santos, D. V., de Oliveria, P. F., Nicolella, H. D., Tavares, D. C., Silva, A. C. A., Dantas, N. O. & Spanó, M. A. (2015). Assessment of the Genotoxic Potential of Two Zinc Oxide Sources (Amorphous and Nanoparticles) Using the in Vitro Micronucleus Test and the in Vivo Wing Somatic Mutation and Recombination Test. *Food and Chemical Toxicology*, 84, 55–63.

Dermatas, D., Mpouras, T., & Panagiotakis, I. (2018). Application of Nanotechnology for Waste Management: Challenges and Limitations. *Waste Management & Research the Journal for a Sustainable Circular Economy*, 36(3), 197–199.

Franco-Molina, M. A., Gamboa, E. M., Sierra-Rivera, C. A., Gómez-Flores, R. A., Zapata-Benavides, P., Castillo-Tello, P., Alcocer-González, J. M., Miranda-Hernández, D. F., Tamez-Guerra, R. S., & Rodríguez-Padilla, C. (2010). Antitumor Activity of Colloidal Silver on MCF-7 Human Breast Cancer Cells. *Journal of Experimental & Clinical Cancer Research*, 29(1), 148.

Gajanan, K., & Tijare, S. N. (2018). Applications of Nanomaterials. *Materials Today Proceedings*, 5(1), 1093–1096.

Gao, X., & Lowry, G. V. (2018). Progress Towards Standardized and Validated Characterizations for Measuring Physicochemical Properties of Manufactured Nanomaterials Relevant to Nano Health and Safety Risks. *NanoImpact*, 9, 14–30.

García, M., Aleixandre, M., Gutiérrez, J., & Horrillo, M. C. (2005). Electronic Nose for Wine Discrimination. *Sensors and Actuators B Chemical*, 113(2), 911–916.

Garimella, R., & Eltorai, A. E. M. (2016). Nanotechnology in Orthopedics. *Journal of Orthopaedics*, 14(1), 30–33.

Ge, S., Sang, D., Zou, L., Yao, Y., Zhou, C., Fu, H., Xi, H., Fan, J., Meng, L., & Wang, C. (2023). A Review on the Progress of Optoelectronic Devices Based on TiO_2 Thin Films and Nanomaterials. *Nanomaterials*, 13(7), 1141.

Gupta, V., Mohapatra, S., Mishra, H., Farooq, U., Kumar, K., Ansari, M., Aldawsari, M., Alalaiwe, A., Mirza, M., & Iqbal, Z. (2022). Nanotechnology in Cosmetics and Cosmeceuticals—A Review of Latest Advancements. *Gels*, 8(3), 173.

Hamzat, A. K., Omisanya, M. I., Sahin, A. Z., Oyetunji, O. R., & Olaitan, N. A. (2022). Application of Nanofluid in Solar Energy Harvesting Devices: A Comprehensive Review. *Energy Conversion and Management*, 266, 115790.

Hashemzadeh, H., Allahverdi, A., Ghorbani, M., Soleymani, H., Kocsis, A., Fischer, M. B., Ertl, P., & Manesh, H. N. (2019). Gold Nanowires/Fibrin Nanostructure as Microfluidics Platforms for Enhancing Stem Cell Differentiation: Bio-AFM Study. *Micromachines*, 11, 50.

Hashemzadeh, H., Javadi, H., & Darvishi, M. H. (2020). Study of Structural Stability and Formation Mechanisms in DSPC and DPSM Liposomes: A Coarse-grained Molecular Dynamics Simulation. *Scientific Reports*, 10(1), 1837.

Hoang, V. V., & Ganguli, D. (2012). Amorphous Nanoparticles — Experiments and Computer Simulations. *Physics Reports*, 518(3), 81–140.

Hobson, D. W., & Guy, R. C. (2014). Nanotoxicology. In *Elsevier eBooks*.

Hu, J. (2024). Nanomaterials in aerospace: Revolutionizing flight and exploration through nanoscale advancements. *Applied and Computational Engineering*, 60, 1–5.

Ibrahim, R. K., Hayyan, M., AlSaadi, M. A., Hayyan, A., & Ibrahim, S. (2016). Environmental Application of Nanotechnology: Air, Soil, and Water. *Environmental Science and Pollution Research*, 23(14), 13754–13788.

Janfaza, S., Nojavani, M. B., Nikkhah, M., Alizadeh, T., Esfandiar, A., & Ganjali, M. R. (2019). A Selective Chemiresistive Sensor for the Cancer-related Volatile Organic Compound Hexanal by Using Molecularly Imprinted Polymers and Multiwalled Carbon Nanotubes. *Microchimica Acta*, 186(3), 137.

Jiang, J., Oberdörster, G., Elder, A., Gelein, R., Mercer, P., & Biswas, P. (2008). Does Nanoparticle Activity Depend Upon Size and Crystal Phase?. *Nanotoxicology*, 2(1), 33–42.

Kanth, A. P., & Soni, A. K. (2022). Application of Nanocomposites for Conservation of Materials of Cultural Heritage. *Journal of Cultural Heritage*, 59, 120–130.

Khan, I., Saeed, K., & Khan, I. (2017). Nanoparticles: Properties, Applications and Toxicities. *Arabian Journal of Chemistry*, 12(7), 908–931.

Kokina, I., Plaksenkova, I., Jankovskis, L., Jermaļonoka, M., & Galek, R. (2024). New Insights on Biosynthesis of Nanoparticles Using Plants Emphasizing the Use of Alfalfa (*Medicago Sativa L.*). *Journal of Nanotechnology*, 2024(1), 9721166.

Kumar, H., Kuča, K., Bhatia, S. K., Saini, K., Kaushal, A., Verma, R., Bhalla, T. K.,& Kumar, D. (1966). Applications of Nanotechnology in Sensor-Based Detection of Foodborne Pathogens. *Sensors*, 20(7), 1966.

Lai, S-M., Tsai, T-Y., Hsu, C-Y., Tsai, J-L., Liao, M-Y., & Lai. P-S. (2012). Bifunctional Silica-Coated Superparamagnetic FePt Nanoparticles for Fluorescence/MR Dual Imaging. *Journal of Nanomaterials*, 2012(1), 631584.

Lamsal, R. P., Hineman, A., Stephan, C., Tahmasebi, S., Baranton, S., Coutanceau, C., Jerkiewicz, G., & Beauchemin, D. (2020). Characterization of Platinum Nanoparticles for Fuel Cell Applications by Single Particle Inductively Coupled Plasma Mass Spectrometry. *Analytica Chimica Acta*, 1139, 36–41.

Liakos, I. L., Grumezescu, A. M., & Holban, A. M. (2018). Consumer's Health and Regulatory Organizations. In *Handbook of Food Engineering*, Academic Press.

Loo, C., Lowery, A., Halas, N., West, J., & Drezek, R. (2005). Immunotargeted Nanoshells for Integrated Cancer Imaging and Therapy. *Nano Letters*, 5(4), 709–711.

McNamara, K., & Tofail, S. A. M. (2016). Nanoparticles in Biomedical Applications. *Advances in Physics: X*, 2(1), 54–88.

Mekuye, B., & Abera, B. (2023). Nanomaterials: An Overview of Synthesis, Classification, Characterization, and Applications. *Nano Select*, 4(8), 486–501.

Meyer, M., & Kuusi, O. (2006). *Nanotechnology: Generalizations in an Interdisciplinary Field of Science and Technology*. World Scientific eBooks.

Mohammad, Z. H., Ahmad, F., Ibrahim, S. A. & Zaidi, S. (2022). Application of Nanotechnology in Different Aspects of the Food Industry. *Discover Food*, 2, 12.

Moreau, L. M., Ha, D. H., Zhang, H., Hovden, R., Muller, D. A., & Robinson, R. D. (2013). Defining Crystalline/Amorphous Phases of Nanoparticles Through X-ray Absorption Spectroscopy and X-ray Diffraction: The Case of Nickel Phosphide. *Chemistry of Materials*, 25(12), 2394–2403.

Phillips, J. D. (2021). Energy Harvesting in Nanosystems: Powering the Next Generation of the Internet of Things. *Frontiers in Nanotechnology*, 3, 633931.

Pitkethly, M. J. (2004). Nanomaterials – The Driving Force. *Materials Today*, 7(12), 20–29.

Razavi, H., Darvishi, M. H., & Janfaza, S. (2018). Silver Sulfadiazine Encapsulated in Lipid-Based Nanocarriers for Burn Treatment. *Journal of Burn Care & Research*, 39(3), 319–325.

SafeWork NSW. (2023). *Nanotechnology*. SafeWork NSW.

Sajid, M. (2021). Nanomaterials: Types, Properties, Recent Advances, and Toxicity Concerns. *Current Opinion in Environmental Science & Health*, 25, 100319.

Sharma, V. K., Filip, J., Zboril, R. & Varma, R. S. (2015). Natural Inorganic Nanoparticles – Formation, Fate, and Toxicity in the Environment. *Chemical Society Reviews*, 44(23), 8410–8423.

Shi, D., Guo, Z., & Bedford, N. (2015). Basic Properties of Nanomaterials. *Nanomaterials and Devices*, 49–82.

Sierra-Fernandez, A., Gomez-Villalba, L. S., Rabanal, M. E., & Fort, R. (2017). New Nanomaterials for Applications in Conservation and Restoration of Stony Materials: A Review. *Materiales De Construcción*.

Singh, T., Shukla, S., Kumar, P., Wahla, V., Bajpai, V. K. & Rather, I. A. (2017). Application of Nanotechnology in Food Science: Perception and Overview. *Frontiers in Microbiology*, 8, 1501.

Spano, M. A., Frei, H., Wurgler, F. E., & Graf, U. (2001). Recombinagenic Activity of Four Compounds in the Standard and High Bioactivation Crosses of Drosophila Melanogaster in the Wing Spot Test. *Mutagenesis*, 16(5), 385–394.

Weerasinghe, J., Prasad, K., Mathew, J., Trifoni, E., Baranov, O., Levchenko, I., & Bazaka, K. (2023). Carbon Nanocomposites in Aerospace Technology: A Way to Protect Low-Orbit Satellites. *Nanomaterials*, 13(11), 1763.

Zhang, D., & Guan, L. (2014). Laser Ablation. In *Elsevier Comprehensive Material Processing*.

Zhao, Y., Zhang, Z., Pan, Z., & Liu, Y. (2021). Advanced Bioactive Nanomaterials for Biomedical Applications. *Exploration*, 1, 20210089.

10 Advanced Applications of Nanomaterials Across Industries
Transforming the Future of Manufacturing

Manu Mehrotra, Manish Baboo Agarwal, Manish Kumar Panday and Pankaj Mittal

10.1 INTRODUCTION

Nanomaterials are materials with dimensions ranging between 1nm and 100nm. The materials tend to geotactically exhibit characteristics of that scale which will be stronger, more conductive, and more reactive. These are also classified into different groups on the basis of size ranging from zero-dimensional (0D) nanomaterials like quantum dots (QDs) to nanoparticles. Simple cylinders or tubes of a nanoscale are called one-dimensional (1D) nanomaterials or more commonly nanotubes and nanowires. Over the past few years, considerable efforts have been directed to the manufacture of nanosized materials on a substrate, almost two-dimensional nanomaterials – graphene or nanosheets. There are also terahertz metamaterial resonators that exhibit smaller replicas of their three-dimensional (3D) nanomaterial structures and patterns.

Nanomaterials can be synthesized using various methodologies, physical and chemical as well as biological approaches (Agarwal et al. 2018, 2020, 2021, 2023). Nanomaterials are widely used in various industries due to their unique properties which very much encourage the use of these materials in a variety of applications.

10.2 APPLICATIONS OF NANOMATERIALS – IN ENERGY AND ENVIRONMENT

10.2.1 For Energy Storage and Conversion

Nanomaterials such as nanoparticles, nanowires, and graphene (Gu et al. 2023) are being incorporated in Lithium-ion batteries, supercapacitors, and fuel cells for the purposes of enhancing the energy density, power density, and life cycle parameters associated with their performance and their operational efficiencies and durability.

 DOI: 10.1201/9781003560234-13

Nanomaterials such as QDs, nanoparticles, and nanostructured materials are being put to applications that enhance the operation of solar cells, including the efficiency, power conversion efficiency, and stability. These materials are being utilized toward the enhancement of the efficiency of thermoelectric devices. These materials are being incorporated in the development of hydrogen storage systems with the aim of achieving improved storage capacity, stability, and safety. Nanomaterials such as carbon nanotubes (CNTs) (Subhedar and Singh 2023), graphene, and metal oxides are being harnessed in the performance enhancement of electrochemical capacitors, enhancing energy density, power density, and cyclability of these devices.

10.2.2 For Energy Harvesting

Energy harvesting, also referred to as micro energy harvesting, involves extracting energy from the environment, creating electric energy out of it (Ramadoss and Sonachalam 2024, Dragoman 2023). This technology holds promise for powering small devices like sensors, wearables, and Internet of Things (IoT) devices without the use of batteries or wired plug points.

***Solar energy harvesting* –** Advanced nanomaterials are being incorporated in the enhancement of the solar cells and increasing the efficiency and stability of power conversion in solar cells (Abdin et al. 2013, Low et al. 2019). The solar cell market is gradually shifting to employ QDs for solar cells. QDs serve to enhance solar cells by increasing the amount of light that is captured and turned into energy. Advanced materials that are made in nanoscale dimensions, such as nanoparticles, nanowires, and so forth, are being incorporated into photovoltaic devices for performance enhancement.

***Thermal energy harvesting* –** Advanced nanomaterials are being incorporated in the enhancement of the thermoelectric materials in which efficiency and power of thermoelectric conversion are focused. These materials include nanoparticles, nanowires, and nanostructured materials whose aim is to enhance the efficiency of thermal energy harvesters and also their power density stability.

***Vibration energy harvesting* –** Advanced nanomaterials are being used to enhance the efficiency of piezoelectric materials, such as improving their piezoelectric coefficient and power density, etc. Nanomaterials like porous nanoparticles, nanowires, and nanostructured materials are being incorporated to enhance the working efficiency of vibration energy harvesters (Jing and Kar-Narayan 2018), including enhancement of power density and stability.

***Wind energy harvesting* –** Nanomaterials are being used to enhance the performance of wind turbine blades, including an increase in their strength, durability, and power generation capability. Wind energy harvesters are made of these materials to enhance their performance.

***Bio-energy harvesting* –** Nanomaterials are being used in places where the efficiency of bio-energy harvesting systems is to be improved (Markandan and Chai 2022) and are being applied to bio-energy harvesting systems to liberate and augment their power density and stability.

Ocean energy harvesting – Nanomaterials are being used in the improvement of systems that harvest ocean energies with an emphasis on internal power density and stability. Triboelectric nanogenerators (TENG) (Nazar et al. 2021) have emerged as a promising new source of ocean energy, and it has shown significant technical advantages in applying low-frequency ocean energy-efficient harvesting.

10.2.3 For Environmental Remediation

Nanomaterials have so far proved to be effective and useful in the field of environmental removal of pollutants (Saleh 2022) with new ideas for water treatment, air management, and waste treatment. The following are some possible uses of nanomaterials in these fields:

Water purification – Filtration performance can be improved by using the nanomaterials such as graphene oxide, CNTs, and metal organic frameworks (MOFs), which are incorporated into water filtration membranes (Zamel et al. 2023) to remove heavy metals, pesticides, and bacteria from the contaminated water. Metal nanoparticles, like titanium dioxide (TiO_2) or zinc oxide (ZnO) (Hussain et al. 2024), act as catalysts against water pollutants like organic dyes, while nanomaterials of carbon, silica, and alumina are used for the removal of heavy metals (Baby et al. 2022), pesticides, and other pollutants from wastewater. Some nanomaterials, particularly gold nanoparticles (Fang et al. 2016) and QDs, are used for the detection of water pollutants and help in the establishment of monitoring and warning systems.

Air pollution control – Nanoparticles such as platinum (Pt) and palladium (Pd) catalysts are also used in (Fang and Guo 2018) the treatment of CO, NOx, and other pollutants converting them into innocuous products. Nanomaterials such as CNTs and MOFs are blended with high-performing air filters in order to eliminate particulate matter (PM) and gaseous pollutants. Titanium dioxide (TiO_2) and zinc oxide (ZnO) nanomaterials are used in the making of self-cleaning surfaces (Padmanabhan and John 2020) and are equally useful to destroy contaminants like NOx. Nanomaterials such as metal oxide nanoparticles and graphene are used for the purpose of sensing the pollutants in the air which helps in smog alarm monitoring in the real time.

Waste management - Iron oxide (Fe_2O_3) and zinc oxide (ZnO) nanoparticles act on the organic wastes like plastics and pesticides. Nanomaterials such as activated carbon, silica, and alumina are used for the adsorption of heavy metals and other wastes that can be safely disposed of and even recycled. Carbon and graphene nanosheet materials are used to fabricate new composite materials (Mazari et al. 2021) for waste management such as use of biodegradable plastics and self-healing materials. To remove pollution from soil and groundwater, zero valent iron (ZVI) and iron oxide (Galdames et al. 2020) nanoparticles are used which dissolves many contaminants including chlorinated solvents and heavy metals.

There are various advantages of using nanomaterials for the remediation of pollution which include, but not limited to, enhancement of reactivity and catalytic activity, higher degree of adsorption or filtration, better sensitivity and selectivity in the detection of pollutants, as well as cheaper and scalable options (Figure 10.1).

10.3 APPLICATIONS OF NANOMATERIALS IN PHOTONICS AND ELECTRONICS

10.3.1 Nanomaterials for Photonics

Nanomaterials in the field of photonics (Mishra et al. 2017, Zhao et al. 2022) offer unique materials with new optical characteristics which can be exploited to develop lasers, LEDs, and optical communication systems. The fictionalization of nanomaterials to enhance performance in photonics includes:

Lasers – Laser designers can also incorporate nanoparticles into the plastic medium in order to improve thermal stability and broadness of tunability of the laser type. QDs are used for better purity and stability at larger modules, while graphenes are used in the manufacture of ultra-fast and compacted lasers of high power density.

LEDs - QD would be ideally used in making highly efficient LEDs that will overcome color, purity, and stability issues. Nanocrystals have been synthesized for LED applications with better color and luminous output. There are others that utilize graphene to achieve flexible transparent LED with high carrier mobility.

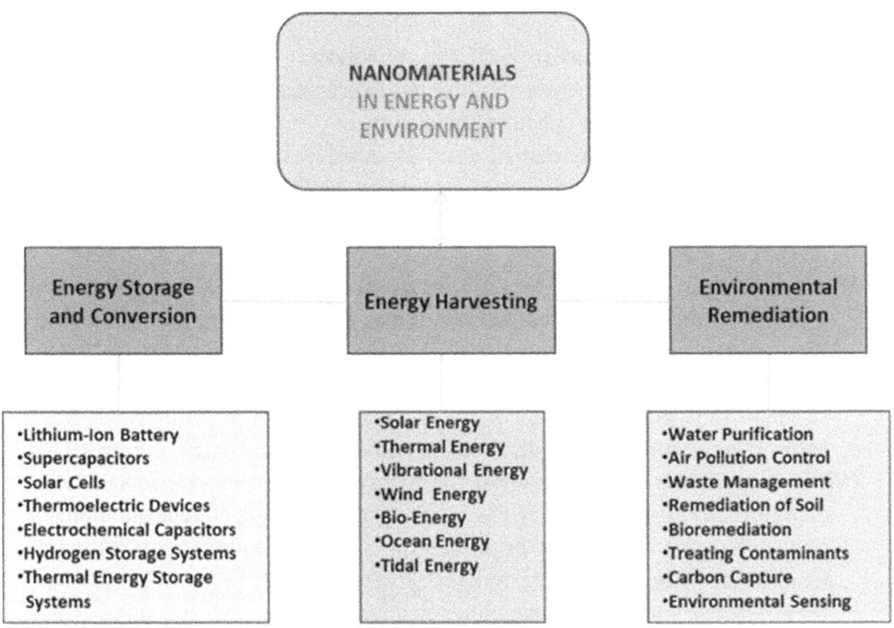

FIGURE 10.1 Nanomaterials (in energy and environment).

Optical communications – Diode lasers and nanophotonic waveguides can be integrated with optical links for more compact and faster interconnects. Ultra-fast compact modulators can be developed using graphene within optical communication systems while optical amplifiers can be developed from nanoparticles with better gain and noise filter performance.

Other applications – Nanomaterials can be effectively used in designing optical sensors with high sensitivity for use in biomedical and environmental purposes. They are employed in the fabrication of advanced imaging systems for applications in medicine and materials. They can also be utilized in making solar cells with higher power optimization.

Nanomaterials cut across all fields such as efficiency and performance, development of miniature and more power-packed photonics devices, cost and new functionalities, and applicability in photonic. Hence, it means that there is room for advancement in photonics due to the presence of new materials that have not been used before in fabricating electronic devices which are able to harness light into electric luminescence such as LEDs, laser systems, and optical communication systems.

10.3.2 Nanomaterials for Electronics

In the area of electronics, they provide high intelligence, better electrical properties which can be used for making high-performance tetrodes, interconnects, and memory devices (Mhetre et al. 2021). The proper utilization of nanomaterials in the field of electronics is in:

Transistors – Utilization of graphene enables achievement of very fast, high-mobility, flexible transistors. Using nanowires, transistor manufacturers improve mobility while minimizing power consumption. Using QDs improves the switching speed of transistors.

Interconnects – Graphene can be used to develop interconnects which are ultra-fast and flexible with carrier mobilities. High speed and low interconnect power are possible due to the CNTs and nanowires coupled with better thermal management systems.

Memory devices – These devices are useful in the making of low-power and high-speed memory devices that can retain data for a long period of time, like resistance random access memory (RRAM) devices (Sarid et al. 2004). Low-power memory devices with better data retention are made from nanocrystals. Low-power and fast memory storage media is synthesized from phase change substances with excellent anti-mist disturbance (Figure 10.2).

Other applications – Nanomaterial fabrication with better electrostatic discharge (ESD) protection devices with improved surge protection subjects is also possible. They may be utilized to construct lightweight, thin, and flexible electromagnetic interference (EMI) shielding materials with better efficiency. Flexible and wearable electronics with better mechanical and electrical properties are manufactured using these materials. They also help in the development of high-density electronics with better performance. They can provide cost-effective solutions for electronics applications.

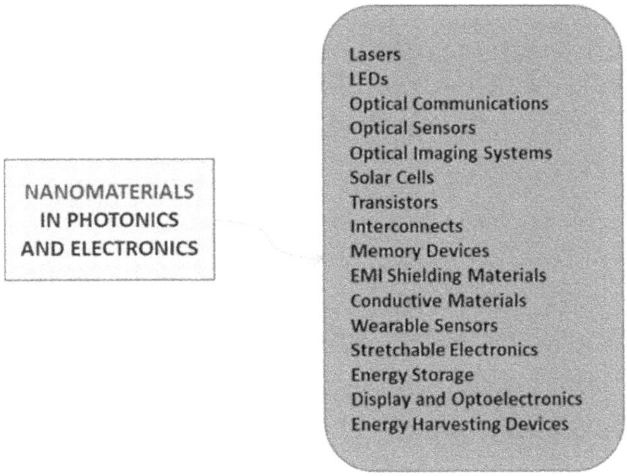

FIGURE 10.2 Applications of nanomaterials in photonics and electronics.

10.3.3 Nanomaterials for Flexible and Wearable Electronics

A review (Sedighi and Montazer 2022) explains innovative materials with incredible electrical, mechanical, and optical properties which can aid in the emergence of efficient, comfortable, and portable devices. Some of its important applications are:

Conductive materials – Composites of graphene and CNTs are conductive and pliable sufficient materials that allow the fabrication of fully transparent electrodes (LeMieux and Bao 2008), interconnects, and other electrodes for flexible and portable devices.

Stretchable electronics – Nanomaterials such as graphene and CNTs can function as stretchable conductors which can find applications in wearable devices. Examples of such nanomaterials include elastomers and hydrogels synthesized to yield substrates suitable for use in wearable devices.

Wearable sensors – Nanowires are ultra-thin structures which can be employed to fabricate ultra-sensitive and flexible sensors incorporated in wearable devices. These sensors are used in glucose monitor and pressure monitor. Nanowires and nanodisks are used to fabricate temperature monitoring devices and humidity sensors which are to be worn by users.

Energy storage – In recent years, there have been growing interest in incorporating nanomaterials into traditional energy storage devices. For instance, graphene oxide sheets (Kim et al. 2021), as an anode polymer composite, are of particular interest in achieving high-density and high-power flexible batteries and supercapacitors.

Display and optoelectronics – Nanomaterials, such as graphene and CNTs, are appropriate for the manufacture of thin stretchable displays of high quality. Furthermore, nanomaterials such as QDs and graphene can be incorporated into optoelectronic devices like wearables, such as LEDs or lasers.

Other applications – Nanomaterials, such as graphene and CNTs, are used for the manufacture of flexible antennas to be embedded in wearable gadgets. Additionally, advanced nanomaterials such as piezoelectric nanowires (Wang et al. 2015) can help in the production of miniaturized and flexible energy harvesting devices. They come up with economical solutions and facilitate the fabrication of lightweight and more comfortable devices that are also more flexible in the range of structures.

10.4 APPLICATIONS OF NANOMATERIALS IN HEALTHCARE AND BIOMEDICAL IMAGING

Nanomaterials used in medicine, healthcare, and bioengineering (Malik et al. 2023) are in a position to change the mode of disease treatment, diagnosis, and even prevention. The exposure provided by them in this sector is a large and informative one. With the advancement of health technology, we will be able to see great improvements in disease evaluation accuracy, treatment outcomes, and patient care.

10.4.1 Nanomaterials for Biomedical Imaging

Nanomaterials have changed the role in the field of bioimaging (Chow 2022) especially in the area of novel contrast agents, site-specific drug delivery, and drug and imaging therapy. The use of nanomaterials in the following ways embodies nanotechnology in biomedical imaging:

Contrast agents **–** Since gold nanoparticles have a high X-ray attenuation coefficient, as well as magnetic and optical properties, they can be used as contrast agents for computed tomography (CT) (Dong et al. 2019) scanning, magnetic resonance imaging (MRI), and photoacoustic imaging. QD contrast agents for fluorescence imaging are the most sensitive and specific for cancer diagnosis and biomarker detection. Lipid nanoparticles are used for contrast enhancement in ultrasound imaging as well as targeted delivery of therapeutics. Dendrimers are used for MRI contrast and enhanced therapeutic delivery with high reflexivity.

Targeted delivery – Nanoparticles like liposomes, polymeric nanoparticles, and gold nanoparticles are used to transport contrast agents, drugs, and even genes to the particular cells or tissues. Liposomes employ targeting ligands like antibodies, peptides, or small molecules to carry therapeutic and contrast agents to specific tissues or cells.

Optimal use of therapeutics, genes, and short-interfering RNA (siRNA) (Ashique et al. 2022) is performed by encapsulating them in nanocarriers like nanoparticles, liposomes, and dendrimers and directing them to the corresponding cells or tissues.

Theranostics **–** Nanoparticles like gold nanoparticles, liposomes, and polymeric nanoparticles are utilized for the treatment, imaging (Handa et al. 2023), and allowing treatment response assessment in real time. Therapeutic techniques are directed using photoacoustic imaging, in which gold nanoparticles and liposomes act as contrast agents, while polymeric nanoparticles, inorganic nanoparticles, and others act as therapeutic agents.

MRI is the guiding modality of therapy, using contrast and therapeutic agents that belong to nanoparticles such as iron oxide nanoparticles and liposomes, while optical imaging is the guiding imaging modality that includes QDs and gold nanoparticles.

Other applications **–** Nanoparticles are employed to fabricate hybrid imaging agents which combine positron-emission tomography with MRIs, and fluorescent imaging remain sequential. These are utilized to create surgical systems for image-guided surgery which allow intraoperative imaging and navigation over the surgical procedure (Paris and Vallet-Regí 2018). They are utilized to establish approaches of personalized medicine targeting the delivery and imaging of therapeutic agents according to patient-based characteristics. The application of nanomaterials in biomedical imaging has certain benefits like improved contrast and sensitivity, selective imaging and delivery, multimodal imaging, real-time monitoring and guidance, as well as personalized medicine approaches.

10.4.2 Nanomaterials for Drug Delivery and Therapy

Nanomaterials have changed the way that drugs are delivered within the body and have given rise to the growth of new therapies (Afzal et al. 2022, Witharana and Napagoda 2022), including those on how to fight cancer, study genomics, and improve tissue reconstruction. Some applications of nanomaterials in these areas include:

Cancer management – Targeted drug delivery includes nanosomes, polymeric nanosomes, and gold nanosomes, which work toward delivering most of the cytotoxic drugs to the site of the cancer cells, thereby reducing the toxicity to the normal patients. They are formulated for specific cancer target cells and tissues by using targeting ligands like antibodies, peptides, and small molecules.

Metal nanomaterials are known to absorb light and generate heat, which is used for photothermal ablation of cancer cells. Tumor-targeting nanomedicine approaches employing gold nanoparticles have also been used to deliver immune stimulatory drugs for cancer therapy (Păduraru et al. 2022, Kashyap et al. 2023).

Genetic eraser – Nanosomes, polymeric nanosomes, and virus vectors enable the insertion of genes into the right tissues and body cells which permit gene therapy of inherited diseases. They are utilized to introduce siRNAs or micro-RNAs (miRNAs) into cells for gene silencing to fight cancers and other genetic diseases. Nanoparticles are also used to transport CRISPR-Cas9 gene editing tools to achieve gene therapy.

Regenerative medicine **–** Nanoparticles, nanofibers, and nano-composites are employed to make scaffolds for tissue engineering to provide appropriate surroundings for cell growth and differentiation. Furthermore, nanoparticles are incorporated with stem cells to enhance their efficiency, whereby growth factors and other biomolecules are loaded into these cells. Other nanomaterials like nanoparticles and nanofibers are employed to prepare wound dressing materials, aiding growth of tissues and healing of wounds. Other nanomaterials such as nanoparticles and nano-composites are employed where porous scaffolds are made for use in application to the bone regeneration process, promoting growth and repair of bone (Alzate-Correa et al. 2022).

***Other applications* –** Nanoparticles are involved in the preparation of vaccines which improve their immunogenicity and effectiveness. These are used to transport protective agents to neurons in the treatment of degenerative diseases of the brain such as Alzheimer and Parkinson. They are also used for targeting the eye with the purpose of treating diseases such as glaucoma and age-related macular degeneration, which are difficult to treat using normal dosage forms. They are also utilized for targeted drug delivery to the skin in the treatment of skin diseases like acne, psoriasis, and skin cancer.

Nanomaterials with regard to use in drug delivery and therapy provide numerous benefits such as less toxicity and side effects, targeted nature of delivery and release, increased effectiveness and specificity, higher bioavailability and pharmacokinetics, and therapy customized medicine.

10.4.3 Nanomaterials for Regenerative Medicine and Tissue Engineering

Nanomaterials have changed the field of regenerative medicine (Khalilov 2023) and tissue engineering by providing new biomaterials, scaffolds, and implants which have the ability to replicate the microenvironment of tissues and facilitate tissue regeneration. The applications include:

***Biomaterials* –** This method is generally referred to as the top-down approach for L-biofabrication of functional constructs using polymeric materials. In recent years, considerable progress has been achieved in the development of nanomedicine, where biomaterials such as silver (Bansal et al. 2020) and gold sodium thiomalate have potential advantages over metallic particles. It can be a polymer nano-composite and nanoparticles with suitable mechanical properties and biocompatibility. Nanobiomaterials are promising candidates for the application of biocompatible titanium alloy and stainless steel.

***Scaffolds* –** Nanofibers produced by electrospinning can serve as cell scaffolds by providing a three-dimensional structure that resembles the natural extracellular matrix (ECM) of tissues. Nanogels are formed by incorporating cells into nanostructured hydrogels that mimic native tissues for efficient cell growth and regeneration. Nanoporous polymers or nanoporous foam for tissue engineering, like Orkotec active scaffold systems, provide a large active surface area upon which cells can attach and proliferate.

Implants – Titanium or hydroxyapatite nano-coatings are able to enhance the osseointegration and biocompatibility of certain implants. Implantation of nanostructured porous titanium or nanostructured stainless-steel structures allows for high integration surfaces in contact with tissues and

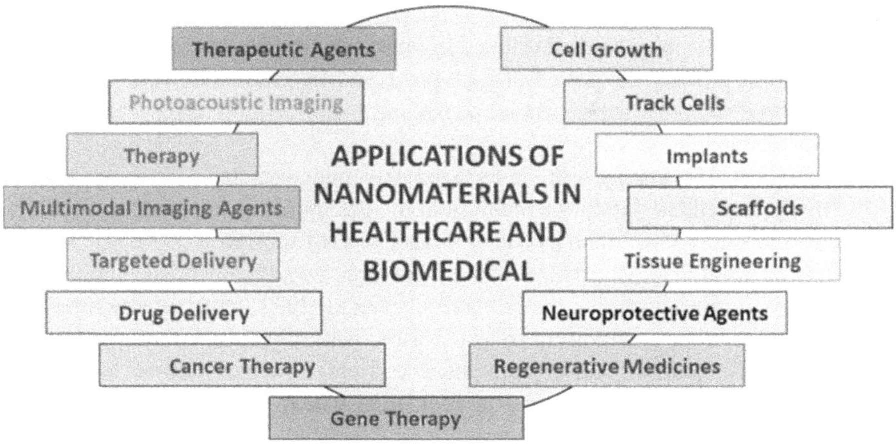

FIGURE 10.3 Applications of nanomaterials in the medical field.

osteogenesis. Nanosilver titanium substrate for implants may enhance their tissue biocompatibility and integration. Silica alumina ceramic nanoparticle composites are indicated for use in biopolymers and bone implants.

Other applications – Nanomaterials can be employed for making adjunct drug delivery systems for tissue engineering and regenerative medicine formulations. Nanomaterials like QDs (Bajwa et al. 2015) or gold nanoparticles can be used to visualize cells in vivo and follow the processes of tissue regeneration. They can be utilized in the development of scaffolds and biomaterials for the tissue engineering of particular organs such as heart, liver, or kidney.

Nanomaterials applied in regenerative medicine and tissue engineering have certain advantages such as targeted and localized drug delivery, improved cell proliferation and differentiation, better biocompatibility, and better mechanical strength (Figure 10.3).

10.5 APPLICATIONS OF NANOMATERIALS IN FOOD AND AGRICULTURE

10.5.1 Nanomaterials in Food Packaging Industry

Nanomaterials possess the capability of developing more efficient and adequate materials that have the possibilities of enhancing the durability, safety, and quality aspects of the food packaging materials (Ashfaq et al. 2022, Kuswandi 2016). The areas of application in the food packaging for using such materials include:

Enhanced barrier properties – Nano-clay has assisted in creating barrier-type packaging material, which has less oxygen, carbon dioxide, and moisture permeability than conventional packaging materials. Graphene is incorporated into mushroom to fabricate thin and less burdening packing films with high barrier properties.

Antimicrobial packaging **–** Antimicrobial packaging materials based on silver nanoparticles have been developed to inhibit the growth of microorganisms and thus prolong the period of food preservation. Zinc oxide nanoparticles are used to create antimicrobial packaging materials with improved safety and efficacy.

Sensing and Monitoring **–** In the food industry, nanoparticles have been effectively embedded within the medium in the design of sensors to detect quality and safety flaws such as contaminants or food spoilage.

Active packaging **–** Amorphous silica nanoparticles are embedded in antimicrobial active packaging materials (Franco et al. 2022) which act by releasing of microbes, killing agents, or antioxidants to prevent food products from deterioration. Aside from these carriers, enzymes can be used on nanoparticles to make "active" e-packaging which will destroy the spoilage-causing compounds, hence prolonging the shelf life of food.

Biodegradable packaging **–** Applications of nanocellulose include development of high-strength and barrier-biodegradable packaging materials, while polylactic acid (PLA) includes the formulation of biodegradable packaging with enhanced mechanical strength and barrier properties.

Nanocoating **–** It is done in order to increase the barrier properties and the antibacterial activity of the food packaging materials.

Nano-composites **–** Nano-composites are used to prepare food packaging material (Sarfraz et al. 2020) which have better strength, barrier properties, and antibacterial activity.

10.5.2 Nanomaterials for Food Processing and Safety

Nanomaterials help to improve food safety (Biswas et al. 2022) by preventing as well as decreasing the chances of any form of food contamination and spoilage. They can also enhance the quality of foods by modifying their texture, flavor, and nutritional value. They help to improve the sustainability of food processing and safety by reducing waste and improving the efficiency of food production. Due to these reasons, they find applications in:

Food processing **–** Nanofiltration technologies may be beneficial for getting rid of undesirable elements, impurities, or even harmful elements inside foodstuffs so that their quality and safety are well improved. Nanocatalysts can be utilized to enhance the effectiveness and environmental friendliness of particular food processing reactions, for instance, enzymatic reactions. They can be used in the production of new food supplements with advanced safety and more ability. They are used to develop new methods of food processing like nano-emulsification and nano-encapsulation.

Food safety **–** Nanoparticle-based sensors can also be utilized to determine the level of contamination and spoilage of food products, which enhances food protectiveness. New methods of testing food for any potential hazard are now possible with nanotechnology, such as nano-based PCR and nano-based

ELISA. It can also be utilized to develop new food wrappers and other packages with greater barrier efficiency and antibacterial effectiveness.

Nanoparticles could also be utilized for manufacturing antimicrobial substances that will help in enhancing the safety of food by inhibiting the maturation of microorganisms. Silver nanoparticles and zinc oxide nanoparticles into food products where they act as antimicrobial agents for better food quality. To enhance the flavor and texture, food additives like titanium dioxide nanoparticles, nanocellulose and graphene are added.

10.5.3 Applications of Nanomaterials in Agriculture

Due to their specific characteristics, the application of nanomaterials in the agricultural sector (Dwivedi et al. 2016) can increase the yield of crops, decrease the amounts of pesticides and fertilizers used, and increase food quality. Systems employing nanomaterials include:

Crop protection – Nanopesticides help to manage insect pests and diseases of crops with a lesser effect on the environment than conventional pesticides, while nanofungicides assist in controlling the fungal pathogens of crops with minimal impact on the environment than traditional fungicides.

Fertilizers and nutrients – Nanofertilizers are capable of enhancing the level of productivity by supplying essential nutrients to the crops while decreasing dependence on chemical fertilizers. They can be utilized in nutrient application, crops enhancing the absorption of nutrients, and minimizing wastage of nutrients.

Food safety – Nano-sensors are useful in sensing contaminants in foods, thus enhancing safety of foods and decreasing the incidence of food-borne diseases.

Water management – Nanomaterials also have an advantage in treating water used for agricultural purposes and thereby decreasing water pollution. They can also help in the improvement of crops through the use of minimal amounts of water in agricultural systems (Figure 10.4).

Soil remediation – Nanoparticles can be employed in eliminating soil contamination (Sathish et al. 2023; Prakash and Smitha 2023), thereby improving the soil structure and benefiting the environment. Efforts have been made to get more benefits from nanotechnology for agricultural biotechnology applications, particularly in enhancing crop production and reducing microbial impact. Nanomaterials used in agriculture are: Nanoparticles of silver are used as nanopesticides for pest and disease control in crops. TiO_2 nanoparticles are used as nanofungicides to combat fungal infections in crops. Zinc oxide nanoparticles are used as functional nanofertilizers that provide crops with the needed nutrients. Nanographene is used for remediation of soils and purification of waters. Nanocellulose is a type of nanomaterial bearing potential uses in agriculture including crop protection and soil remediation.

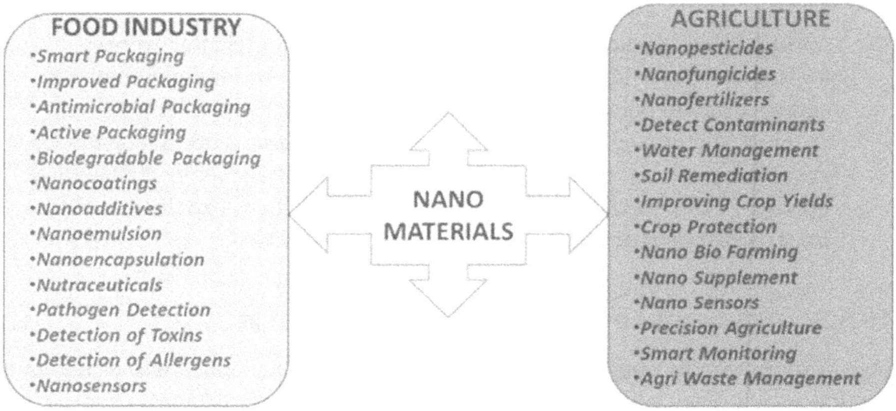

FIGURE 10.4 Applications of nanomaterials in food and agriculture.

10.6 APPLICATIONS OF NANOMATERIALS IN CONSTRUCTION AND INFRASTRUCTURE

Nanomaterials are also being employed in the synthesis of high-performance construction and (Jones et al. 2016) other infrastructural materials with better mechanical properties, durability, green performance, and energy efficiency. They are being harnessed in the development of internal materials and sensors that are intelligent in the sense that they can detect and respond to changes in internal parameters. Some of them used or under study for building materials are nano-silica (SiO_2), nano-alumina (Al_2O_3), nano-titanium (TiO_2), nano-zinc oxide (ZnO), CNT, graphene, nano-clay, nano-cement, nano-fibers (Daniyal et al. 2017).

10.6.1 Nanomaterials for Construction Materials

High-performance concrete **–** Different types of nanoparticles such as silica, alumina, and calcium carbonate stand to enhance properties such as strength, durability, and sustainability of concrete. The application of nano-cement leads to the reduction of cement, which provides an economic advantage and an environmental benefit due to its low content (Zhang et al. 2023).

Self-healing materials **-** Nanomaterials such as bacteria-based self-healing concrete smartly self-repairs its cracks and damages; therefore, maintenance costs will be minimized and structures' life span can be increased.

Insulating and energy efficiency measures – Nanomaterials such as graphene, nanocellulose, and nano-silica polymers also serve as thermal insulation materials to save energy and carbon emissions. Nano-coatings can improve energy efficiency of buildings by minimizing heat transfer and improving the performance of windows.

Sustainable and durable coatings **–** Nanomaterials such as titanium dioxide and zinc oxide can help to make buildings and structures durable by using self-cleaning, anti-glare, and ultraviolet (UV) protective coatings. These coatings make the construction materials' durability and sustainability more efficient by cutting down on repairs and replacements.

Reinforced composite materials – Nanomaterials such as CNTs, graphene, and nano-clay have the effect of improving the composite materials' mechanical properties which include strength, stiffness, and toughness. These nanocomposites can then be used in roofing, cladding and structural parts.

Smart materials and sensors – Sensors that respond to environmental factors, such as temperature, humidity, etc. can be embedded into piezoelectric nanoparticles and are referred to as smart materials. Health condition and performance of building structures can be diagnosed using nano-sensors which will help in constant observation and forecasting of maintenance of the structure. QDs used in construction helps to reduce energy loss by acting as efficient thermal insulators.

Water repellency and waterproofing – Nanomaterials such as nano-silica and nano-titanium dioxide develop coating over construction materials that are non-wettable and non-permeable, ensuring further decrease in maintenance cost.

Fire-resistant materials – Nanomaterials such as nano-clay and nano-silica effectively enhance the fire-retardant properties of construction materials, reducing the risk of fire incidents and enhancing safety.

Advanced ceramics – Advanced ceramics with enhanced mechanical properties, thermal stability, and resistance to corrosion can be produced from nano-alumina and nano-zirconia. In construction, these nano-ceramics find applications in tiles, bricks, and refractory materials, among others.

3D printing and additive manufacturing – In the last few years, even more development has been made in nanomaterials for construction such as nano-clay and nano-silica that are added to, or naturally occur in, 3D printing materials to enhance the mechanical properties of the fabricated complex structures and components.

10.6.2 Nanomaterials for Infrastructure Development

Smart roads and highways – In constructing vibrant structures today, the advanced use of nanomaterials such as graphene and CNTs, which have been used in building and construction, assists in improving the mechanical properties of asphalt and concrete, thus increasing safety and cutting down the cost of maintenance. Nano-sensors help in observing road conditions, the amount of traffic, and climate which makes it possible to manage and maintain the structure in advance.

Advanced bridge construction – Carbon fiber and graphene applied in bridge structures enhance strength, durability, and sustainability, while nano-coatings improve the resistance to corrosion and the durability of bridge elements.

Sustainable and durable buildings – Nanomaterials such as nano-silica and nano-alumina serve as thermal reliefs and conserve energy measures within the buildings, whereas nano-coatings protect and enhance buildings' façade and windows. In view of sustainable features of these nanomaterials, it is possible to make up construction materials, whose manufacture gives less damage to nature and involves less energy and carbon release.

TABLE 10.1
Nanomaterials in Building Construction

Nanomaterial	Applications
Nano-silica	Self-healing concrete, fire-resistant material
Nano-clay	Advanced composites, bricks
Nano-cement	High-strength concrete
Nano-alumina	Fire-resistant materials
Nano-carbon	High-performance Concrete
Nano-titanium dioxide	Photocatalytic concrete
Nanoparticles-iron oxide	Cement
Nano-sensors	Structural health monitoring, water quality monitoring
Nano-coatings	Corrosion protection, low-E windows
Nanoparticles	Water treatment, phase change materials
Nano-ceramics	Advanced pavement materials
Nano-composites	Advanced bridge construction, advanced building envelopes, advanced water pipes
Nano-fibers	Advanced geotextiles
Nano-foams	Advanced insulation materials
Nano-insulation	Advanced thermal insulation

Water infrastructure – Nanomaterials such as nano-ceramics and nano-membranes, being the new generation of water purification technology, also favorably affect desalination processes (Table 10.1).

Smart buildings – Nanomaterials like graphene and CNTs (Pruthviraj et al. 2021) are likely to further the structural enhancement of tunnel linings and building structures. Nano-sensors are able to help in tracking such things as the buildings' condition, energy being used, and environmental factors, enabling and predicting maintenance.

Geotechnical engineering – Nanomaterials such as nano-clay and nano-silica measure up to improve the mechanical property and stability of soils and foundations while exploiting nano-coatings to enhance the performance and safety of geotechnical structures.

Disaster-resistant construction – Nanomaterials in the form of nano-ceramics and nano-membranes can enhance the performance and use of building materials in areas prone to disasters like earthquakes and tornados.

10.7 APPLICATIONS OF NANOMATERIALS IN AEROSPACE AND DEFENSE

The aerospace as well as defense sector has been ever more demanding for new materials and efficient technologies that enhance performance, cut down on weight, and improve efficiency. Nanomaterials provide very good solutions to these concerns.

10.7.1 CNTs, Graphene, and Nano-Clay-Based Composites

The aerospace and defense sectors have been looking into CNTs, graphene, and nano-clay-based composites (Kausar et al. 2017) due to their various advantages, including improved mechanical, thermal, and electrical properties, which are very useful in these sectors. Some of the most prominent uses of these nanomaterial-based composites in creation of aerospace and defense components are as listed below:

CNT composites – CNT polymer composites are a potential replacement of traditional materials employed in the structures of aircraft and spacecraft and hence minimize weight and allow better fuel efficiency. They can also be used for effective EMI shielding of electronic components for aerospace and defense applications. Owing to high thermal conductivity, they can be employed in thermal management applications including heat sinks and thermal interfaces. As substitute materials for capacitors and batteries, they are applied in aerospace and defense applications for energy storage. They are employed in enhancing the functioning of propulsion systems including rocket engines and fuel cell units. Alternative to traditional materials, the introduction of CNT-based composites may provide "smart" skins to aircraft and spacecraft that will enable monitoring of structural health and damage as it occurs. They are employed in the development of materials for the exploration of space, such as planetary protection materials and materials for radiation shielding.

Graphene composites – Graphene-reinforced polymeric materials are lightweight yet contribute to high weight-to-strength ratio constructions for aircraft and spacecraft. Because of its excellent electrical conductivity and mechanical strength, graphene can be incorporated into electronics and sensors for aerospace and defense applications. It has applications in thermal management systems in electronics including heat sinks and heat spreaders. An electrochemical storage system that includes batteries and supercapacitors made from graphene-based composites are suitable for use in aerospace and defense applications. They are employed in the development of advanced radar-absorbing materials for use in stealth technologies. Their applications include flexible electronic screens and wearable displays, especially for aerospace and defense systems (Figure 10.5).

Nano-clay composites – Nano-clay-reinforced composites have a potential application advantage in the areas of aerospace and defense, where improved flame retardancy is required. They include improved barrier properties which minimize gas permeability and enhance the efficiency of the fuel. They can enhance factors like tensile strength and toughness that are necessary in the aerospace and defense domains. They also enhance the durability of foreign metal parts, associated with aviation and military use, due to increased resistance to corrosion. They are employed in the manufacture of high-performance coatings in aerospace and military sectors, offering better resistance to wear and corrosion. Nano-clay-based composites can also be used in making self-repairing materials in aerospace and military fields, providing damage tolerance and lowering maintenance costs.

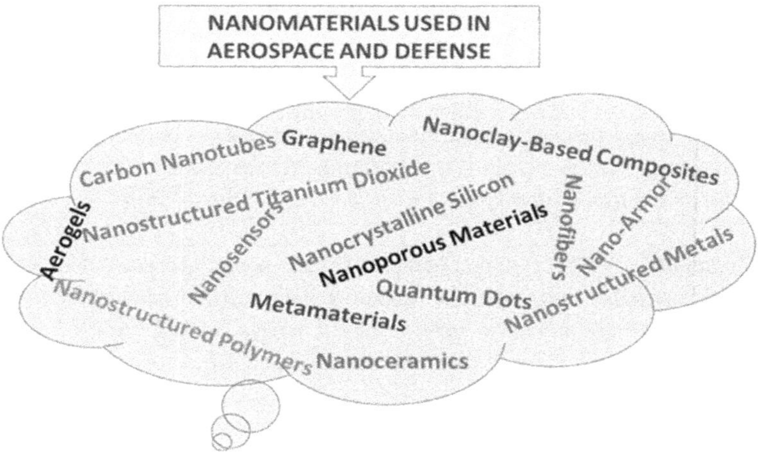

FIGURE 10.5 Nanomaterials used in aerospace, defense, and space exploration.

10.7.2 Nanostructured Metals, Nano-Ceramics, Aerogels, and Metamaterials

Nanostructured metals – These materials can be used in the fabrication of lightweight, strong structural parts for aircraft and spacecraft (Pu and Chen 2022), facilitating weight reduction and increasing the efficiency of fuel consumption. They can also be used in propulsion systems for aerospace, such as in rocket engines and fuel cells and in sensors for aerospace and military applications.

Nano-ceramics – The aerospace industry makes use of nanoceramic (Ming et al. 2022) materials in the manufacture of next-generation thermal protection systems such as orbiting thermal protection application systems, mounting devices, and heat shield devices. These materials have high technology sensing elements for aerospace, defense, and control systems for electricity storage and generation, batteries, and fuel cells.

Aerogels – They serve in aircraft and spacecraft as thermal barriers inhibiting heat transfer and enhancing the efficiency of energy use. These materials find application in acoustical protection of aircraft and spacecraft by sound absorption and in spacecraft and satellites for radiation shielding. These materials are used in body and vehicular armors for augmenting the ballistic protection levels. These materials are employed against thermal attack, like missile attacks from heat-seeking missiles. Radar absorbent materials are used in such a way that they help to shield the materials due to radar waves, thus making them less visible. These materials also are being used to make lighter-weight structural parts and advanced sensors and help designing advanced electric power storage and conversion modules for aerospace and defense industry (Hebalkar et al. 2020), including batteries and fuel cells. NASA's Environmental Control and Life Support System (ECLSS) employs such materials to clean air and water in spacecraft and habitats, for example, aerogel-based materials.

Metamaterials – The emphasis can be given in order to fabricate next-generation electromagnetic materials like materials for negative index refraction and perfect absorption (Mei et al. 2021) which could be utilized in developing materials for stealth technology, advanced sensing technologies, energy harvesting systems for aerospace and defense sectors, enhancement of power generation, and reduction of energy usage.

10.7.3 Nanomaterials for Space Exploration

CNTs, graphene, metal nanostructures (aluminum, copper), polymeric nanostructures (polyimides, polyurethanes), titanium dioxide nanofibers (e.g., polymeric, ceramic) (Erol et al. 2023) are used for the development of ultra-lightweight, high-strength materials for spacecraft structures, next-generation thermal protection systems, energy storage, generators (supercapacitors, batteries), space exploration for thermal protection, and radiation shielding materials. They are also used as advanced materials for water purification systems.

QDs are utilized in advanced sensor and detector systems, specially designed imaging systems, as well as in the fields of quantum computation and communication. Nano-crystalline silica is used in advanced solar cells and energy-harvesting systems along with active electronics resistant to radiation.

Nanostructured silver (Ag) and copper oxide (CuO) materials are used in spacecraft (Ghidini 2018) (e.g., biomedical and vehicle applications, atmospheric safety), water filtration systems, and thermal control.

10.8 CONCLUSION

Emerging trends in using nanomaterials for various purposes are likely to change the conventional process of production in many different sectors of industry. Due to the properties of nanomaterials such as their high tensile strength, conductivity, and reactivity, they are meant for a great number of uses ranging from energy storage and generation to biomedical products and advanced composites. Further enhancements in product performance, efficiency, and the sustainability of the product will lead to invention of new nanomaterials and modes of manufacturing.

Nanomaterials provide means for creating new products and processes that one could only imagine before. Everything from self-repairable materials and changeable structures to next-generation water purifying devices and drug formulations targeting diseases can be made; the range is huge and diverse. Furthermore, it is projected that incorporating nanomaterials in the already established methods of production will be a catalyst for economic development and increase employment opportunities.

Like any other new technology, there are also potential risks and uncertainties in the field of nanotech. As the field of nanomaterials is emerging, it is essential that the use of nanomaterials is done safely and responsibly to reach its full potential. This would imply not only a continued search for solutions to the existing problems, but also the building of efficient regimes and standards of regulation.

In spite of these obstacles, prospects for manufacturing using nanomaterials are very bright. Many industries will benefit from these evolving materials as we begin to maximize on the use of nanomaterials in aerospace, automotive, healthcare, and energy industries. Numerous innovations can come from nanomaterials, which will improve our lives and environment.

REFERENCES

Abdin, Z., M. A. Alim, R. Saidur, Islam, W. Rashmi, S. Mekhilef, and A. Wadi. "Solar Energy Harvesting with the Application of Nanotechnology." *Renewable and Sustainable Energy Reviews* 26 (July 18, 2013): 837–52. https://doi.org/10.1016/j.rser.2013.06.023

Afzal, Obaid, Abdulmalik S. A. Altamimi, Muhammad Shahid Nadeem, Sami I. Alzarea, Waleed Hassan Almalki, Aqsa Tariq, Bismillah Mubeen, et al. "Nanoparticles in Drug Delivery: From History to Therapeutic Applications." *Nanomaterials* 12, no. 24 (December 19, 2022): 4494. https://doi.org/10.3390/nano12244494

Agarwal, Manish Baboo, Shahjad Ali, Manu Mehrotra, Manish Kumar Panday, Neeraj Gupta, and Seema Agarwal. 2023. "Magnetic Anisotropy in the Ferromagnetic Mn and N Doped ZnO Nanocrystals." *Journal of Physics Conference Series* 2570 (1): 012008. https://doi.org/10.1088/1742-6596/2570/1/012008

Agarwal, Manish Baboo, M. Malaidurai, Sumit Sahoo, and R. Thangavel. 2021. "Fabrication and Characterization of Mn Doped ZnO Nanoarrays Prepared by Two Step Process." *AIP Conference Proceedings* 2327 (January): 020010. https://doi.org/10.1063/5.0040024

Agarwal, Manish Baboo, M. Malaidurai, Akash Sharma, and R. Thangavel. 2020. "Effect of Al Doping on Hydrothermal Growth and Physical Properties of Doped ZnO Nanoarrays for Optoelectronic Applications." *Materials Today Proceedings* 21 (January): 1781–86. https://doi.org/10.1016/j.matpr.2020.01.231

Agarwal, Manish Baboo, Akash Sharma, M. Malaidurai, and R. Thangavel. 2018. "Effect of Sn Doping on Structural, Mechanical, Optical and Electrical Properties of ZnO Nanoarrays Prepared by Sol-gel and Hydrothermal Process." *Superlattices and Microstructures* 117 (March): 342–50. https://doi.org/10.1016/j.spmi.2018.03.061

Alzate-Correa, D., W. R. Lawrence, A. Salazar-Puerta, N. Higuita-Castro, and D. Gallego-Perez. "Nanotechnology-Driven Cell-Based Therapies in Regenerative Medicine." *The AAPS Journal* 24, no. 2 (March 15, 2022). https://doi.org/10.1208/s12248-022-00692-3

Ashfaq, Alweera, Nazia Khursheed, Samra Fatima, Zayeema Anjum, and Kaiser Younis. "Application of Nanotechnology in Food Packaging: Pros and Cons." *Journal of Agriculture and Food Research* 7 (January 10, 2022): 100270. https://doi.org/10.1016/j.jafr.2022.100270

Ashique, Sumel, Basmah Almohaywi, Nazima Haider, Sabina Yasmin, Afzal Hussain, Neeraj Mishra, and Ashish Garg. "siRNA-based Nanocarriers for Targeted Drug Delivery to Control Breast Cancer." *Advances in Cancer Biology - Metastasis* 4 (May 25, 2022): 100047. https://doi.org/10.1016/j.adcanc.2022.100047

Baby, Rabia, Mohd Zobir Hussein, Abdul Halim Abdullah, and Zulkarnain Zainal. "Nanomaterials for the Treatment of Heavy Metal Contaminated Water." *Polymers* 14, no. 3 (January 31, 2022): 583. https://doi.org/10.3390/polym14030583

Bajwa, Neha, Neelesh K. Mehra, Keerti Jain, and Narendra K. Jain. "Pharmaceutical and Biomedical Applications of Quantum Dots." *Artificial Cells Nanomedicine and Biotechnology*, June 10, 2015, 1–11. https://doi.org/10.3109/21691401.2015.1052468

Bansal, Suneev Anil, Vanish Kumar, Javad Karimi, Amrinder Pal Singh, and Suresh Kumar. "Role of Gold Nanoparticles in Advanced Biomedical Applications." *Nanoscale Advances* 2, no. 9 (January 1, 2020): 3764–87. https://doi.org/10.1039/d0na00472c

Biswas, Rahul, Mahabub Alam, Animesh Sarkar, Md Ismail Haque, Md Moinul Hasan, and Mominul Hoque. "Application of Nanotechnology in Food: Processing, Preservation, Packaging and Safety Assessment." *Heliyon* 8, no. 11 (November 1, 2022): e11795. https://doi.org/10.1016/j.heliyon.2022.e11795

Chow, James C. L. "Special Issue: Application of Nanomaterials in Biomedical Imaging and Cancer Therapy." *Nanomaterials* 12, no. 5 (February 22, 2022): 726. https://doi.org/10.3390/nano12050726

Daniyal, Md, Ameer Azam, and Sabih Akhtar. "Application of Nanomaterials in Civil Engineering." In *Advanced Structured Materials*, 169–89, 2017. https://doi.org/10.1007/978-981-10-6214-8_6

Dong, Yuxi C., Maryam Hajfathalian, Portia S. N. Maidment, Jessica C. Hsu, Pratap C. Naha, Salim Si-Mohamed, Marine Breuilly, et al. "Effect of Gold Nanoparticle Size on Their Properties as Contrast Agents for Computed Tomography." *Scientific Reports* 9, no. 1 (October 17, 2019). https://doi.org/10.1038/s41598-019-50332-8

Dragoman, Daniela. "Nanomaterials for Energy Harvesting." *Nanomaterials* 13, no. 7 (March 24, 2023): 1154. https://doi.org/10.3390/nano13071154

Dwivedi, Sourabh, Quaiser Saquib, Abdulaziz A. Al-Khedhairy, and Javed Musarrat. "Understanding the Role of Nanomaterials in Agriculture." In *Springer eBooks*, 271–88, 2016. https://doi.org/10.1007/978-81-322-2644-4_17

Erol, Mustafa, Sibel Demiroğlu Mustafov, Salih Alper Akalın, Begüm Uzunbayır, Erdem Tevfik Özdemir, Berkay Coşkun, and Zeynep Ertekin. "Carbon Nanostructures for Automotive and Aerospace Applications." In *Springer eBooks*, 1–29, 2023. https://doi.org/10.1007/978-3-031-14955-9_67-1

Fang, Cheng, Rajarathnam Dharmarajan, Mallavarapu Megharaj, and Ravi Naidu. "Gold Nanoparticle-based Optical Sensors for Selected Anionic Contaminants." *TrAC Trends in Analytical Chemistry* 86 (November 8, 2016): 143–54. https://doi.org/10.1016/j.trac.2016.10.008

Fang, Yarong, and Yanbing Guo. "Copper-based Non-precious Metal Heterogeneous Catalysts for Environmental Remediation." *Chinese Journal of Catalysis (Chinese Version)* 39, no. 4 (April 1, 2018): 566–82. https://doi.org/10.1016/s1872-2067(17)62996-6

Franco, Domenico, Giovanna Calabrese, Salvatore Pietro Paolo Guglielmino, and Sabrina Conoci. "Metal-Based Nanoparticles: Antibacterial Mechanisms and Biomedical Application." *Microorganisms* 10, no. 9 (September 3, 2022): 1778. https://doi.org/10.3390/microorganisms10091778

Galdames, Alazne, Leire Ruiz-Rubio, Maider Orueta, Miguel Sánchez-Arzalluz, and José Luis Vilas-Vilela. "Zero-Valent Iron Nanoparticles for Soil and Groundwater Remediation." *International Journal of Environmental Research and Public Health* 17, no. 16 (August 11, 2020): 5817. https://doi.org/10.3390/ijerph17165817

Ghidini, Tommaso. "Materials for Space Exploration and Settlement." *Nature Materials* 17, no. 10 (September 17, 2018): 846–50. https://doi.org/10.1038/s41563-018-0184-4

Gu, Yanling, Zhongzhu Yang, Junwu Zhou, and Zhaoyong Chen. "Application of Graphene/LDH in Energy Storage and Conversion." *Sustainable Materials and Technologies* 37 (August 15, 2023): e00695. https://doi.org/10.1016/j.susmat.2023.e00695

Handa, Mayank, Vanshul Saini, Shourya Tripathi, Awesh Yadav, and Rahul Shukla. "Hybrid Nanomaterials as Novel Therapeutic and Imaging Modalities." In *Springer eBooks*, 147–73, 2023. https://doi.org/10.1007/978-3-031-30529-0_5

Hebalkar, Neha, Keerthi Sanghamitra Kollipara, Yamini Ananthan, and Murali Krishna Sudha. "Nanoporous Aerogels for Defense and Aerospace Applications." In *Springer eBooks*, 121–63, 2020. https://doi.org/10.1007/978-3-030-16347-1_5

Hussain, Rabia Tasaduq, Md Sanower Hossain, and Jun Haslinda Shariffuddin. "Green Synthesis and Photocatalytic Insights: A Review of Zinc Oxide Nanoparticles in Wastewater Treatment." *Materials Today Sustainability* 26 (March 23, 2024): 100764. https://doi.org/10.1016/j.mtsust.2024.100764

Jing, Qingshen, and Sohini Kar-Narayan. "Nanostructured Polymer-based Piezoelectric and Triboelectric Materials and Devices for Energy Harvesting Applications." *Journal of Physics D Applied Physics* 51, no. 30 (May 26, 2018): 303001. https://doi.org/10.1088/1361-6463/aac827

Jones, Wendy, Alistair Gibb, Chris Goodier, Phil Bust, Mo Song, and Jie Jin. "Nanomaterials in Construction – What Is Being Used, and Where?" *Proceedings of the Institution of Civil Engineers - Construction Materials* 172, no. 2 (July 22, 2016): 49–62. https://doi.org/10.1680/jcoma.16.00011

Kashyap, Brijendra Kumar, Virendra Vikram Singh, Manoj Kumar Solanki, Anil Kumar, Janne Ruokolainen, and Kavindra Kumar Kesari. "Smart Nanomaterials in Cancer Theranostics: Challenges and Opportunities." *ACS Omega* 8, no. 16 (April 10, 2023): 14290–320. https://doi.org/10.1021/acsomega.2c07840

Kausar, Ayesha, Irum Rafique, and Bakhtiar Muhammad. "Aerospace Application of Polymer Nanocomposite with Carbon Nanotube, Graphite, Graphene Oxide, and Nanoclay." *Polymer-Plastics Technology and Engineering* 56, no. 13 (January 3, 2017): 1438–56. https://doi.org/10.1080/03602559.2016.1276594

Khalilov, Rovshan. "A Comprehensive Review of Advanced Nano-Biomaterials in Regenerative Medicine and Drug Delivery." *Advances in Biology & Earth Sciences*, 8(1), (2023): 5–18.

Kim, Seong Dae, Arijit Sarkar, and Jong-Hyun Ahn. "Graphene-Based Nanomaterials for Flexible and Stretchable Batteries." *Small* 17, no. 48 (March 7, 2021). https://doi.org/10.1002/smll.202006262

Kuswandi, Bambang. "Nanotechnology in Food Packaging." In *Sustainable Agriculture Reviews*, 151–83, 2016. https://doi.org/10.1007/978-3-319-39303-2_6

LeMieux, Melburne C., and Zhenan Bao. "Stretching Our Imagination." *Nature Nanotechnology* 3, no. 10 (October 1, 2008): 585–86. https://doi.org/10.1038/nnano.2008.296

Low, Foo Wah, Chin Wei Lai, Sieh Kiong Tiong, and Nowshad Amin. "Graphene-Based Nanocomposites for Renewable Energy Application." In *Springer eBooks*, 1–36, 2019. https://doi.org/10.1007/978-3-030-10614-0_26-1

Malik, Shiza, Khalid Muhammad, and Yasir Waheed. "Emerging Applications of Nanotechnology in Healthcare and Medicine." *Molecules* 28, no. 18 (September 14, 2023): 6624. https://doi.org/10.3390/molecules28186624

Markandan, Kalaimani, and Wai Siong Chai. "Perspectives on Nanomaterials and Nanotechnology for Sustainable Bioenergy Generation." *Materials* 15, no. 21 (November 3, 2022): 7769. https://doi.org/10.3390/ma15217769

Mazari, Shaukat Ali, Esfandyar Ali, Rashid Abro, Fahad Saleem Ahmed Khan, Israr Ahmed, Mushtaq Ahmed, Sabzoi Nizamuddin, et al. "Nanomaterials: Applications, Waste-Handling, Environmental Toxicities, and Future Challenges – A Review." *Journal of Environmental Chemical Engineering* 9, no. 2 (January 10, 2021): 105028. https://doi.org/10.1016/j.jece.2021.105028

Mei, Hui, Wenqiang Yang, Dou Yang, Li Yao, Yongtao Yao, Chao Chen, and Laifei Cheng. "Metamaterial Absorbers Towards Broadband, Polarization Insensitivity and Tunability." *Optics & Laser Technology* 147 (November 8, 2021): 107627. https://doi.org/10.1016/j.optlastec.2021.107627

Mhetre, Harshada V., Yuvraj K. Kanse, and Suhas S. Patil. "Nanomaterials: Applications in Electronics." *International Journal of Advanced Engineering and Nano Technology* 4, no. 6 (September 30, 2021): 7–19. https://doi.org/10.35940/ijaent.d0464.094621

Ming, Wuyi, Zhiwen Jiang, Guofu Luo, Yingjie Xu, Wenbin He, Zhuobin Xie, Dili Shen, and Liwei Li. "Progress in Transparent Nano-Ceramics and Their Potential Applications." *Nanomaterials* 12, no. 9 (April 27, 2022): 1491. https://doi.org/10.3390/nano12091491

Mishra, Yogendra Kumar, N. Arul Murugan, Jani Kotakoski, and Jost Adam. "Progress in Electronics and Photonics with Nanomaterials." *Vacuum* 146 (September 28, 2017): 304–7. https://doi.org/10.1016/j.vacuum.2017.09.035

Nazar, Ali Matin, King-James Idala Egbe, Azam Abdollahi, and Mohammad Amin Hariri-Ardebili. "Triboelectric Nanogenerators for Energy Harvesting in Ocean: A Review on Application and Hybridization." *Energies* 14, no. 18 (September 7, 2021): 5600. https://doi.org/10.3390/en14185600

Padmanabhan, Nisha T., and Honey John. "Titanium Dioxide Based Self-Cleaning Smart Surfaces: A Short Review." *Journal of Environmental Chemical Engineering* 8, no. 5 (July 2, 2020): 104211. https://doi.org/10.1016/j.jece.2020.104211

Păduraru, Dan Nicolae, Daniel Ion, Adelina-Gabriela Niculescu, Florentina Mușat, Octavian Andronic, Alexandru Mihai Grumezescu, and Alexandra Bolocan. "Recent Developments in Metallic Nanomaterials for Cancer Therapy, Diagnosing and Imaging Applications." *Pharmaceutics* 14, no. 2 (February 17, 2022): 435. https://doi.org/10.3390/pharmaceutics14020435

Paris, Juan Luis, and María Vallet-Regí. "Nanostructures for Imaging, Medical Diagnostics and Therapy." In *Elsevier eBooks*, 1–28, 2018. https://doi.org/10.1016/b978-0-323-51255-8.00001-x

Prakash, Priyanka, and Chandran S. Smitha. "Nano-Phytoremediation of Heavy Metals from Soil: A Critical Review." *Pollutants* 3, no. 3 (August 1, 2023): 360–80. https://doi.org/10.3390/pollutants3030025

Pruthviraj, R. D., N. Gangadhara, A. A. Jahagirdar, and Santhosh A S Kumar. "A Review on - Hybrid Nanomaterials for Future Applications in Construction." *Bulletin of Pure and Applied Sciences-Chemistry* 40, no. 1 (2021): 39–47. https://doi.org/10.48165

Pu, Pengpeng, and Tijun Chen. "Nanostructured Metals with an Excellent Synergy of Strength and Ductility: A Review." *Materials* 15, no. 19 (September 23, 2022): 6617. https://doi.org/10.3390/ma15196617

Ramadoss, Jagadeesh, and Arumugam Sonachalam. "Energy Harvesting and Storage Applications for Hybrid Nanomaterials." In *Advances in Chemical and Materials Engineering Book Series*, 119–37, 2024. https://doi.org/10.4018/979-8-3693-1261-2.ch005

Saleh, Tawfik A. "Applications of Nanomaterials to Environmental Remediation." In *Interface Science and Technology*, 291–315, 2022. https://doi.org/10.1016/b978-0-12-849876-7.00005-1

Sarfraz, Jawad, Tina Gulin-Sarfraz, Julie Nilsen-Nygaard, and Marit Kvalvåg Pettersen. "Nanocomposites for Food Packaging Applications: An Overview." *Nanomaterials* 11, no. 1 (December 23, 2020): 10. https://doi.org/10.3390/nano11010010

Sarid, Dror, Brendan McCarthy, and Ghassan E. Jabbour. "Nanotechnology for Data Storage Applications." In *Springer eBooks*, 899–920, 2004. https://doi.org/10.1007/3-540-29838-x_30

Sathish, T., N. Ahalya, M. Thirunavukkarasu, T. S. Senthil, Zakir Hussain, Md Irfanul Haque Siddiqui, Hitesh Panchal, and Kishor Kumar Sadasivuni. "A Comprehensive Review on the Novel Approaches Using Nanomaterials for the Remediation of Soil and Water Pollution." *Alexandria Engineering Journal* 86 (December 2, 2023): 373–85. https://doi.org/10.1016/j.aej.2023.10.038

Sedighi, Ali, and Majid Montazer. "Nanomaterials for Wearable, Flexible, and Stretchable Strain/Pressure Sensors." In *Nanotechnology in Electronics: Materials, Properties, Devices*, 155–206, 2022. https://doi.org/10.1002/9783527824229.ch6

Subhedar, K. M., and B. P. Singh (2023). Introduction to Carbon Nanotubes and Graphene and their Emerging Applications. In *Emerging Applications of Carbon Nanotubes and Graphene*, ed. K. M. Subhedar and B. P. Singh, 1–22. CRC Press.

Wang, Zhao, Xumin Pan, Yahua He, Yongming Hu, Haoshuang Gu, and Yu Wang. "Piezoelectric Nanowires in Energy Harvesting Applications." *Advances in Materials Science and Engineering* 2015 (January 1, 2015): 1–21. https://doi.org/10.1155/2015/165631

Witharana, Sanjeeva, and M. T. Napagoda. *Nanotechnology in Modern Medicine*, 2022. https://doi.org/10.1007/978-981-19-8050-3

Zamel, Doaa, Atta Ullah Khan, Abdul Waris, Asmaa Ebrahim, and Nour E. A. Abd El-Sattar. "Nanomaterials Advancements for Enhanced Contaminant Removal in Wastewater Treatment: Nanoparticles, Nanofibers, and Metal-Organic Frameworks (MOFs)." *Results in Chemistry* 6 (September 3, 2023): 101092. https://doi.org/10.1016/j.rechem.2023.101092

Zhang, Peng, Jia Su, Jinjun Guo, and Shaowei Hu. "Influence of Carbon Nanotube on Properties of Concrete: A Review." *Construction and Building Materials* 369 (February 1, 2023): 130388. https://doi.org/10.1016/j.conbuildmat.2023.130388

Zhao, Xiaoli, Hong Jin, Jiayao Liu, Jiale Chao, Tingyi Liu, Han Zhang, Gang Wang, et al. "Integration and Applications of Nanomaterials for Ultrafast Photonics." *Laser & Photonics Review* 16, no. 11 (September 12, 2022). https://doi.org/10.1002/lpor.202200386

11 Biomedical Nanomanufacturing
Implants, Diagnostics, and Drug Delivery Systems

Corey Glickman, Bharti Raina, Vijay Sharma and Belhi Guerira

11.1 INTRODUCTION

Biomedical nanomanufacturing is an emerging field that integrates nanoscale science with advanced fabrication techniques to produce materials, devices, and systems intended for healthcare applications. This domain leverages the precise control of matter at nanometer scales to engineer structures and functionalities that directly interface with biological systems. From bone implants and neural probes to biosensors and nanocarriers, the landscape of modern medicine is being reshaped by innovations in nanomanufacturing. The convergence of materials science, tissue engineering, and nanotechnology has enabled researchers to bridge the gap between synthetic structures and living tissues, providing unprecedented opportunities in personalized and regenerative medicine as shown in Figure 11.1.

In the past, medical products and systems have been designed through the traditional manufacturing processes, which are, in most cases, not able to mimic the complex geometrical structure of biological systems (Zadpoor 2017). But all that changed in the late 20th century with the emergence of nanotechnology in the field of biomedicine. Carbon nanotubes (CNTs), quantum dots, gold nanoparticles, and polymeric nanostructures are the most studied nanomaterials in regard to their physicochemical and biological properties. At the same time, other techniques like electron beam lithography, nanoimprint lithography, electrospinning, self-assembly, etc., are becoming sufficiently developed to enable large-scale fabrication of nanoscale components. Such devices have been used to create nanostructured surfaces on implants, nano-enabled diagnostic sensors, and smart drug delivery sensors which are responsive to the environmental triggers within the body (Gaur et al. 2021). Table 11.1 is the biomedical applications of different nanomaterials.

Biomedical relevance of nanoscale engineering is accentuated by the fact that biological events are naturally transpiring on the nanoscale. All the essential components of life, such as proteins, enzymes DNA, and cellular receptors, are on the nanoscale range. This comparative scale enables nanostructures to better and

DOI: 10.1201/9781003560234-14

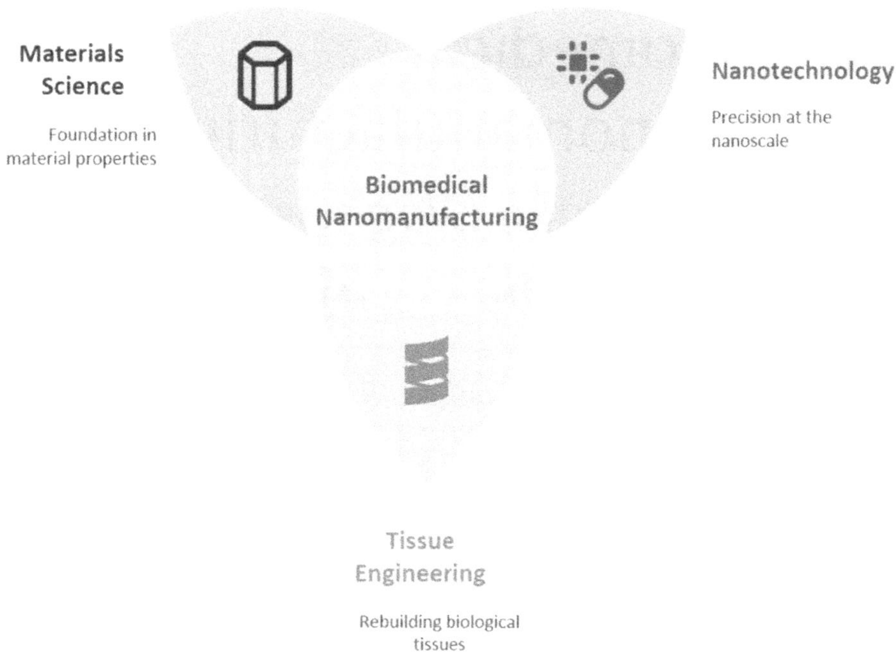

FIGURE 11.1 Synergy driving the biomedical nanomanufacturing.

TABLE 11.1
Examples of Nanomaterials in Biomedical Applications

Nanomaterial	Key Properties	Biomedical Use Case
CNTs (Wang et al. 2018)	High tensile strength, electrical conductivity, high aspect ratio	Neural interfaces, biosensors, drug delivery conduits
Gold nanoparticles (AuNPs) (Chen et al. 2022)	Optical tunability (SPR), inertness, biocompatibility	Targeted cancer therapy, imaging, lateral flow assays
Quantum dots (Mashinchian et al. 2017)	Narrow emission spectrum, high brightness, photostability	Cellular imaging, molecular diagnostics, tracking agents
Polymeric nanoparticles (e.g., PLGA, PEGylated systems) (Niculescu and Grumezescu 2021)	Biodegradability, tunable drug release, flexibility	Sustained drug delivery, vaccine platforms
Silica nanoparticles (Huq et al. 2025)	Porosity, surface modifiability, chemical stability	Gene delivery, theranostics

selectively interact with biomolecules and cells, and this scale has a degree of control and specificity not possible with conventional microscale devices. As an example, nanopatterned surfaces can be harnessed to greatly maximize osteointegration through increasing adhesion and proliferation of bone cells on orthopedic implants (Karazisis et al. 2017). Along the same line, the resolution and sensitivity of imaging techniques, including MRI and PET, have been enhanced with nanoparticle-based contrast agents.

There are three categories of core applications in biomedical nanomanufacturing: their design and manufacture of functional systems in implants, diagnostics, and drug delivery. Nanomanufacturing in implant devices has made it possible to construct surfaces that act as an extracellular matrix (ECM), providing superior tissue integration, minimized inflammation, and superior mechanical compatibility in the device (Shekaran and Garcia 2011). In diagnostics, nano-enabled platforms enable sensitive, multiplex-based detection of biomarkers within a rapid and feasible timeframe and can be used in the early detection and tracking of diseases. The use of nanoscale-based drug delivery systems through direct targeting, selectivity, and site-responsiveness has enabled the development of nanoscale drug delivery systems such as liposomes, dendrimers, and polymeric nanoparticles.

The driving force behind the need to conduct research in the field of biomedical nanomanufacturing is the high demand for patient-specific and minimally invasive solutions. Conventional implants do not succeed to a great extent either because of lack of biocompatibility or mechanical incompatibility with the surrounding tissues. In addition, off-target effect and poor treatment efficacy are caused by systemic drug delivery. Nanomanufacturing provides an opportunity to customize the properties of materials (physical, chemical, biological) to the requirements of individual patients. As an illustration, titanium implants have been shown to decrease bacterial adhesion and stimulate cell differentiation when provided with nanoscale grooves or pits. Likewise, the targeted delivery that can minimize systemic toxicity can be realized by designing nanomaterials that target tumor microenvironments in terms of pH or enzymatic activity.

The synergy between materials science, tissue engineering, and nanotechnology is central to the success of biomedical nanomanufacturing. Materials science contributes insights into the structure–property relationships of nanomaterials, guiding the design of biocompatible and mechanically robust systems. Tissue engineering introduces biological requirements, such as scaffold porosity, degradation rate, and cellular signaling. Nanotechnology, on the other hand, provides tools for precise manipulation and functionalization at the molecular level. This triad enables the development of hybrid systems that are structurally sound, biologically active, and functionally dynamic. Table 11.2 shows the core contributions of various disciplines in biomedical nanomanufacturing.

As an example, the use of nanofibrous scaffolds that have been prepared via electrospinning has the potential to mimic the natural fibrous structure of the ECM, thus enabling cellular attachment and growth. The scaffolds may be administered with growth factors or drugs in nanoparticles, having the dual purposes of generating tissue and localized treatment (De Witte et al. 2018). Equally, transistor organic-based field-effect transistors (FETs) have shown the ultra-sensitive detectability of DNA and protein translation with prospects of early cancer detection (Darwish et al. 2024). Antibodies conjugated to gold nanoparticles can also be used as colorimetric biosensors, as they can give a quick detection of the pathogen at the point of care (Chauhan et al. 2025).

Even with such progress, there are still obstacles in the move to translate laboratory-to-scale innovations into clinically viable products. The big challenges are

TABLE 11.2
Disciplinary Contributions to Biomedical Nanomanufacturing

Discipline	Core Contributions
Materials science	Understanding nanomaterial properties, ensuring biocompatibility and mechanical robustness
Tissue engineering	Scaffold design to mimic ECM, promoting cellular integration and regeneration
Nanotechnology	Nanoscale fabrication (top-down and bottom-up), surface functionalization, and responsiveness
Biomedical engineering	Integration into systems, ensuring clinical relevance and usability
Pharmaceutical sciences	Encapsulation, controlled release, pharmacokinetics optimization

reproducibility, scalability, and regulatory permission. Nanomaterials have the property of batch-to-batch variability, and the patterns of their interactions with biological systems are wicked and unpredictable. Besides, long-term biocompatibility, biodegradability, and even cytotoxicity of some nanomaterials are to be investigated in detail. The difficulties have led to joint initiatives between academia, industry, and regulatory entities to design common methods to characterize and test nanomaterials with standardized tests.

There are some distinguished research studies that have led to biomedical nanomanufacturing. A nano-scaled coating of titanium implants has been devised that promotes the bone-regenerative capabilities of the implants in animal models (Di Spirito et al. 2024). A graphene-based biosensor to detect cancer biomarkers at femtomolar concentrations has been proposed (Alsharabi et al. 2023). Lipid-based nanoparticle has been FDA approved as an siRNA delivery agent, as demonstrated in Patisiran, a drug against hereditary transthyretin-mediated amyloidosis in the drug delivery field (Moazzam et al. 2024). These incidences underscore the translational potential of nanomanufacturing in combination with high scientific validation and smart regulatory navigation.

In the future, the discipline is likely to adopt the use of multifunctional devices, such as those that can reproduce and sense at the same time with therapeutic functions. Bioanalytical nanomanufacturing together with artificial intelligence (AI) and digital health systems have the potential to transform the patient care continuum with real-time information as a basis and autonomous treatment. Besides, the theme of sustainability and green nanomanufacturing is on the rise, and efforts have been made to come up with biodegradable nanomaterials or energy-efficient methods of fabrication.

To sum up, biomedical nanomanufacturing is a transformational and powerful force that will remake healthcare by creating accurate, personalized, and proactive medical interventions. Based on the special characteristics of nanomaterials and combined with biological knowledge and engineering concepts, a new age of healthcare gadgets and treatments can be created. In the following parts of this chapter, more details of the particular applications in implants, diagnostics, and drug delivery will be provided, and current technologies, manufacturing processes, key metrics, and perspectives will be given. This historical development forms a backdrop of understanding of the intricacies and futures of biomedical nanomanufacturing in the improvement of human health.

11.2 FUNDAMENTALS OF BIOMEDICAL NANOMANUFACTURING

Biomedical nanomanufacturing is a very fast-developing field that combines the accuracy of nanotechnology together with the biological complexity of healthcare. It lies within the overlap of engineering, materials science, and medicine, to provide revolutionary methods to develop new sophisticated medical systems and equipment that could be used at the same level as the human body at the molecular level. In this section, the main ideas will be explored in detail, the distinction between nanomanufacturing and its counterpart on a microscale will be made, regulatory and biocompatibility issues will be discussed, and the major classes of nanomaterials used in this area will be introduced.

In its most basic definition, biomedical nanomanufacturing focuses on producing manmade structures and systems that are engineered to have features in a range of 1 to 100 nanometers that have the capacity to interact with in terms of functionality within a biological environment. Among these, one can design drug carriers that can traverse intracellular pathways, biosensors that can sense biomarkers as low as femtomolar, and materials used as a scaffold that can resemble an extracellular-like matrix at the molecular scale (Cicha et al. 2022).

Nanomanufacturing approaches are broadly classified into two categories: top-down and bottom-up methods (Bora et al. 2025; Ahmed 2020). Top-down methods include lithographic patterning, etching, and mechanical milling, where bulk materials are systematically reduced to nanoscale dimensions as shown in Figure 11.2. Bottom-up methods such as self-assembly, chemical vapor deposition, and nanoprecipitation build nanostructures from atoms and molecules, allowing for superior control over surface chemistry and morphology as shown in Figure 11.3.

The central tenets of nanomanufacturing are: Atomic-scale precision that allows the fabrication of functional materials and devices at the molecular level. Surface functionalization, which tailors materials for selective biological interaction, is often achieved via covalent bonding of ligands or bioactive molecules. System-level integration, where nanostructures are embedded into macroscale systems, enabling controlled responses to biological cues or environmental stimuli.

11.2.1 Differences Between Micro- and Nanomanufacturing in Biomedical Contexts

Although both micro- and nanomanufacturing encompass the process of miniaturization and provide the improved course of action of devices, these two processes have entirely dissimilar scope when it comes to choosing the magnitude of operation and the level of interaction with the biological system. Devices of microscales are usually between 1 micrometer and 1000 micrometers and can be used in microfluidic chips and implantable stents. Nevertheless, they are not as resolute or biologically concrete as nanostructures.

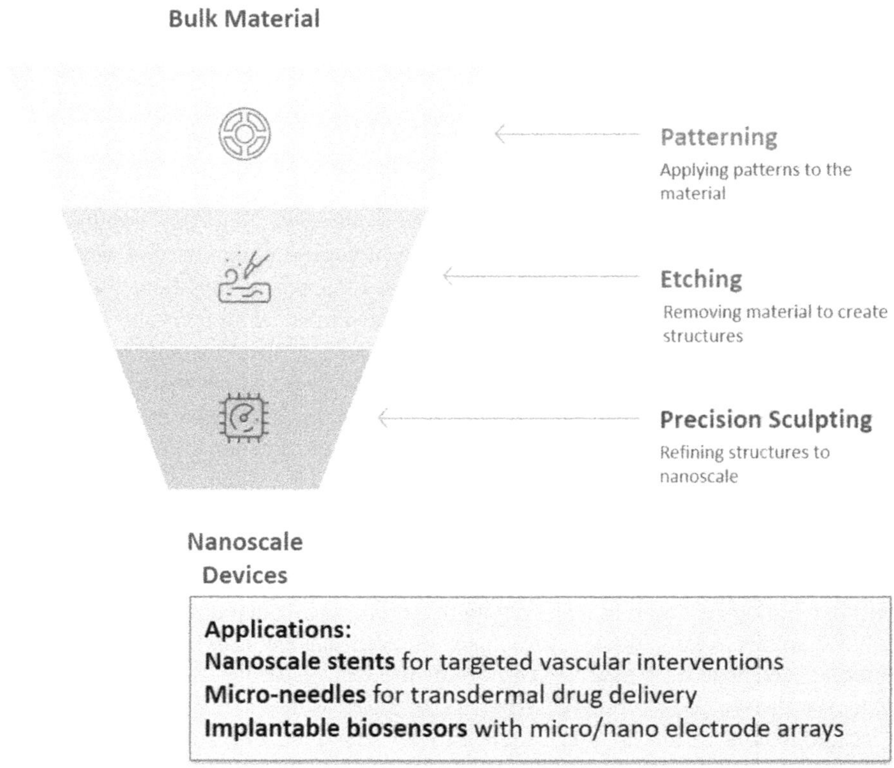

FIGURE 11.2 Top-down nanofabrication process.

Nanomanufacturing is at a level equivalent to that of biological molecules, proteins, DNA, and viruses; hence, direct biochemical interaction is assured. As an example, nanoparticles have the ability to cross blood–brain barrier, which would not be possible at the microscale level. Nanostructures are also highly useful in the magnitude of their surface area-to-volume ratios that are significant in drug loading, cell signaling, and catalytic reactions as illustrated in Figure 11.4.

11.2.2 Biocompatibility, Sterility, and Regulatory Constraints

Biocompatibility is a critical requirement for any biomedical material. At the nanoscale, unique interactions emerge between nanomaterials and cellular systems. While this enables higher efficacy, it also raises concerns about cytotoxicity and immunogenic responses. Therefore, materials must undergo rigorous in vitro and in vivo evaluations. Surface modifications, such as PEGylation or the addition of bioactive ligands, can enhance compatibility and reduce recognition by the immune system (Meyers and Grinstaff 2012).

Sterilization techniques commonly used for traditional medical devices, like autoclaving or gamma radiation, can significantly alter the properties of nanomaterials. Autoclaving may deform thermally sensitive polymers, while gamma radiation can

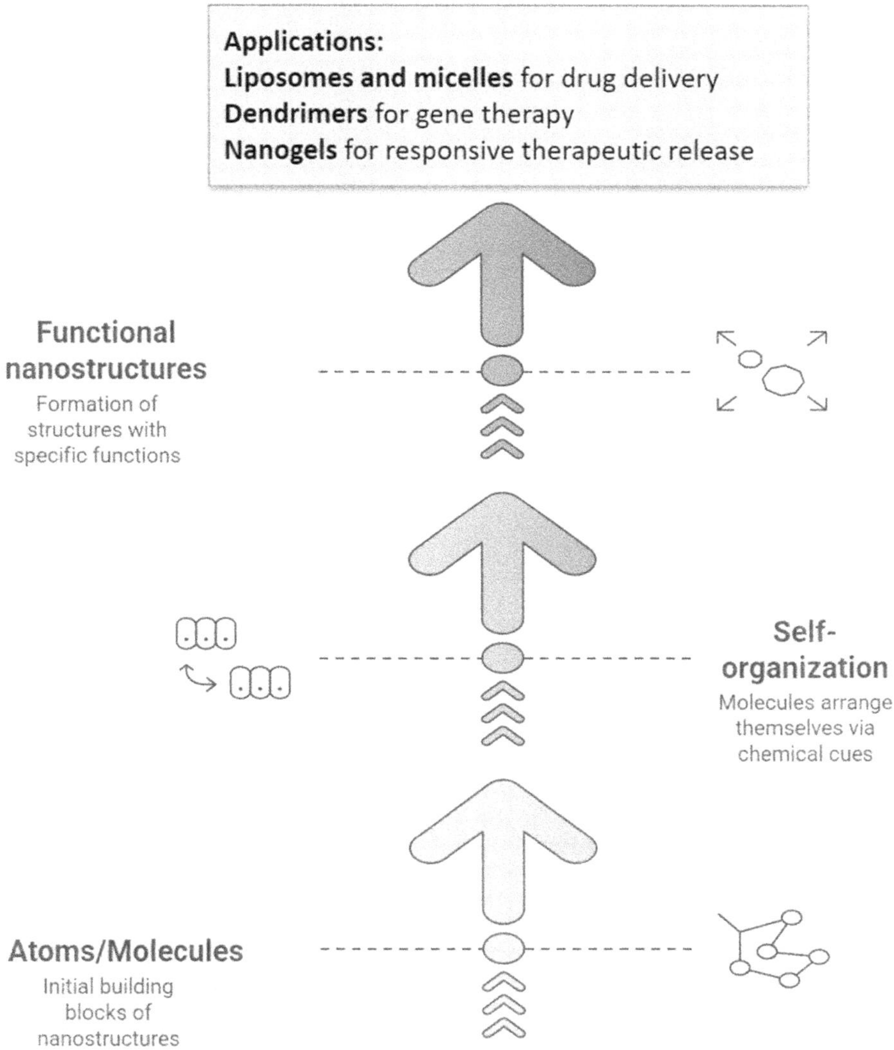

FIGURE 11.3 Bottom-up nanofabrication process.

break chemical bonds and induce reactive oxygen species. Ethylene oxide gas is often preferred for sterilizing nanostructures, although it requires careful removal of residues. Table 11.3 shows the compatibilities and limitations of various sterilization methods.

The regulatory landscape for nanomedicine is still evolving. Agencies like the FDA (United States) and EMA (Europe) emphasize comprehensive characterization, including size distribution, surface charge (zeta potential), morphology, and interaction with biological components. Additionally, long-term pharmacokinetic behavior and potential accumulation in organs must be addressed in preclinical studies. Standardization efforts by ISO and ASTM are underway to establish testing protocols specific to nanoscale biomedical devices (Desai et al. 2025).

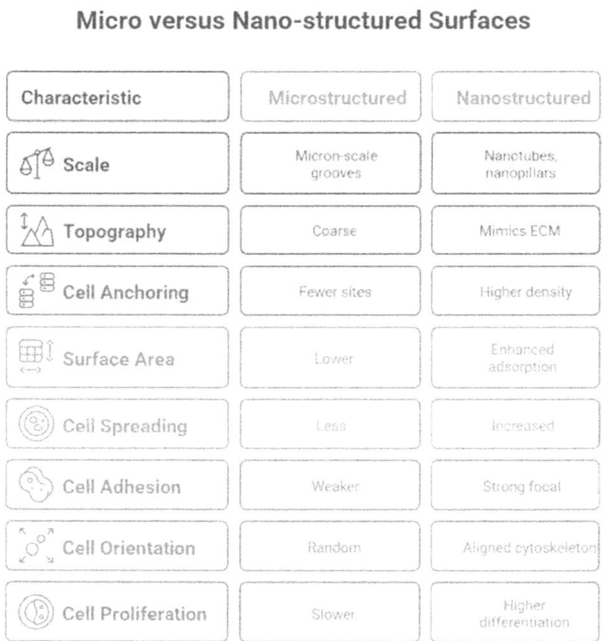

FIGURE 11.4 Comparative illustration of micro- versus nanostructured implant surfaces and their influence on cell adhesion and proliferation.

TABLE 11.3
Sterilization Techniques and Their Suitability for Nanomaterials

Method	Compatibility	Limitations
Gamma irradiation (Kolanthai et al. 2015)	High	Potential ROS generation and bond breakage
Autoclaving (Vetten et al. 2014)	Low for most nanomaterials	High heat and pressure may damage nanoscale integrity
Ethylene oxide (da Cunha Mendes, da Silva Brandão, and Miranda Silva 2008)	High	Residual toxicity must be thoroughly removed

11.2.3 Material Classes in Biomedical Nanomanufacturing

Biomedical nanomanufacturing exploits a very wide range of nanomaterials, each having special advantages in particular applications of medicine. Such materials consist of nanoparticles, nanotubes, nanocomposites, and nanogels as illustrated in Figure 11.5, each designed to provide specific capability with regard to targeted therapy, regenerative scaffolds, or sensitive diagnostics.

Nanoparticles can be defined as either inorganic (e.g., gold, iron oxide), organic (e.g., PLGA, chitosan), or lipid-based (e.g., liposomes, solid lipid nanoparticles). They have many uses in the controlled delivery of drugs, imaging, and immunomodulation.

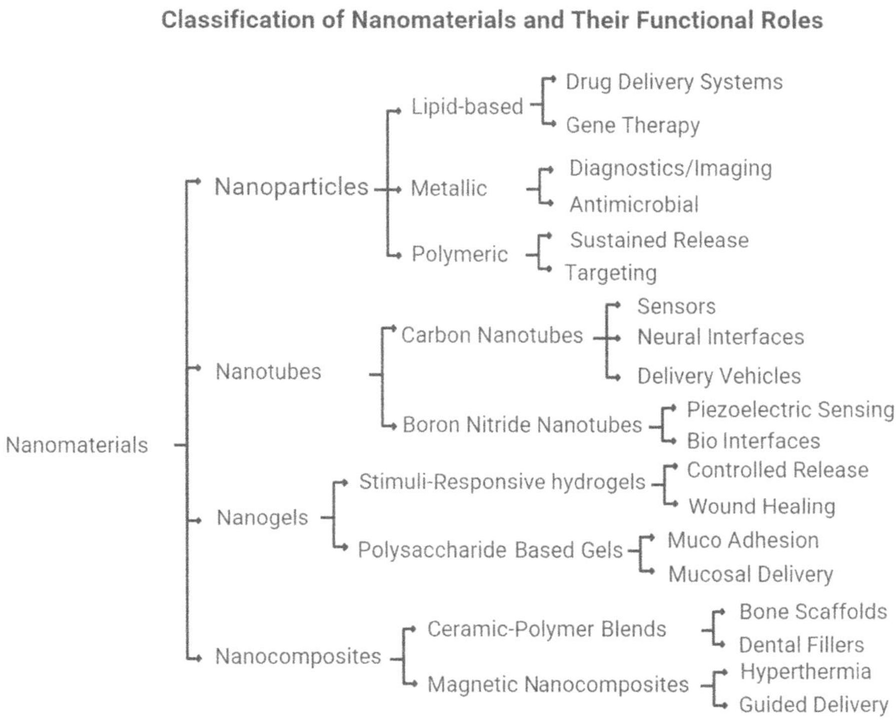

FIGURE 11.5 Classification of nanomaterials (nanoparticles, nanotubes, nanogels, nanocomposites) and their functional roles.

As an example, gold nanoparticles are utilized in photothermal therapy, whereas the polymeric nanoparticles, e.g., PLGA, provide biodegradable drug carriers in chemotherapy (Sun et al. 2017).

CNTs and Boron nitride nanotubes (BNNTs) are one of the cylindrical nanostructures that have been identified to be strong and electrical owing to their mechanical strength. The use of CNTs as bioelectronics or neural probes is common because they are both conductive and flexible. Instead, BNNTs are insulating and piezoelectric, thus capable of adapting unique functions for a pressure-based biosensing platform.

Nanocomposites are a mixture of nanoparticles and polymers, ceramics, or even metals to obtain better mechanical strength, bioactivity, or electric conductivity. As an example, nanoparticles of hydroxyapatite within a collagen matrix resemble bone structure and help in osteointegration during the use of orthopedic implants (Sobczak-Kupiec et al. 2021).

Nanogels are crosslinked, hydrophilic polymer networks that swell in response to environmental stimuli such as pH, temperature, or enzymatic activity. Their high-water content and tunable porosity make them excellent candidates for targeted, stimuli-responsive drug delivery systems, especially in oncology and wound healing applications (H. Zhang et al. 2016).

TABLE 11.4
Nanomaterial Classes and Representative Biomedical Applications

Class	Examples	Key Properties	Applications
Nanoparticles (Alkilany, Abulateefeh, and Murphy 2019)	AuNPs, PLGA, liposomes	Small size, tunable surface chemistry	Drug delivery, imaging
Nanotubes (Feng et al. 2021)	CNTs, BNNTs	High mechanical strength, conductivity	Biosensing, neural interfaces
Nanocomposites (Kumar et al. 2020)	Hydroxyapatite + Collagen	Enhanced strength, osteoconductivity	Orthopedic and dental implants
Nanogels (Delgado-Pujol et al. 2025)	PEG, PNIPAM	Swelling, biodegradability	Smart therapeutic delivery, regenerative aids

The applications and key properties of various nanomaterials are shown in Table 11.4. Biomedical nanomanufacturing builds on principles of molecular engineering, biological safety, and functional performance to create highly sophisticated medical solutions. By operating at a scale that mirrors biology, nanomanufactured systems promise better integration with the human body, from smarter drug delivery to real-time diagnostic interfaces. Understanding the distinctions between micro- and nanomanufacturing, as well as the implications of biocompatibility and regulatory oversight, is crucial for advancing these technologies from lab-scale to clinical settings.

11.3 NANOMANUFACTURED IMPLANTS

Nanostructured orthopedic implants employ surface modifications and coatings engineered at the nanoscale to improve osseointegration and longevity. Traditional titanium implants, while mechanically robust, often require enhanced biomimetic surfaces to encourage bone integration and reduce infection risk. One emerging strategy involves anodized titania (TiO_2) nanotube arrays on titanium substrates. These nanotubes, with pore diameters of 15–100 nm, dramatically increase surface area and support the adhesion, proliferation, and differentiation of osteoblasts. Additionally, the incorporation of nano-hydroxyapatite (nHA) onto TiO_2 nanotube arrays, via plasma spraying or electrodeposition, has shown synergistic effects: improved bone cell activity and increased bone-to-implant contact in vivo. Animal models with TiO_2–nHA–Ti composite implants demonstrate enhanced early osseointegration compared to smooth Ti (Radulescu et al. 2023).

Surface functionalization extends beyond osteoconduction to infection prevention. Coatings of silver-loaded fluorhydroxyapatite (FAgHA) on TiO_2 nanotubes have achieved strong antibacterial properties while preserving cell compatibility. Overall, these nanocoatings not only mimic natural bone architecture but also offer active biological control. Further, direct studies show that nanoscale roughness alone significantly improves bone fixation by promoting protein adsorption and integrin engagement (Coman et al. 2021).

Additive manufacturing (AM) adds another layer of customization. Powder bed fusion of Ti alloys enables the fabrication of patient-specific implants with integrated nanostructured surfaces. Post-processing steps, such as electrochemical etching and hydrothermal growth of TiO_2 nanotubes, confer nanoscale texture onto complex, porous titanium scaffolds. These scaffolds replicate trabecular bone structures and support vascularized bone in-growth, reducing stress shielding and improving fixation longevity.

The mouth environment poses additional challenges, including high bacterial load, fluctuating pH, and mechanical loading. Dental and craniofacial implants must therefore combine mechanical resilience with antimicrobial protection. Nanoscale coatings on Ti surfaces are being actively explored. Nano-spray-dried antibacterial coatings, such as PMMA films embedded with ZnO or Ag nanoparticles, have demonstrated strong activity against oral pathogens while maintaining mechanical durability. Likewise, TiO_2 nanotube arrays loaded with silver nanoparticles and growth factors (e.g., IGF-1) have significantly improved osseointegration in diabetic animal models, simultaneously resisting polymicrobial biofilms (Liu et al. 2020).

Bioinspired approaches further enhance integration by mimicking tooth microtopographies. Nanofeatured surfaces that replicate dentin tubule diameter (~1–2 μm) promote early calcium phosphate deposition and bacterial resistance. Clinical evaluations of Ti implants with such customized nano surfaces indicate reduced peri-implantitis rates and stronger bone anchorage (Choi et al. 2020).

Nanostructured coatings of cardiovascular stents are also receiving attention to minimize the incidence of thrombosis and restenosis. Although the current state of the market is occupied by drug-eluting polymer stents, nanotextured surfaces such as TiO_2 nanotubes and CNT coatings are under development to immobilize heparin and to control endothelialization. Equally, surface nanostructuring (e.g., platinum nanoparticles, films of iridium oxide) is applied in pacemaker leads and electrodes of neurostimulation, where it raises charge storage capability and biocompatibility. Such alterations lower the impedance and enhance electrical coupling (Udrişte et al. 2024).

CNT-coated or graphene-coated neural electrodes have a large surface area and conductivity, which in turn improves the signal transduction. Preclinical experiments demonstrate that CNT-modified microelectrode arrays (MEAs) show several impedance characteristics and assist in adhering the neurons and targeting the axons. Also, hybrid composites, i.e., CNTs combined with iridium oxide or PEDOT (poly(3,4-ethylenedioxythiophene)), are further optimized, also providing low impedance and tissue compliance. The presence of stent-mounted electrodes in the form of the Stentrode, which is shown to be feasible using early clinical data, although not nano-coated at this point, can be viewed as the first step toward the long-term goal of nanoscale-enabled devices being used in the future (S. Gupta et al. 2022).

Glial scaring and immune activity must be overcome in nanostructured neural interfaces over chronic implants. Protein-coating or hydrogel decreases foreign body responses (promoting close neuron-electrode coupling) substantially, and laminin-derived proteins are effective in this regard. Such materials as CNTs and TiO_2, which show good biocompatibility and have relatively lower tissue activation, are beneficial. There is also a reinforcement of conductive polymers (e.g., PEDOT) which adds to mechanical compliance, increasing performance over time (Huang 2020).

It is clear that the current development of nanostructured implants in orthopedics, dentistry, cardiovascular, and neural fields follows a common theme: the nanostructure of the implant surface determines the biological response. These technologies hold promise in boosting osteointegration, biofilm resistance, or electrical signal transduction and are also far greater than flat surfaces or micro-rough surfaces. Critical mass occurs when surface changes are combined with systemic models, e.g., implanting an implant with a certain anatomy with the patient via 3D printing, and making the nanocoating as an implant, controlled drug release, and antimicrobial and electrical characteristics.

Nevertheless, regulatory translation requires long-term preclinical data, standardized testing, and compliance with ISO/FDA guidance on nanomaterial characterization and safety. As microbiological, immunological, and biomechanical models evolve, the field moves closer to routine clinical adoption of nanomanufactured implants.

11.4 NANOMANUFACTURING FOR BIOMEDICAL DIAGNOSTICS

Biomedical diagnostics have been revolutionized by nanomanufacturing, enabling unprecedented sensitivity, portability, and integration at the point of need. By leveraging nanomaterials such as CNTs, graphene, quantum dots, and plasmonic nanoparticles, modern diagnostic platforms achieve precise biomolecule detection, real-time monitoring, and compatibility with microfluidic systems.

11.4.1 Lab-on-a-Chip and Nanosensors

Lab-on-a-chip (LOC) devices are the diminutive version of complex laboratory workflows (sample preparation, reaction, separation, and detection) on micro- or nanofluidic chips. To enable rapid diagnostics, incorporating common types of functional nanomaterials such as CNTs and graphene into LOC systems increases the efficiency of the capture and the level of signal, and real-time information can be read.

High-sensitivity CNTs and graphene are transducers of choice in FET-based biosensors based on their single-molecule sensitivity, biocompatibility, and high conductivity. Functionalized CNT-FET arrays of DNA-RNA or antibody-antigens have demonstrated femtomolar real-time detection of nucleic acids-label free sensitivity with limit of detection surpassing ELISA sensitivity. Equally, graphene FETs (GFETs) incorporated in the Cardea BPU 1M device have high multiplexing capabilities because they can identify a variety of cancerous biomarkers simultaneously, making them applicable to multiplex diagnostics (Parthasarathy and Venkatesh 2025).

Microfluidic chips that go hand in hand with nanomaterials have been applied to maximize the concentration and signal transduction of the analyte. The increase in electron transfer rates of the walls of the microchannel covered with CNT provides low electron transfer limits that make small-volume assays possible. Quantum dot conjugates with Droplet-based microfluidics provide ultrafast, multiplexed nucleic acids (e.g., DNA/RNA) detection with time and concentration control of reactions. In addition, the requirements of nanomaterial-enhanced LOCs have drastically reduced the use of reagents, which significantly cuts down expenditure and increases accessibility to resource-poor environments.

The enhanced LOC systems based on nanomaterial sensors allow point-of-care testing (POCT) to be performed in non-traditional laboratories. There are portable microchips based on CNT and GFETs capable of measuring cardiac biomarkers, infectious diseases, and metabolic abnormalities in serum samples in minutes. Other businesses such as Grapheal and Cardea have commercially developed wearable GFET platforms to detect SARS-CoV-2 and wound biomarkers in a few minutes, with potential readouts on smartphones. These trends highlight the intersection of nanomaterial sensitivity, microfluidic efficiency, and the user-friendly interfaces (Krishnan et al. 2023).

11.4.2 Nano-Optical and Electrochemical Sensors

In conjunction with LOC platforms, there are nano-optical and electrochemical sensors, which have broad-spectrum sensing properties with the ability to detect proteins in the analyte to metabolic biomarkers with amplification properties that use plasmonic and electrochemical properties.

SPR sensors take advantage of changes in refractive index in the vicinity of metal nanostructures as a result of molecular binding. An excellent reproducibility of as low as 1 nM limit of detection was obtained with gold nanostars functionalized with DNA to detect DNA. Most recently, wearable SPR sensors made of nanowire arrays (e.g., silver-coated silicon) have been shown to monitor glucose in human sweat with detection limits of ~0.12 mM non-invasively. Also, in fluorescence-based assays, quantum dots (QDs) are employed: QDs conjugated to an oligonucleotide can be used in a microdroplet to multiplex DNA assays on a nanosecond turnaround (Kant et al. 2024).

Electrochemical transducers are used to convert the biochemical bound events into a measurable change in current or voltage. Nanomaterial-coated transistors used in FET systems are fast, label-free, as sensing switch feeds in micro/nano fluidic devices. CNT FETs that are functionalized with enzymes, aptamers, or antibodies detect glucose, insulin, or cytokines in a very specific way and within a few seconds. The functional tag-supported improvement in GFETs supplies point-of-care multiplex biomarker detectivity (Sengupta and Hussain 2025).

In addition, electrochemical sensors can be applied to wearables. Gold nanomesh surface-enhanced Raman spectroscopy (SERS) sensors were proved to be able to be used as sweat-based sensors to detect in situ sweat biomarkers, including drugs, metabolites, and chemicals. In the meantime, lactate, pH, and electrolytes can be long-term monitored with the electrochemical-based wearable patches (Khonina and Kazanskiy 2025).

The integration of nanomaterials with flexible substrates and electronic systems has facilitated the advancement of real-time wearable diagnostic technologies. Bandages and patches are covered with sensors of CNT, graphene, or nanoparticle that continuously monitor physiological parameters, including glucose, electrolytes, fatigue characteristics, or environmental toxins. These sensors send data to smartphones, which provide off-site analysis and warnings to the cloud, and this can be used to monitor individual health without visiting hospitals.

Nanomanufacturing has helped in making a paradigm shift in diagnostics, where large machinery used in the laboratory has now become small-scale, high-capacity, easy-to-use mechanisms. When applied to microfluidic chips, FETs based on CNT/

graphene enable highly sensitive and fast assays capable of being designed in acute care and international health situations. Similarly, nano-optical sensors such as SPR, QDs, and SERS platforms are displayed as having a superior multiplexing ability for wearable monitoring and diagnostics.

Such a combination of nanotechnology, electronics, microfluidics, and cloud connectivity is manifesting itself in a new age in truly personal diagnostics, so that real-time measures can be made and fed directly into the decision-making of therapeutic interventions that may be taken outside of conventional healthcare contexts.

11.5 NANOMANUFACTURED DRUG DELIVERY SYSTEMS

Nanomanufacturing has had a sweeping effect on the way drugs are delivered as it allows drugs to be deployed with precision over pharmacokinetics, biodistribution, and therapeutic payload discharge. Using direct engineering on the nanoscale, there can be a designed drug carrier moving through biological barriers, precise targeting of diseased tissues, and drug release under controlled or stimulus-responsive conditions.

11.5.1 Principles of Controlled Drug Release

Nano-level control of drug release means that the drug release is carefully controlled, including the diffusion of drugs out of the carriers with time, the degradation of the carriers through time, and external triggering conditions.

Diffusion is the process by which small drug molecules migrate within a carrier matrix (e.g., polymeric nanoparticles, hydrogels) in response to concentration gradients. The rate of release is dictated by the porosity of the matrix, the solubility of the drug, and rate of diffusion. A universal analog is the case of chemotherapeutics-loaded PLGA nanoparticles, in which reducing the polymer molecular weight or changing how the particles were prepared allows regulation of pace, whether it be a quick burst or a longer, several-hour progression (Banstola et al. 2019).

The role of controlling degradation-controlled systems is matrix disintegration by hydrolysis or through an enzyme. Liposome and dendrimer carriers are degraded slowly in physiological conditions, which makes them appropriate for chronic applications like analgesia or hormone substitution therapy, which gives them a sustained release profile.

Figure 11.6 presents the relative release kinetics of PLGA nanoparticles (arylactosyl-controlled diffusion) and of liposomes/dendrimers (arylactosyl-controlled degradation). After an initial burst release, typical of PLGA-based systems, In many cases, sustained diffusion is observed with liposome and dendrimer carriers, where minimal or no initial release occurs until carrier degradation, after which drug release proceeds relatively rapidly.

Stimulus-responsive systems add a layer of control, reacting to external cues such as pH, temperature, light, or magnetic fields. These "smart" carriers remain inert during circulation but release their payload upon encountering specific signals.

a. ***pH-sensitive carriers:*** Kidney and tumor tissues are more acidic than blood; polymer-lipid nanocarriers that destabilize in acidic environments can selectively release chemotherapy agents in tumor microenvironments.

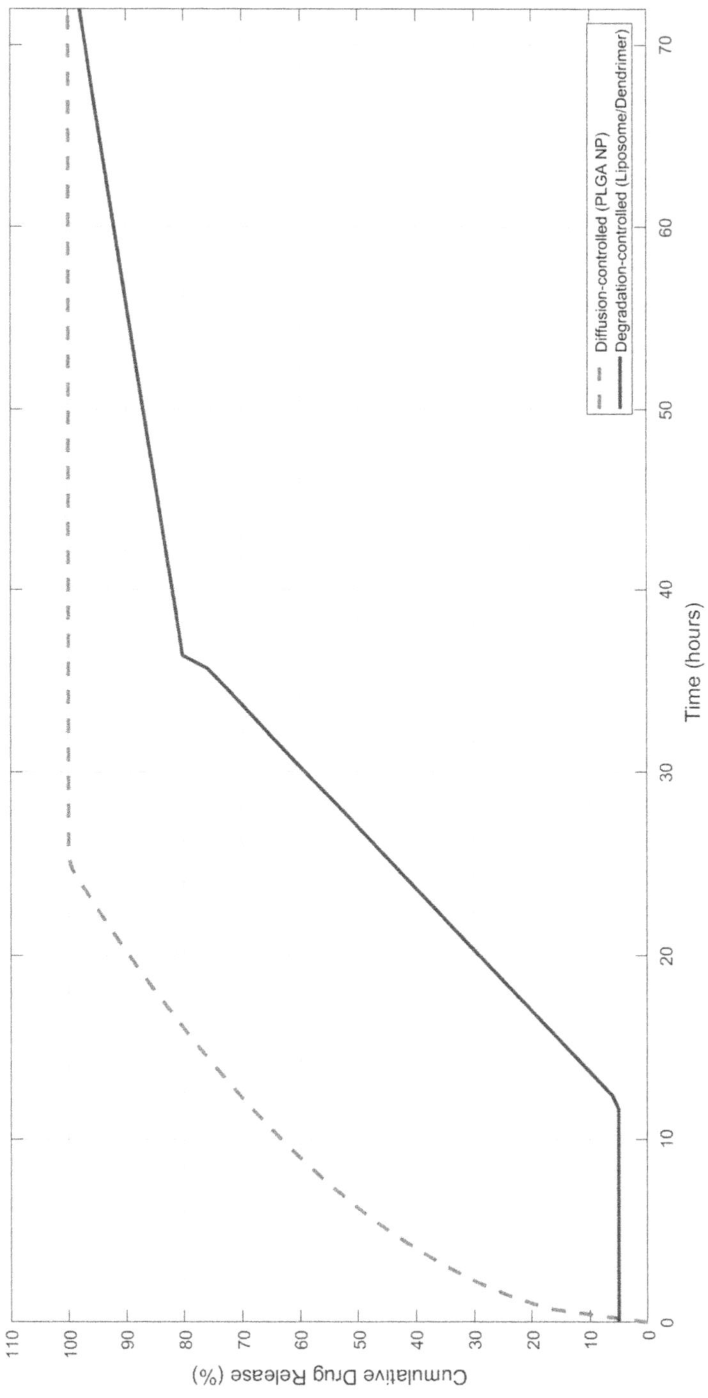

FIGURE 11.6 Drug release profiles of diffusion versus degradation-based nanocarriers.

b. ***Thermo- and photo-responsive systems***: Gold-coated nanogels can serve as on/off thermal switches activated by infrared light exposure, enabling localized bursts of drug release.
c. ***Magnetically triggered carriers:*** Superparamagnetic iron oxide nanoparticles (SPIONs) within thermoresponsive matrices allow pulsatile, magneto-thermal drug release on demand.

These mechanisms collectively form a versatile toolkit for achieving precise temporal and spatial delivery of therapeutics.

11.5.2 Targeted Delivery and Site-Specific Therapy

To enhance the therapeutic efficacy and decreasing side effects systemically, nanocarriers are motivated to become tissue-specific via passive or active targeting mechanism.

Nanoparticles of 10–200 nm are able to permeate tumor vasculature through a phenomenon known as enhanced permeability and retention (EPR) leading to queued accumulation because of poor lymphatic drainage. This effect is also used to produce PEGylated liposomes and polymeric nanoparticles to localize chemotherapy drugs around cancerous targets but does little to counter tumor heterogeneity and microenvironment signature heterogeneity (Nel, Ruoslahti, and Meng 2017).

Active targeting enhances specificity by functionalizing nanoparticle with a molecule that binds to host receptors of diseases:

a. Targeting carriers can be done via antibodies (e.g., anti-HER2 on dendrimer platforms); besides enhancing uptake, the off-target activity is inhibited in this way (Sitia et al. 2022).
b. Other ligands to attach to integrins, including RGD or YIGSR, can be used to help target integrin-expressing cells and enhance internalization in solid tumors or inflammation sites (Avgoustakis and Angelopoulou 2024).
c. Aptamers: Ligands composed of nucleic acid that can be synthetically bound to gold nanoparticles exhibit high specificity in binding therapeutic agents and delivering them to prostate or breast cancer cells. (Graczyk et al. 2020).

11.5.3 Responsive and Smart Drug Systems

The integration of the smart materials and AI-enabled design leads to the creation of nanocarriers that respond to changes in physiology and change to fit the specific condition of the patient.

Such systems are programmed to only release drugs under certain conditions: a pH-sensitive system will only release in an acidic tumor microenvironment or in an endosomal compartment; however, the release will remain intact in the bloodstream (Pandey and Pandey 2025). By heating thermoresponsive carriers with external light or ultrasound, such as a gold nanogel-based composite, pulsatile delivery of drugs can be controlled remotely (J. Gupta and Sharma 2025). On the one hand, magneto-responsive ones, in which the drug is released under the influence of alternating electromagnetic fields in SPION-containing membranes (Vangijzegem et al. 2023).

It is also supported by the multifunctionality that is achieved through a combination of the delivery with imaging to produce theranostic systems that can be used in treating cancer.

An enhanced use of AI, machine learning, and genetic algorithms can further be used in nanoparticle design: Genetic algorithms are used to optimize the size, shape, and ligand density to maximize tumor cell uptake and minimize immune clearance. The neural networks are used to predict pharmacokinetic behavior (e.g., circulation half-life, biodistribution) to lower the cost of formulation development. The responsive systems are incorporating reinforcement learning paradigms to adapt to the triggering of doses in real-time with respect to patient biomarkers.

Such combination of nanomaterials/AI represents a breakthrough—personalized nanomedicine—where a drug release mechanism can be adjusted to the unique biology of a patient in real time as shown in Figure 11.7.

Nanomaterials offer immense flexibility in drug delivery, enabling controlled kinetics, tissue targeting, and smart release. Table 11.5 consolidates the principal classes of nanocarriers and their features.

Next generation in mind, future plans are focused on: Creation of hybrid carriers capable of carriage of multiple payloads, responsive features, and imagery. Scaling of AI-based design to optimized designs that can be reproduced in manufacturing processes. The achievement of regulatory standards through the incorporation of predictable profiles of safety and efficacy.

The fact that controlled release, targeting, sensory responsiveness, and AI integration have been making strides suggests that in the near future, smart nanocarriers will ensure personalized, accurate therapy delivery with low collateral effects.

The general use of AI has resulted in the development of nanoinformatics as a field that employs the computational modeling approach to invigorate nanoparticle design, toxicity screening, and therapeutic compatibility. This specifically remains crucial in personalized medicine with a specific genetic and physiological profile of the patient that can be coupled with a personalized nanotherapeutic design as depicted in Figure 11.8.

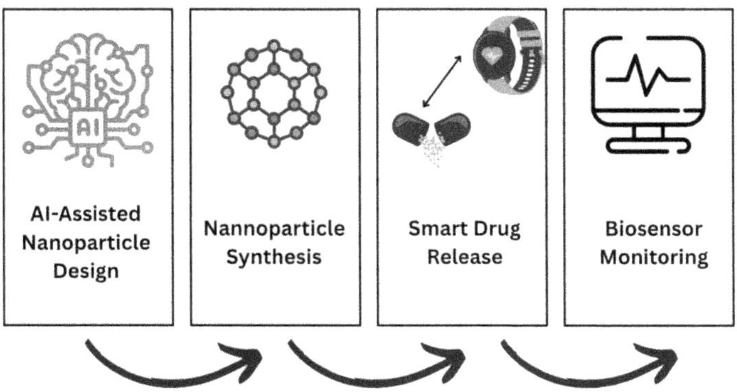

FIGURE 11.7 Workflow from AI-assisted nanoparticle design to stimulus-responsive drug release and real-time patient monitoring.

TABLE 11.5
Summary of Nanocarriers and Drug Delivery Characteristics

Nanocarrier	Release Mode	Targeting Strategy	Stimulus Response	Clinical Status
Liposomes (Jain et al. 2021)	Diffusion/ degradation	Passive/active (PEG, antibodies)	pH, Thermo	Multiple licensed drugs (e.g., Doxil®)
Dendrimers (Zhernov et al. 2024)	Surface-conjugated releases	Active via ligands/ antibodies	pH, enzymatic	Preclinical/early trials
PLGA nanoparticles (Xu et al. 2015)	Diffuse and degrade	Passive via EPR	Release kinetics tuning	Multiple in trials
CNT carriers (Alfei, Reggio, and Zuccari 2025)	Diffusion through tube walls	Targeting peptides	NIR/light, magnetism	Preclinical
Magnetic nanogels (L. Zhang, Wang, and Hou 2025)	Triggered via external fields	Passive, surface ligands	Magnetic/ thermal	Prototype stage

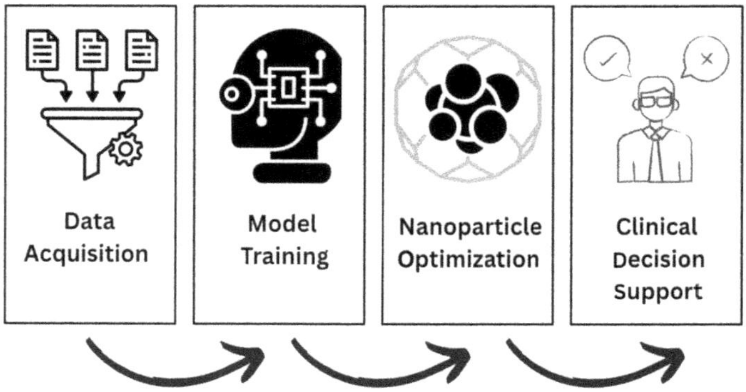

FIGURE 11.8 Workflow showing AI-assisted design of personalized nanomedicine.

11.6 CHALLENGES IN BIOMEDICAL NANOMANUFACTURING

Despite the remarkable promise and growing implementation of nanomanufactured biomedical systems, several persistent challenges hinder their full-scale translation from laboratory innovation to clinical adoption. These challenges span technical, regulatory, biological, and ethical domains, and addressing them is crucial for the sustained growth and integration of nanotechnologies into modern healthcare.

11.6.1 Scalability and Reproducibility

Reproducibility and scalability of biomedical nanomanufacturing are among the biggest problems in biomedical nanomanufacturing. Although laboratory synthesis methods can be directly nanoprecipitated, sol-gel and chemical vapor deposition processes enable fine-tuning of the particle morphology and particle surface functionality, and industrial-scale manufacturing of nanoparticles based on these methods tends to increase the nanoparticle size heterogeneity, purity, as well as drug loading efficiency.

Previous studies remarked that more than 60 percent of promising nanocarriers were abandoned because they were not able to reproduce laboratory-scale properties at batch-scale manufacturing. Such preoccupations as the delay in product development, inconsistency at the batch-to-batch level, and agglomeration at the storage phase as well as the collapse of bioactivity is the result of such concerns. Another huge discrepancy exists between innovation in academia and production-ready under Good Manufacturing Practice (GMP) (Shegokar and Nakach 2020).

In addition, the techniques of characterization have to be standardized within the industry. This further compounds the difficulty of process validation, where the same material gives different dynamic light scattering (DLS), zeta potential, and TEM imaging results in different labs.

11.6.2 Regulatory and Ethical Considerations

One of the toughest obstacles that can present themselves in the process of clinical translation is regulatory navigability. Whereas traditional drugs or even the medical devices are categorized as drugs or medical devices or perhaps even medical machines, the nanomedicine products usually exist in subsequent regulations gray areas, and they are in any case categorized as a product or a drug, a biologic, or even a combination product.

In the United States, the FDA would need a lot of pharmacokinetics and immunogenicity and biodistribution data and needs the nanoscale materials to comply with GMP. Nonetheless, there are hardly any established practices as far as nanotechnology applied in therapies is concerned. Nanotechnology Task Force of the FDA is still in the process of updating the standard, but the turnaround to approving nanotherapeutics takes longer compared to conventional small molecules (Souto et al. 2020).

Nanomedical products in the EU have to acquire CE marking as a result of conformity assessment procedures in the EU, with specific reference made to the new Medical Devices Regulation (MDR 2017/745) and In Vitro Diagnostic Regulation (IVDR). Such regulations require risk-based classification and clinical examination of any substance whose dimension is less than 100 nm. Table 11.6 shows the comparative analysis of FDA and CE mark requirements for various nanomedical devices.

The questions of morality are related to the disclosure of data (at least in case of AI-equipped gadgets), the lack of long-term health effects understanding, and equal provision of nanotherapy at high prices. Safer improvement was promoted with the demand of better regulatory harmonization and open databases of proper nanomaterials by the stakeholders.

11.6.3 Biodegradation, Immune Response, and Long-Term Safety

Due to their close interactions with biology, nanoparticles raise issues of biodegradability, immune response, and chronic toxicity. Although some materials are considered to be biocompatible on the macroscale, on the nanoscale their mechanism

TABLE 11.6
Comparison of FDA and CE Mark Requirements for Nanomedical Devices (Arch-Douglas et al. 2024)

Criteria	FDA (USA)	CE Mark (EU)
Oversight agency	CDER (drugs), CBER (biologics), CDRH (devices)	EMA and national competent authorities
Risk classification	Based on intended use, invasiveness, and material composition	Based on risk: Class I (low risk) to Class III (high risk)
GMP compliance	Mandatory under 21 CFR Part 210/211/820	Mandatory under MDR/IVDR regulatory framework
Nanotech-specific guidance	Limited but evolving (FDA Nanotechnology Task Force Reports)	Addressed in MDR Annex I, with references to ISO/TR 10993-22 and ISO 22442
Clinical evaluation requirements	Extensive non-clinical and clinical data, pharmacokinetics	Mandatory for Class IIa and above; equivalence data allowed for legacy devices
Post-market surveillance	Required with risk-based periodic reporting	Required with Periodic Safety Update Reports (PSURs) and trend analysis

of interaction with cells can be toxic and related to their more reactive surface and changed biodistribution.

As an illustration, the non-degradable materials such as gold or silica nanoparticles have been found to accumulate in liver, spleen, and kidneys over a long period of animal experiments (Paris, Baeza, and Vallet-Regí 2019). Alternatively, safer clearance mechanisms are done through the kidney or by enzyme degradation by the biodegradable systems such as PLGA and chitosan where they show acidic PMCA degradation products causing local inflammation (Puricelli et al. 2023).

The other concern is immune system activation. Experiments conducted at the Nanotechnology Characterization Laboratory (NCL) emphasize that the creation of the protein corona and the surface charge affects immune cell recognition critically. The positively charged particles have a higher possibility to cause cytokine storms or activate the complement (Souto et al. 2020).

Therefore, in vivo testing and uniform measures of toxicity are needed. According to the NCL, all the new nanomedical candidates should be investigated using multitiered studies, including hematological, histopathological, and immunotoxicity tests.

11.6.4 Interfacing with Biological Tissues and Fluids

The other important limiting factor is the interfacing of the nanostructure with complex biological systems such as the tissues, blood, and extracellular matrices. Functional biomedical nanomaterials have to move through active, enzyme-filled, immunity-sensitive environments keeping intact as far as possible their functional performance.

An example is that oral nanotherapeutics should not be degraded by gastric acid and proteases whereas injected systems should not be prematurely cleared out by reticuloendothelial system (RES). Also, the brain tissue requires an action that penetrates the blood–brain barrier, which is the most selective and impermeable biological filtration system (Z. Yang et al. 2022).

Site-specific improvement of bioavailability with minor surface interactions (i.e., through PEGylation, zwitterionic coating, or ligand functionalization) is an appropriate change, though it may cause unexpected pharmacokinetics or interactions. And in addition, many tissues (such as mucosal linings) may need mucoadhesive polymers or charge effects to adhere to the tissue, making the design more complex still.

These interactions are being examined in vitro through advance forms of technology such as organ-on-chip platforms, 3D bioprinted tissue models, and more before deployment into clinical practice. The latter can be studied in real time using these models exposed to environments that represent lung, liver, or tumor tissues (Carvalho et al. 2021).

To sum up, the field of biomedical nanomanufacturing is promising, but it also faces some serious problems of technical, regulatory, and biological nature. The tools of interdisciplinary cooperation, standardized tests systems, adaptive regulatory mechanisms, and further investments in scalable, safe, and ethically oriented nanotechnologies will need to be employed when dealing with these issues.

11.7 FUTURE TRENDS AND OPPORTUNITIES

Biomedical nanomanufacturing is not only evolving rapidly but is poised to reshape how healthcare systems approach disease prevention, diagnosis, and therapy. As foundational technologies like nanofabrication, bioprinting, and bioelectronics mature, several emerging frontiers are opening new avenues for clinical impact. This section explores transformative opportunities, including 4D bioprinting, regenerative nanomedicine, implantable nanobioelectronics, and sustainability in nanomanufacturing.

11.7.1 4D Bioprinting with Nanomaterials

Based on 3D bioprinting, 4D bioprinting adds the factor of time by designing biomaterials that are able to change dynamically after fabrication in order to respond to physiological changes like pH, temperature, hydration, or enzyme movement. To achieve such responsive behavior, nanomaterials are now being added to the bioinks, including nanosilicates, graphene oxide, and nanocellulose.

These nanomaterials endow the printed constructs with important properties such as reinforcement of the structure, conductivity, and augmented cell adhesion. On the same note, the use of gold nanoparticles within gelatin hydrogels incorporates photothermal responsiveness, which is conducive to remote-controlled scaffold remodeling (Yadav et al. 2024).

These applications are self-folding stents, dynamic physiological models, organ-on-chip models, and smart tissue scaffolds.

11.7.2 Regenerative Medicine and Stem Cell Guidance Using Nanostructures

Nano-topography is essential in determining the morphology of the stem cells, their fate, attachment, growth, and differentiation. Mechano-transduction pathways can also be modulated by nanopatterned surfaces with grooves, pits, or ridges, and thus play a role in influencing lineage specification.

According to the previous research, TiO2 nanotube arrays enhance osteogenic differentiation of mesenchymal stem cells (MSCs). Chondrogenic lineage commitment is improved by Nanoscale Collagen-mimetic fibers. Neurogenesis can be carried out with the help of the electrical conductivity of nanocomposites (e.g., polypyrrole-CNT) to stimulate neuronal stem cells (Chang et al. 2019).

Conjugation of such nanostructures to biocompatible hydrogels has facilitated the development of injectable systems capable of tissue regeneration at the site of the injection in musculoskeletal, neural and cardiac tissues. Additionally, tissue-specific morphogens (e.g., BMP-2) are loaded into nanocarriers, which will be localized to target sites in an attempt to generate microenvironments permissive to regeneration with reduced systemic side effects (Kowalczewski and Saul 2018).

11.7.3 Implantable Nanobioelectronics

Nanoscale implantable electronics are also redesigning personalized medicine. The long-lifetime physiological monitoring and treatment systems are developed where the sensing, stimulation, and wireless transmission are integrated.

Nanoscale FET sensors have been integrated into neural tissues to enable real-time electrophysiological monitoring. Bendable graphene-based brain–machine interface electrodes further enhance biocompatibility and signal fidelity. In addition, real-time closed-loop biosensors and implantable devices capable of monitoring glucose, lactate, and cytokine levels are being developed to provide continuous physiological assessment (S. Yang et al. 2025).

Such platforms employ extremely thin substrates such as parylene-C or silk fibroin to promote better tissue compatibility and biodegradability. These issues are power sourcing (that can be addressed through wireless energy harvesting), biofouling, and reliable tissue-electronics integration. Nevertheless, these obstacles are quickly being surmounted by the convergence of nanomanufacturing and soft electronics.

11.7.4 Sustainable and Green Nanomanufacturing Approaches

With the scaling up of nanomanufacturing across the world, the issue of environmental footprint, along with the toxicity of nanomaterial production, is coming to the limelight. The conventional procedures usually use toxic solvents, heavy metals, and energy-consuming synthesis pathways.

Anthocyanin-based green nanomanufacturing is concerned with: Nanoparticle synthesis using tea, neem, or algae extracts. Aqueous-based processing and solvent-free fabrication. The reusable nanomaterials (nanocellulose and silk proteins) (Gharanjig et al.

2024). Closely related, chitosan- and alginate-based biopolymer-based nanoparticle-based drug delivery and tissue engineering applications are building popularity.

Nanoproducts now have life cycle assessments (LCAs) which are included in the design phase in order to calculate carbon footprint, disposal of the waste, and the effect on the ecosystem. Some agencies such as OECD and EPA are in the process of developing safer nanomanufacturing guidelines (Nizam, Hanafiah, and Woon 2021).

To conclude, biomedical nanomanufacturing is the area which holds its future in the convergence not only with biology but also with electronics, sustainability, and intelligent systems. Such cross-faculty initiatives will help in taking healthcare beyond the reactive framework to that of predictive, responsive, and regenerative model.

11.8 CONCLUSION

Biomedical nanomanufacturing is at the top of technological advancements, a science that combines engineering at the nanoscale with the life sciences to transform the way diseases in human beings are diagnosed, treated, and monitored. The evidenced abilities of this sphere throughout implants, diagnostics, drug delivery systems, and bioelectronic devices demonstrate the sphere's enormous potential. Manipulation of biological interactions has never been pre-controlled. New nanostructured materials have already opened the way to greater precision in terms of targeting, responsivity to cues, and compatibility with digital health technologies.

Nevertheless, there are still some limitations. The translation of laboratory successes into commercial solutions in the world of healthcare takes time due to the existence of challenges in measures of reproducibility, scale, biocompatibility, long-term safety, and regulatory pathways. Moreover, nanodevices embedded into dynamic systems and diverse biologies require further interface engineering, computational modeling, and material science developments. Deployment is complicated further by the fact that there is a need to have standardized protocols, ethical governance, and sustainable practices.

Beyond the horizon, the vision of long-term biomedical nanomanufacturing is one where we move beyond generalized treatments, and toward precision medicine that considers a variety of data of an individual (genetic, physiological, and environmental data). Such a vision consists of 4D-bioprinted scaffolds taking shape after implantation, AI-optimized drug carriers with response capabilities in real time, and the use of nanosensors to monitor health on IoT platforms continuously. Nanotechnology, convergent with digital health, regenerative biology, and systems medicine, promises that we can realize the Holy Grail of truly personalized medicine, predictive, or preventive medicine.

This possibility will necessitate long-term interdisciplinary efforts between materials scientists on one hand and bioengineers, clinicians, data scientists, and regulators on the other hand. Universities, research institutes, industry, and policy organisms must cooperate to provide the safety, efficacy, accessibility, and ethical embedding. Such an approach of collaboration and inclusiveness in research and innovation is the only way through which biomedical nanomanufacturing can go beyond being a promising research frontier and become a clinical mainstay of 21st-century medicine.

REFERENCES

Ahmed, Marya. 2020. "Nanomaterial Synthesis." In *Polymer Science and Nanotechnology*, 361–99. Elsevier. https://doi.org/10.1016/B978-0-12-816806-6.00016-9

Alfei, Silvana, Caterina Reggio, and Guendalina Zuccari. 2025. "CNTs as Excellent Adjuvants for Anticancer Therapeutic Strategies and Cancer Diagnosis: A Plethora of Laboratory Studies versus Few Clinical Trials and Still No Clinical Application." https://doi.org/10.20944/preprints202505.2363.v1

Alkilany, Alaaldin M., Samer R. Abulateefeh, and Catherine J. Murphy. 2019. "Facile Functionalization of Gold Nanoparticles with PLGA Polymer Brushes and Efficient Encapsulation into PLGA Nanoparticles: Toward Spatially Precise Bioimaging of Polymeric Nanoparticles." *Particle & Particle Systems Characterization* 36 (2). https://doi.org/10.1002/ppsc.201800414

Alsharabi, Rim M, Suyash Rai, Hamed Y. Mohammed, Maamon A Farea, Sesha Srinivasan, Preeti S. Saxena, and Anchal Srivastava. 2023. "A Comprehensive Review on Graphene-Based Materials as Biosensors for Cancer Detection." *Oxford Open Materials Science* 3 (1). https://doi.org/10.1093/oxfmat/itac013

Arch-Douglas, Katherine, Nathalie Dubois, Stephen Mayer, Rich Pelt, and Andrew Nelson. 2024. "Overview of Complexities of Global CMC Regulatory Affairs." In *Bioprocessing, Bioengineering and Process Chemistry in the Biopharmaceutical Industry*, 457–71. Cham: Springer Nature Switzerland. https://doi.org/10.1007/978-3-031-62007-2_16

Avgoustakis, Konstantinos, and Athina Angelopoulou. 2024. "Biomaterial-Based Responsive Nanomedicines for Targeting Solid Tumor Microenvironments." *Pharmaceutics* 16 (2): 179. https://doi.org/10.3390/pharmaceutics16020179

Banstola, Asmita, Tung Thanh Pham, Jee-Heon Jeong, and Simmyung Yook. 2019. "Polydopamine-Tailored Paclitaxel-Loaded Polymeric Microspheres with Adhered NIR-Controllable Gold Nanoparticles for Chemo-Phototherapy of Pancreatic Cancer." *Drug Delivery* 26 (1): 629–40. https://doi.org/10.1080/10717544.2019.1628118

Bora, Tonmoy J., Subir Biswas, Arpita Devi, Kusum K. Bania, and Magdi E. A. Zaki. 2025. "Top-Down Approaches of Nanomaterial Synthesis." In *Nanomaterial Green Synthesis*, 425–52. https://doi.org/10.1007/978-3-031-84643-4_14

Carvalho, Violeta, Inês Gonçalves, Teresa Lage, Raquel O. Rodrigues, Graça Minas, Senhorinha F. C. F. Teixeira, Ana S. Moita, Takeshi Hori, Hirokazu Kaji, and Rui A. Lima. 2021. "3D Printing Techniques and Their Applications to Organ-on-a-Chip Platforms: A Systematic Review." *Sensors* 21 (9): 3304. https://doi.org/10.3390/s21093304

Chang, Yongyun, Yongke Shao, Yanchang Liu, Runzhi Xia, Zhicheng Tong, Jingwei Zhang, Zanjing Zhai, Wendan Cheng, and Huiwu Li. 2019. "Mechanical Strain Promotes Osteogenic Differentiation of Mesenchymal Stem Cells on TiO2 Nanotubes Substrate." *Biochemical and Biophysical Research Communications* 511 (4): 840–46. https://doi.org/10.1016/j.bbrc.2019.02.145

Chauhan, Sakshi, Roopal Mittal, Manish Kumar, Amit Mittal, and Ajay Singh Kushwah. 2025. "Gold Nanoparticle-Based Biosensors for Point-of-Care Diagnostics: A Review of Sensing Nanoparticle Applications and Future Prospects." *Combinatorial Chemistry & High Throughput Screening* 28 (3): 417–34. https://doi.org/10.2174/0113862073293557240320065128

Chen, Xirui, Lu Ding, Xiaolin Huang, and Yonghua Xiong. 2022. "Tailoring Noble Metal Nanoparticle Designs to Enable Sensitive Lateral Flow Immunoassay." *Theranostics* 12 (2): 574–602. https://doi.org/10.7150/thno.67184

Choi, Yang-Jung, Moon-Kyoung Bae, Yong-Il Kim, Jeong-Kil Park, and Sung-Ae Son. 2020. "Effects of Microsurface Structure of Bioactive Nanoparticles on Dentinal Tubules as a Dentin Desensitizer." *Plos One* 15 (8): e0237726. https://doi.org/10.1371/journal.pone.0237726

Cicha, Iwona, Ronny Priefer, Patrícia Severino, Eliana B. Souto, and Sona Jain. 2022. "Biosensor-Integrated Drug Delivery Systems as New Materials for Biomedical Applications." *Biomolecules* 12 (9): 1198. https://doi.org/10.3390/biom12091198

Coman, Alina Năstaca, Anca Mare, Corneliu Tanase, Eugen Bud, and Aura Rusu. 2021. "Silver-Deposited Nanoparticles on the Titanium Nanotubes Surface as a Promising Antibacterial Material into Implants." *Metals* 11 (1): 92. https://doi.org/10.3390/met11010092

Cunha Mendes, Gisela da Cristina, Teresa Ribeiro da Silva Brandão, and Cristina Luisa Miranda Silva. 2008. "Ethylene Oxide Potential Toxicity." *Expert Review of Medical Devices* 5 (3): 323–28. https://doi.org/10.1586/17434440.5.3.323

Darwish, Moustafa A., Walaa Abd-Elaziem, Ammar Elsheikh, and Abdelhameed A. Zayed. 2024. "Advancements in Nanomaterials for Nanosensors: A Comprehensive Review." *Nanoscale Advances* 6 (16): 4015–46. https://doi.org/10.1039/D4NA00214H

Delgado-Pujol, Ernesto J., Guillermo Martínez, David Casado-Jurado, Juan Vázquez, Jesús León-Barberena, David Rodríguez-Lucena, Yadir Torres, Ana Alcudia, and Belén Begines. 2025. "Hydrogels and Nanogels: Pioneering the Future of Advanced Drug Delivery Systems." *Pharmaceutics* 17 (2): 215. https://doi.org/10.3390/pharmaceutics17020215

Desai, Nimeet, Dhwani Rana, Mitali Patel, Neha Bajwa, Rajendra Prasad, and Lalitkumar K. Vora. 2025. "Nanoparticle Therapeutics in Clinical Perspective: Classification, Marketed Products, and Regulatory Landscape." *Small* 21 (29). https://doi.org/10.1002/smll.202502315

Feng, Jianyou, Chuanrui Chen, Xuemei Sun, and Huisheng Peng. 2021. "Implantable Fiber Biosensors Based on Carbon Nanotubes." *Accounts of Materials Research* 2 (3): 138–46. https://doi.org/10.1021/accountsmr.0c00109

Gaur, Manish, Charu Misra, Awadh Bihari Yadav, Shiv Swaroop, Fionn Ó. Maolmhuaidh, Mikhael Bechelany, and Ahmed Barhoum. 2021. "Biomedical Applications of Carbon Nanomaterials: Fullerenes, Quantum Dots, Nanotubes, Nanofibers, and Graphene." *Materials* 14 (20): 5978. https://doi.org/10.3390/ma14205978

Gharanjig, Kamaladin, Hamid Gharanjig, Hooman Imani, Siavash Goudarzi, Alireza Khosravi, and Azita Noshirvani. 2024. "Impact of Nanotechnology on Extraction and Application of Carotene and Anthocyanin-Based Natural Colorants." In *Nanoscale Textile Coatings for Enhanced Performance*, 345–86. https://doi.org/10.1007/978-981-97-5922-4_14

Graczyk, Anna, Roza Pawlowska, Dominika Jedrzejczyk, and Arkadiusz Chworos. 2020. "Gold Nanoparticles in Conjunction with Nucleic Acids as a Modern Molecular System for Cellular Delivery." *Molecules* 25 (1): 204. https://doi.org/10.3390/molecules25010204

Gupta, Jitendra, and Gaurang Sharma. 2025. "Nanogel: A Versatile Drug Delivery System for the Treatment of Various Diseases and Their Future Perspective." *Drug Delivery and Translational Research* 15 (2): 455–82. https://doi.org/10.1007/s13346-024-01684-w

Gupta, Sonal, Ram Datt, Anamika Mishra, Wing Chung Tsoi, Asit Patra, and Patrycja Bober. 2022. "Poly(3,4-ethylenedioxythiophene):Poly(Styrene Sulfonate) in Antibacterial, Tissue Engineering and Biosensors Applications: Progress, Challenges and Perspectives." *Journal of Applied Polymer Science* 139 (30). https://doi.org/10.1002/app.52663

Huang, Boyang. 2020. "Carbon Nanotubes and Their Polymeric Composites: The Applications in Tissue Engineering." *Biomanufacturing Reviews* 5 (1): 3. https://doi.org/10.1007/s40898-020-00009-x

Huq, Tamanna Binte, Punnya Anil Kumar Jeeja, Sudip Kumar Dam, Sabya Sachi Das, and Juan L. Vivero-Escoto. 2025. "Recent Applications of Mesoporous Silica Nanoparticles in Gene Therapy." *Advanced Healthcare Materials*. https://doi.org/10.1002/adhm.202404781

Jain, Sanyog, Sandip Vikram Deore, Rohan Ghadi, Dasharath Chaudhari, Kaushik Kuche, and Sameer S. Katiyar. 2021. "Tumor Microenvironment Responsive VEGF-Antibody Functionalized PH Sensitive Liposomes of Docetaxel for Augmented Breast Cancer Therapy." *Materials Science and Engineering: C* 121 (February):111832. https://doi.org/10.1016/j.msec.2020.111832

Kant, Krishna, Reshma Beeram, Yi Cao, Paulo S. S. dos Santos, Lara González-Cabaleiro, Daniel García-Lojo, Heng Guo, et al. 2024. "Plasmonic Nanoparticle Sensors: Current Progress, Challenges, and Future Prospects." *Nanoscale Horizons* 9 (12): 2085–2166. https://doi.org/10.1039/D4NH00226A

Karazisis, Dimitrios, Sarunas Petronis, Hossein Agheli, Lena Emanuelsson, Birgitta Norlindh, Anna Johansson, Lars Rasmusson, Peter Thomsen, and Omar Omar. 2017. "The Influence of Controlled Surface Nanotopography on the Early Biological Events of Osseointegration." *Acta Biomaterialia* 53 (April):559–71. https://doi.org/10.1016/j.actbio.2017.02.026

Khonina, Svetlana N., and Nikolay L. Kazanskiy. 2025. "Trends and Advances in Wearable Plasmonic Sensors Utilizing Surface-Enhanced Raman Spectroscopy (SERS): A Comprehensive Review." *Sensors* 25 (5): 1367. https://doi.org/10.3390/s25051367

Kolanthai, Elayaraja, Suryasarathi Bose, K. S. Bhagyashree, S. V. Bhat, K. Asokan, D. Kanjilal, and Kaushik Chatterjee. 2015. "Graphene Scavenges Free Radicals to Synergistically Enhance Structural Properties in a Gamma-Irradiated Polyethylene Composite through Enhanced Interfacial Interactions." *Physical Chemistry Chemical Physics* 17 (35): 22900–910. https://doi.org/10.1039/C5CP02609A

Kowalczewski, Christine J., and Justin M. Saul. 2018. "Biomaterials for the Delivery of Growth Factors and Other Therapeutic Agents in Tissue Engineering Approaches to Bone Regeneration." *Frontiers in Pharmacology* 9 (May). https://doi.org/10.3389/fphar.2018.00513

Krishnan, Siva Kumar, Nandini Nataraj, M. Meyyappan, and Umapada Pal. 2023. "Graphene-Based Field-Effect Transistors in Biosensing and Neural Interfacing Applications: Recent Advances and Prospects." *Analytical Chemistry* 95 (5): 2590–2622. https://doi.org/10.1021/acs.analchem.2c03399

Kumar, Naresh, Shahid Ali, Bharat Kumar, Muhammad Sohail Zafar, and Zohaib Khurshid. 2020. "Hydroxyapatite and Nanocomposite Implant Coatings." In *Dental Implants*, 69–92. Elsevier. https://doi.org/10.1016/B978-0-12-819586-4.00005-6

Liu, Jianqiao, Jia Liu, Shokouh Attarilar, Chong Wang, Maryam Tamaddon, Chengliang Yang, Kegong Xie, et al. 2020. "Nano-Modified Titanium Implant Materials: A Way Toward Improved Antibacterial Properties." *Frontiers in Bioengineering and Biotechnology* 8 (November). https://doi.org/10.3389/fbioe.2020.576969

Mashinchian, Omid, Mohammad Johari-Ahar, Behnaz Ghaemi, Mohammad Rashidi, Jaleh Barar, and Yadollah Omidi. 2017. "Impacts of Quantum Dots in Molecular Detection and Bioimaging of Cancer." *BioImpacts* 4 (3): 149–66. https://doi.org/10.15171/bi.2014.008

Meyers, Steven R., and Mark W. Grinstaff. 2012. "Biocompatible and Bioactive Surface Modifications for Prolonged In Vivo Efficacy." *Chemical Reviews* 112 (3): 1615–32. https://doi.org/10.1021/cr2000916

Moazzam, Muhammad, Mengjie Zhang, Abid Hussain, Xiaotong Yu, Jia Huang, and Yuanyu Huang. 2024. "The Landscape of Nanoparticle-Based SiRNA Delivery and Therapeutic Development." *Molecular Therapy* 32 (2): 284–312. https://doi.org/10.1016/j.ymthe.2024.01.005

Nel, Andre, Erkki Ruoslahti, and Huan Meng. 2017. "New Insights into 'Permeability' as in the Enhanced Permeability and Retention Effect of Cancer Nanotherapeutics." *ACS Nano* 11 (10): 9567–69. https://doi.org/10.1021/acsnano.7b07214

Niculescu, Adelina-Gabriela, and Alexandru Mihai Grumezescu. 2021. "Polymer-Based Nanosystems—A Versatile Delivery Approach." *Materials* 14 (22): 6812. https://doi.org/10.3390/ma14226812

Nizam, Nurul Umairah M., Marlia M. Hanafiah, and Kok Sin Woon. 2021. "A Content Review of Life Cycle Assessment of Nanomaterials: Current Practices, Challenges, and Future Prospects." *Nanomaterials* 11 (12): 3324. https://doi.org/10.3390/nano11123324

Pandey, Vivek, and Tejasvi Pandey. 2025. "Mechanistic Understanding of PH as a Driving Force in Cancer Therapeutics." *Journal of Materials Chemistry B* 13 (8): 2640–57. https://doi.org/10.1039/D4TB02083A

Paris, Juan L., Alejandro Baeza, and María Vallet-Regí. 2019. "Overcoming the Stability, Toxicity, and Biodegradation Challenges of Tumor Stimuli-Responsive Inorganic Nanoparticles for Delivery of Cancer Therapeutics." *Expert Opinion on Drug Delivery* 16 (10): 1095–1112. https://doi.org/10.1080/17425247.2019.1662786

Parthasarathy, P., and M. Venkatesh. 2025. "Developments in Microfluidic Integrated Lab-on-Chip Devices for DNA Biosensing Towards Unifying the Emergence of Nanobiosensor." *Biomedical Materials & Devices* 3 (2): 1104–24. https://doi.org/10.1007/s44174-024-00257-2

Puricelli, Chiara, Casimiro Luca Gigliotti, Ian Stoppa, Sara Sacchetti, Deepika Pantham, Anna Scomparin, Roberta Rolla, et al. 2023. "Use of Poly Lactic-Co-Glycolic Acid Nano and Micro Particles in the Delivery of Drugs Modulating Different Phases of Inflammation." *Pharmaceutics* 15 (6): 1772. https://doi.org/10.3390/pharmaceutics15061772

Radulescu, Diana-Elena, Otilia Ruxandra Vasile, Ecaterina Andronescu, and Anton Ficai. 2023. "Latest Research of Doped Hydroxyapatite for Bone Tissue Engineering." *International Journal of Molecular Sciences* 24 (17): 13157. https://doi.org/10.3390/ijms241713157

Sengupta, Joydip, and Chaudhery Mustansar Hussain. 2025. "Carbon Nanotube-Based Field-Effect Transistor Biosensors for Biomedical Applications: Decadal Developments and Advancements (2016–2025)." *Biosensors* 15 (5): 296. https://doi.org/10.3390/bios15050296

Shegokar, Ranjita, and Mostafa Nakach. 2020. "Large-Scale Manufacturing of Nanoparticles—An Industrial Outlook." In *Drug Delivery Aspects*, 57–77. Elsevier. https://doi.org/10.1016/B978-0-12-821222-6.00004-X

Shekaran, Asha, and Andres J. Garcia. 2011. "Nanoscale Engineering of Extracellular Matrix-Mimetic Bioadhesive Surfaces and Implants for Tissue Engineering." *Biochimica et Biophysica Acta (BBA) - General Subjects* 1810 (3): 350–60. https://doi.org/10.1016/j.bbagen.2010.04.006

Sitia, Leopoldo, Marta Sevieri, Lorena Signati, Arianna Bonizzi, Arianna Chesi, Francesco Mainini, Fabio Corsi, and Serena Mazzucchelli. 2022. "HER-2-Targeted Nanoparticles for Breast Cancer Diagnosis and Treatment." *Cancers* 14 (10): 2424. https://doi.org/10.3390/cancers14102424

Sobczak-Kupiec, Agnieszka, Anna Drabczyk, Wioletta Florkiewicz, Magdalena Głąb, Sonia Kudłacik-Kramarczyk, Dagmara Słota, Agnieszka Tomala, and Bożena Tyliszczak. 2021. "Review of the Applications of Biomedical Compositions Containing Hydroxyapatite and Collagen Modified by Bioactive Components." *Materials* 14 (9): 2096. https://doi.org/10.3390/ma14092096

Souto, Eliana B., Gabriela F. Silva, João Dias-Ferreira, Aleksandra Zielinska, Fátima Ventura, Alessandra Durazzo, Massimo Lucarini, Ettore Novellino, and Antonello Santini. 2020. "Nanopharmaceutics: Part I—Clinical Trials Legislation and Good Manufacturing Practices (GMP) of Nanotherapeutics in the EU." *Pharmaceutics* 12 (2): 146. https://doi.org/10.3390/pharmaceutics12020146

Spirito, Federica Di, Francesco Giordano, Maria Pia Di Palo, Cosimo Ferraro, Luigi Cecere, Eugenio Frucci, Mario Caggiano, and Roberto Lo Giudice. 2024. "Customized 3D-Printed Mesh, Membrane, Bone Substitute, and Dental Implant Applied to Guided Bone Regeneration in Oral Implantology: A Narrative Review." *Dentistry Journal* 12 (10): 303. https://doi.org/10.3390/dj12100303

Sun, Ying, Qi Wang, Jianhua Chen, Lei Liu, Li Ding, Ming Shen, Jin Li, Baoshan Han, and Yourong Duan. 2017. "Temperature-Sensitive Gold Nanoparticle-Coated Pluronic-PLL Nanoparticles for Drug Delivery and Chemo-Photothermal Therapy." *Theranostics* 7 (18): 4424–44. https://doi.org/10.7150/thno.18832

Udriște, Alexandru Scafa, Alexandra Cristina Burdușel, Adelina-Gabriela Niculescu, Marius Rădulescu, and Alexandru Mihai Grumezescu. 2024. "Coatings for Cardiovascular Stents—An Up-to-Date Review." *International Journal of Molecular Sciences* 25 (2): 1078. https://doi.org/10.3390/ijms25021078

Vangijzegem, Thomas, Valentin Lecomte, Indiana Ternad, Levy Van Leuven, Robert N. Muller, Dimitri Stanicki, and Sophie Laurent. 2023. "Superparamagnetic Iron Oxide Nanoparticles (SPION): From Fundamentals to State-of-the-Art Innovative Applications for Cancer Therapy." *Pharmaceutics* 15 (1): 236. https://doi.org/10.3390/pharmaceutics15010236

Vetten, Melissa A., Clarence S. Yah, Tanusha Singh, and Mary Gulumian. 2014. "Challenges Facing Sterilization and Depyrogenation of Nanoparticles: Effects on Structural Stability and Biomedical Applications." *Nanomedicine: Nanotechnology, Biology and Medicine* 10 (7): 1391–99. https://doi.org/10.1016/j.nano.2014.03.017

Wang, Mian, Gujie Mi, Di Shi, Nicole Bassous, Daniel Hickey, and Thomas J. Webster. 2018. "Nanotechnology and Nanomaterials for Improving Neural Interfaces." *Advanced Functional Materials* 28 (12). https://doi.org/10.1002/adfm.201700905

Witte, Tinke-Marie De, Lidy E Fratila-Apachitei, Amir A Zadpoor, and Nicholas A Peppas. 2018. "Bone Tissue Engineering via Growth Factor Delivery: From Scaffolds to Complex Matrices." *Regenerative Biomaterials* 5 (4): 197–211. https://doi.org/10.1093/rb/rby013

Xu, Helin, Dan Yang, Cuifang Cai, Jingxin Gou, Yu Zhang, Lihui Wang, Haijun Zhong, and Xing Tang. 2015. "Dual-Responsive MPEG-PLGA-PGlu Hybrid-Core Nanoparticles with a High Drug Loading to Reverse the Multidrug Resistance of Breast Cancer: An in Vitro and in Vivo Evaluation." *Acta Biomaterialia* 16 (April):156–68. https://doi.org/10.1016/j.actbio.2015.01.039

Yadav, Deepika, Pramod Kumar Sharma, Rishabha Malviya, Prem Shankar Mishra, Amareswarapu V. Surendra, G.S.N. Koteswara Rao, and Budha Roja Rani. 2024. "Stimuli-Responsive Biomaterials for Tissue Engineering Applications." *Current Pharmaceutical Biotechnology* 25 (8): 981–99. https://doi.org/10.2174/1389201024666230818121821

Yang, Siyi, Xiujuan Qiao, Junlong Ma, Zhihao Yang, Xiliang Luo, and Zhanhong Du. 2025. "Recent Advances in Flexible Sensors for Neural Interfaces: Multimodal Sensing, Signal Integration, and Closed-Loop Feedback." *Biosensors* 15 (7): 424. https://doi.org/10.3390/bios15070424

Yang, Zhengyang, Wei Deng, Xiao Zhang, Yongbo An, Yishan Liu, Hongwei Yao, and Zhongtao Zhang. 2022. "Opportunities and Challenges of Nanoparticles in Digestive Tumours as Anti-Angiogenic Therapies." *Frontiers in Oncology* 11 (January). https://doi.org/10.3389/fonc.2021.789330

Zadpoor, Amir. 2017. "Design for Additive Bio-Manufacturing: From Patient-Specific Medical Devices to Rationally Designed Meta-Biomaterials." *International Journal of Molecular Sciences* 18 (8): 1607. https://doi.org/10.3390/ijms18081607

Zhang, Hui, Yingjie Zhai, Juan Wang, and Guangxi Zhai. 2016. "New Progress and Prospects: The Application of Nanogel in Drug Delivery." *Materials Science and Engineering: C* 60 (March):560–68. https://doi.org/10.1016/j.msec.2015.11.041

Zhang, Lin, Shuren Wang, and Yanglong Hou. 2025. "Magnetic Micro/Nanorobots in Cancer Theranostics: From Designed Fabrication to Diverse Applications." *ACS Nano* 19 (8): 7444–81. https://doi.org/10.1021/acsnano.4c10382

Zhernov, Yury V., Vladislava O. Petrova, Mark Y. Simanduyev, Denis V. Shcherbakov, Roman V. Polibin, Oleg V. Mitrokhin, Artem A. Basov, et al. 2024. "Microbicides for Topical HIV Immunoprophylaxis: Current Status and Future Prospects." *Pharmaceuticals* 17 (6): 668. https://doi.org/10.3390/ph17060668

12 Advancements in Nanomaterials and Nanofluids for Enhanced Tribological and Thermal Applications

Deepika Singh Singraur, Chayan Das, Asad Habeeb, Pushpinder Kumar and Sandeep Bansal

12.1 INTRODUCTION

The cutting tool and workpiece are in contact with each other to perform machining operations, resulting in large friction which causes quite high temperatures at the tool–workpiece contact area. The principal function in the grinding process is due to excess heat which causes tool wear, lessens tool life, and degrades the quality of the surface of the processed material (Grzesik and Rech 2019). Such effects not only deteriorate machining performance but also increase production costs while increasing machining times (Ogedengbe et al., 2019; Devillez et al. 2007). Lubricants are used in order to address these problems by reducing the heat generation and minimizing the mechanical stresses in materials processing (Alimuzzaman et al. 2023). In addition to their primary functions of lubrication and cooling, lubricants offer several other advantages, including anticorrosion protection, facilitating chip removal, and improving the surface quality of the end product (Sen et al. 2023). In the context of sustainable production, lubricants should be recyclable, free of toxic substances, and environmentally compatible (Kabave Kilincarslan and Cetin 2020). Specific heat capacity and thermal conductivity are thermal properties that are critical factors in assessing the efficiency of lubricants to diffuse heat during high-friction behaviors (Wang et al. 2023). The innovative use of nanotechnology is one of the most successful methods to improve lubricant performance (Uysal and Korkmaz 2018), with investigations revealing that lubricant properties can be greatly improved via the addition of nano-sized additives to the base oils. According to their chemical composition, lubricants are divided into four main types—mineral-based, vegetable-based, synthetic, and semi-synthetic. Among these types, mineral-based oils are still the most widely used type in the metal industry because of their availability and cost-effectiveness (Perumal et al. 2022). Lubricants, to improve the performance further,

DOI: 10.1201/9781003560234-15

are commonly mixed with nanoparticles of high thermal conductivity materials like gold, silver, and copper. These nanoparticles improve heat dissipation, minimize thermal damage in high-friction areas, and prolong tool life (Ukueje, Abam, and Obi 2022). Mechanical engineers aim to protect the cooling channels with more efficient heat transfer by replacing traditional micron-sized particle-based fluids with nanoparticles (Tarafdar et al. 2021). Nanofluids (NFs), defined as fluids filled with nanoparticles at certain concentrations, have shown outstanding thermal and tribological properties, which display their tremendous potential in industries such as electronic, automotive, aerospace, and solar energy (Xu et al. 2022, Ghodbane et al. 2022). NFs show better performance due to the fact that the surface-area-to-volume ratio of the nanoscale grains is larger, leading to brokerage with the base fluid; as a result, heat transfer and lubrication properties are greatly improved. Rigoberto Barba, Patricia Monroy, and Keira Avendaño reported that suspension of heat-generating nanoparticles can improve thermal conductivity by a higher viscosity, in addition to static size, shape, and crystal structure, and by providing anti-wear and anti-friction effects through an action mechanism such as the ball effect, film formation, and polishing (Liñeira del Río, López, and Fernández 2021). The utilization of nanomaterials and NFs has hence turned out to be a promising solution for long-standing issues in the field of tribology and thermal management. Nanostructured materials, classified as having one or more dimensions at the nanoscale, are of particular interest due to their unique properties including high surface-area-to-volume ratios, superior strength, and selective chemical reactivity. With these properties, they achieve significant improvements in friction reduction, wear resistance, and thermal stability—key considerations in the automotive, aerospace, and manufacturing industries. NFs and nanolubricants have found applications in various engineering fields, including their use in automotive systems to increase fuel efficiency by minimizing frictional losses, in aerospace components to enhance thermal stability, and in manufacturing

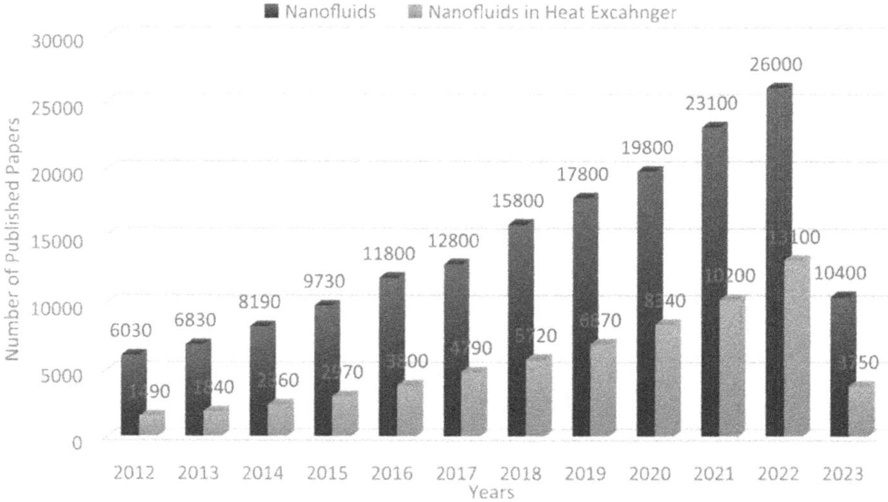

FIGURE 12.1 The research landscape on NFs over the past decade (Kalsi et al. 2023).

processes to prolong the service life of tools. Throughout the establishment of the field, most studies in this domain targeted optimizing the nanoparticle formulations, eco-friendly base fluids, and the long-term performance of nanolubricants under different conditions, as depicted in Figure 12.1. For practically all engineering domains, these materials will have transformational influence, and therefore, this chapter will cover key principles, mechanisms, and applications of the same.

12.2 TYPES OF LUBRICANTS

Lubricants play a crucial role in mitigating the adverse effects on machining processes caused by other physical and chemical phenomena (N. S. Ross et al. 2023). These molecules are aimed to intrude into the interface of the cutting tool with the workpiece and create a film layer that has an earmarked role for both lubrication and cooling (Kuntoğlu, Kaya, and Binali 2023). This resulting film of lubricant drastically cuts that friction between the surfaces, extending the life of the cutting tool. In addition to extending the life of the tool, lubrication can also help improve the surface finish of the workpiece by reducing wear. They also improve overall production quality through increased machining efficiency and lowered production cost (Ross and Manimaran 2020). Despite their benefits, lubricants often possess toxic properties, which pose serious risks to both human health and the environment (Sivalingam et al. 2021). The effectiveness of metal lubricants is largely determined by their physical and chemical properties, which are optimized for their specific application. There are many thermophysical, chemical, and mechanical properties that influence the effectiveness of lubricants (Şirin 2022), which are listed in the following:

i. Combustion and flashpoint are critical for determining the safety and stability of lubricants under high temperatures.
ii. Drop size affects the distribution and coverage of the lubricant, influencing its efficiency in reducing friction.
iii. Transparency plays a role in monitoring and assessing the lubricant's condition and performance during use.
iv. pH value is important as it affects the lubricant's chemical stability and its interaction with materials.
v. Density impacts the lubricant's ability to form an effective film layer between surfaces.
vi. Concentration rate determines the proportion of active ingredients in the lubricant, which affects its overall performance.
vii. Viscosity is crucial for controlling the flow and film formation of the lubricant, especially under varying operational conditions.
viii. Thermal conductivity ensures the lubricant can effectively transfer heat away from the cutting interface, preventing overheating.

The primary purpose of lubrication is to control and minimize friction and wear by creating a protective film that reduces the friction between moving surfaces in contact. Beyond reducing friction and preventing wear, lubricants perform several other essential functions, such as removing heat and contaminants from the interface and protecting

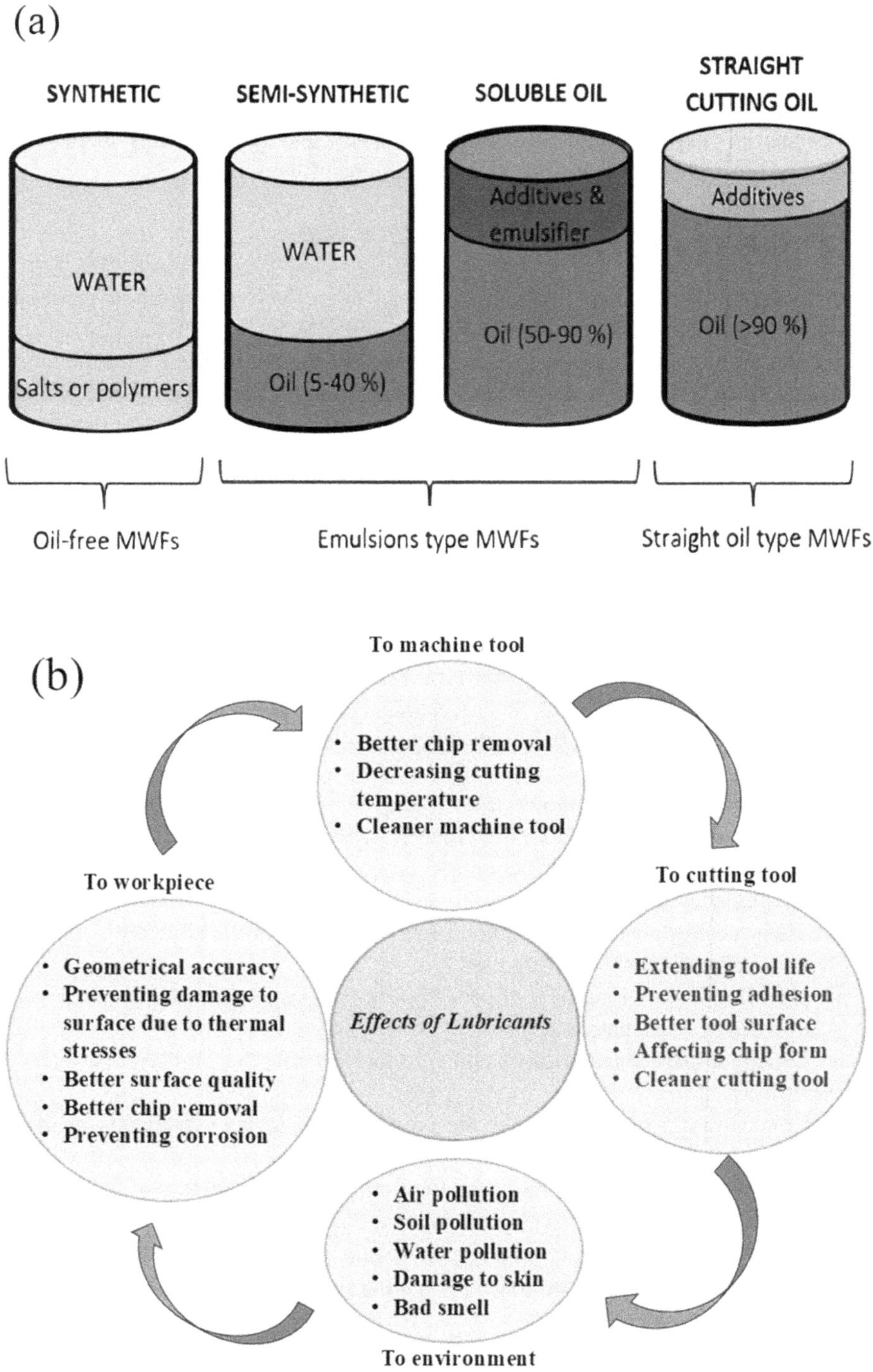

FIGURE 12.2 (a) Types of lubricants (Afonso et al. 2023); (b) effect of lubricants (Erdi Korkmaz and Kumar Gupta 2024).

machinery from corrosion. Figure 12.2(a) categorizes lubricants into four types: solid, liquid, gaseous, and semi-solid lubricants (Farfan-Cabrera et al. 2023). Liquid lubricants, commonly used in machining and tribology applications, can be classified into four major groups based on their chemical structure: synthetic, semi-synthetic, mineral, and vegetable-based oils. Of these, mineral-based oils are the most widely used (Afonso et al. 2023). Figure 12.2(b) illustrates the impact of lubricants on the cutting tool, workpiece, environment, and machine tool. In situations where lubricants are insufficient, various additives may be introduced to improve their performance.

12.3 PROPERTIES AND ADVANTAGES OF NANOMATERIALS AND NFS

Nanomaterials, including nanoparticles and NFs, have become integral components in advancing the performance of various engineering systems, particularly in thermal management and tribology. These advanced materials exhibit exceptional thermophysical properties that offer enhanced heat transfer, friction reduction, and wear resistance compared to conventional materials.

12.3.1 Thermophysical Enhancements

NFs, the colloidal suspensions of nanoparticles in base liquids (solvent), are popular thermal conductive fluids as they have shown very high thermal conductivity other than conventional coolants and heat transfer fluids. Nanoparticles like copper, silver, alumina, or carbon nanotubes are mixed with fluid and increase the surface area and heat rejection properties. This property enables NFs to remove heat from important components in cooling systems including radiators and heat exchangers. There are several mechanisms through which the heat transfer efficiency of NFs is increased. The first reason is the high surface-area-to-volume ratio of the nanoparticles leading to better interaction with the fluid, which helps in the conduction of heat. Second, nanoparticles can cause micro-convection and disturb the boundary layer of the fluid, which enhances the heat transfer coefficient as a whole. Also, the effective viscosity of NFs can be adjusted according to the application, which is necessary to improve the hydrodynamic and thermal performance of the working fluid. This adjustability makes NFs versatile in a range of engineering applications, from industrial cooling systems to electronic heat management. Figure 12.3(a) and (b) compare thermal conductivity with temperature for different heat exchangers and NFs, including various hybrids like ND-Fe3O4, CNT, and GO-Si. Figure 12.3(c) and (d) show that viscosity decreases as temperature increases. Figure 12.3(d) also highlights that TiO_2/water NF achieves the highest viscosity of 1.2 at 10°C among the nanofluids.

12.3.2 Tribological Benefits

Tribology, the science of friction, wear, and lubrication, has also benefited greatly from the introduction of nanomaterials. The use of nanolubricants and nanocoatings has revolutionized the performance of mechanical systems, particularly in

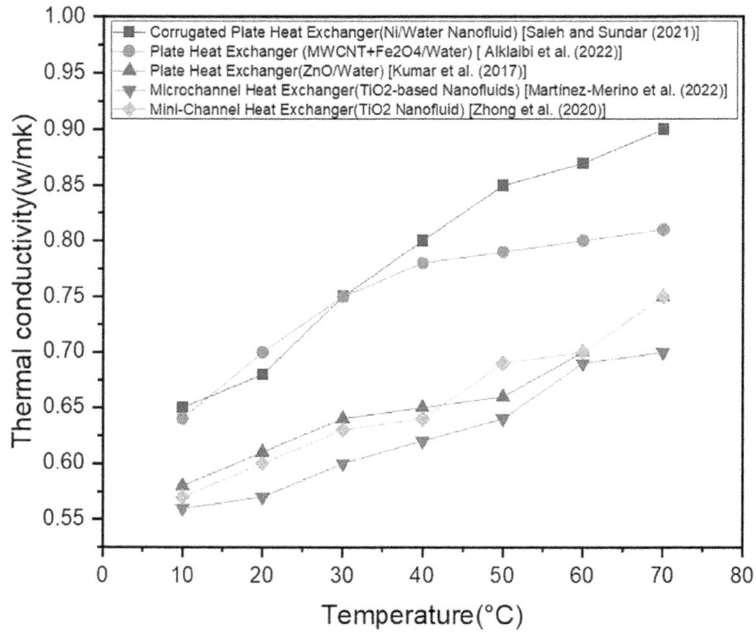

(a)

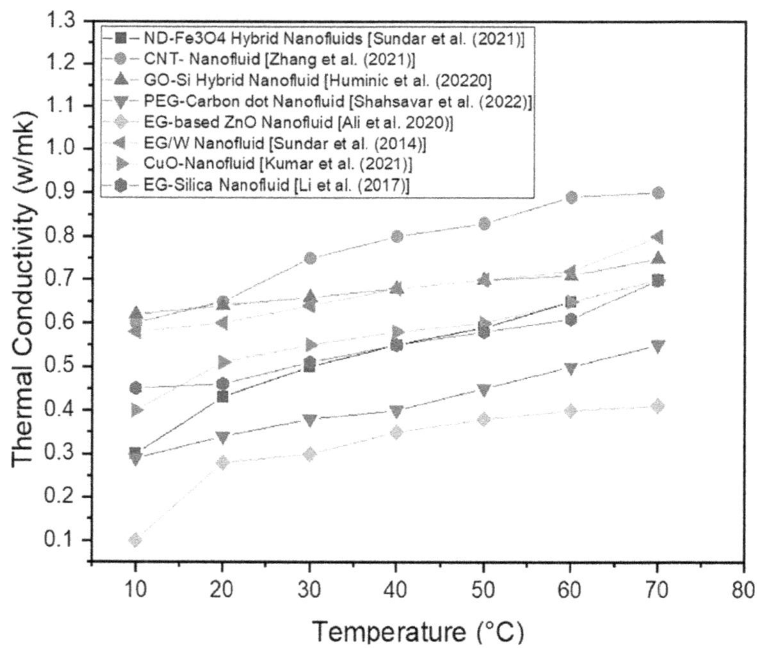

(b)

FIGURE 12.3 Comparison of thermal conductivity and viscosity with temperature for different heat exchangers and different NFs (Kalsi et al. 2023).

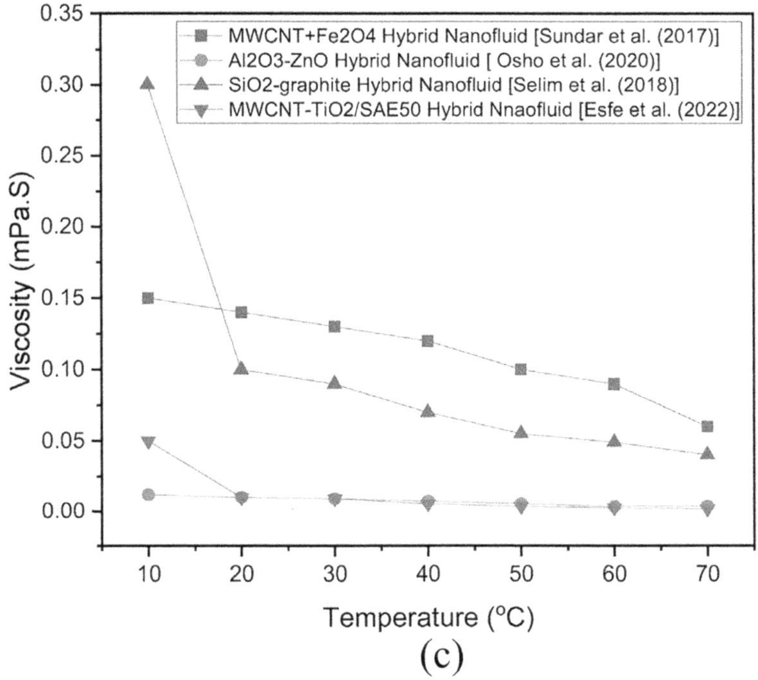

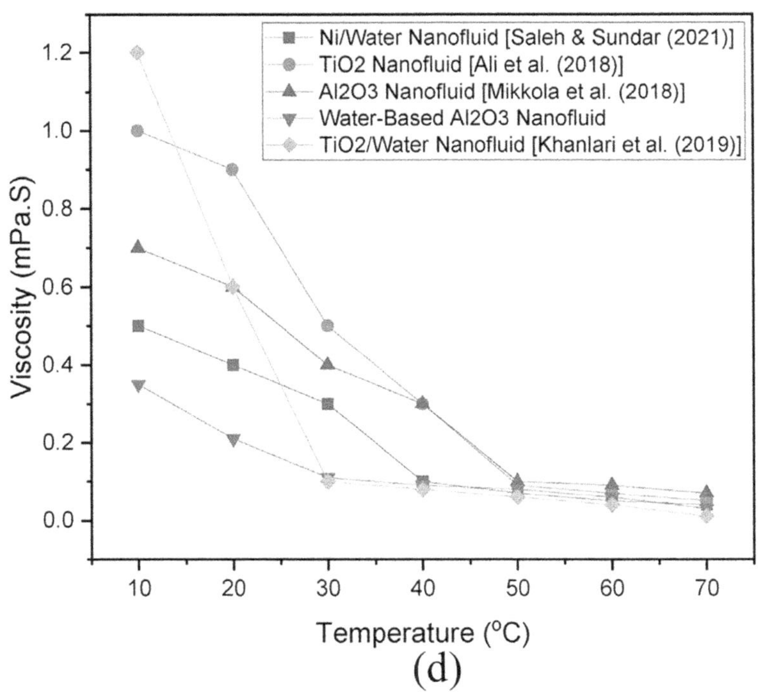

FIGURE 12.3 (Continued)

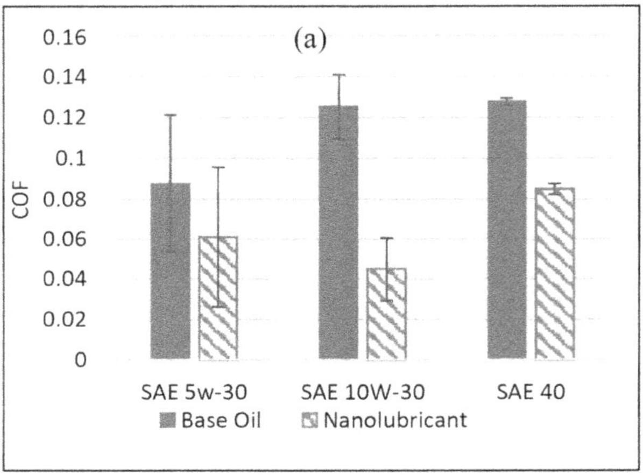

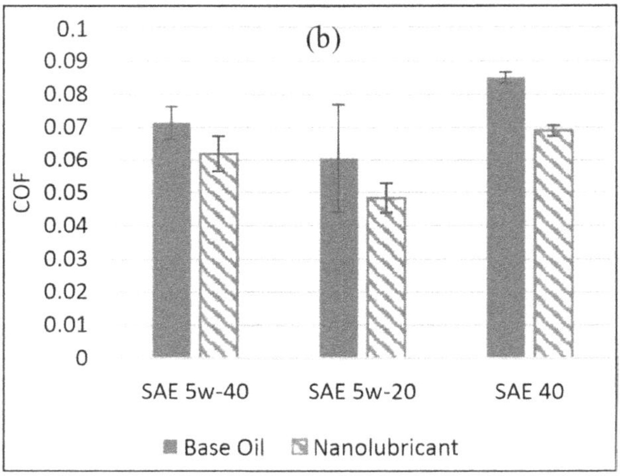

FIGURE 12.4 (a) Effect of TiO2 NPs on the friction performance of synthetic lubricants. (b) Effect of Cu NPs on the friction performance of synthetic lubricants (Ahmed Rashed and Ahmed Nabhan 2018; Waqas et al. 2021).

reducing friction and minimizing wear. Nanoparticles possess unique properties at the nanoscale, including high hardness, reduced friction coefficient, and the ability to form protective layers that minimize direct contact between moving parts. These protective nanolayers penetrate surface asperities and form a lubricating film, reducing the friction between surfaces in contact. This results in a substantial decrease in wear and increases the service life of mechanical parts like gears, bearings, and pistons. Also, the lightweight nature of nanocomposites, combined with superior load-bearing capacities and thermal stability of the composite, makes them ideal for use in extreme environments which may be encountered in aerospace applications or during high-speed operations of machines. These characteristics provide components with corrosion resistance, an area where ceramic nanoparticles, in their nanoscale form, play a crucial role. As shown in Figure 12.4, the addition of nanoparticles could adjust the friction performance of different lubricants. As shown in Figure 12.4(b), the use of nanoparticles in some synthetic lubricants had a significant effect on the friction. Studies have found that the application of copper nanoparticles as additive to SAE 5W-40 oil at an optimal concentration of 0.8 wt% reduces the COF by about 13% in four-ball tribotester tests. Furthermore, in another study, 1.6 wt% copper nanoparticles in SAE 5W-20 oil resulted in COF reduction of 28.5% (Li et al. 2018).

12.4 APPLICATIONS IN KEY INDUSTRIES

Such improvements in fuel economy are driven by new-generation nanomaterials and NFs, which are applied in the automotive, machinery, aerospace, and shipbuilding industries to enhance performance and maximize energy efficiency. NFs are employed for engine cooling, battery cooling, and cooling of electronic components to improve heat dissipation. NFs can provide superior thermal conductivity, enabling efficient heat transfer. This not only helps mitigate the risk of overheating but also improves overall powertrain efficiency and strengthens the performance of thermal management system. Nanocoatings are further used for improving the durability of engine components in a high-stress environment. Nanocoatings on engine pistons, valves, or turbine blades dramatically reduce wear and tear, extending the service life of those parts. In aerospace, nanomaterials combined with 3D printing have led to lighter structural elements that retain their solidity and reliability. Nanocomposites that are lightweight and possess high strength enable the aircraft and spacecraft to fly much faster and also help save fuel.

Figure 12.5 highlights various applications of nanotechnology in automobiles. Nanoadditives in fuel and lubricants enhance efficiency and reduce friction, while nanoparticles in tires and nanocoatings improve durability. Additionally, nanosensors, catalyst nanomaterials, and polymeric nanocomposites contribute to advanced

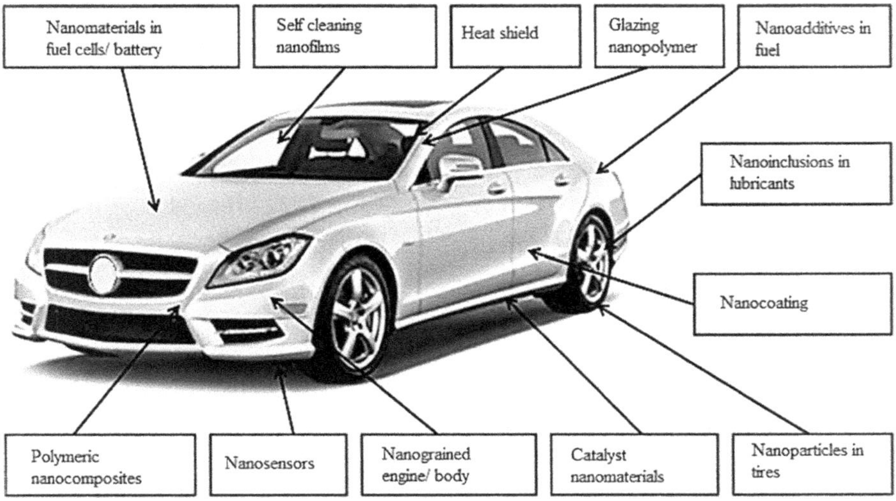

FIGURE 12.5 Various parts of automobile in which nanotechnology is applied (not representing the actual vehicle) (Asmatulu, Nguyen, and Asmatulu 2013).

monitoring, emissions control, and lightweight, robust vehicle components. These innovations enhance vehicle performance, efficiency, and sustainability.

Table 12.1 highlights the significant advancements in various modes of transportation achieved through nanoengineering. Nanotechnology simultaneously enhances comfort, safety, and efficiency—three key factors that influence our choice of transportation.

12.4.1 Manufacturing and Machining

The integration of NFs in machining and manufacturing processes has led to significant advancements in cooling efficiency and tool performance. NFs, when used as coolants in machining, improve heat transfer at the cutting interface, reducing the temperature in the machining zone. This temperature reduction results in improved surface finishes and prolonged tool life, as the cutting tools experience less thermal degradation. Furthermore, the incorporation of nanolubricants in the machining process reduces friction between the tool and the workpiece, leading to lower energy consumption and higher operational efficiency. These lubricants also minimize wear on the machine components, reducing maintenance costs and enhancing productivity. As a result, NFs and nanolubricants have become integral to industries such as automotive manufacturing, metalworking, and precision engineering. Table 12.2 represents a comparison between biochemical methods and cementation processes as foundational chemical steps in the synthesis of nano-cutting fluids.

12.4.2 Biomedical Applications

The biomedical field has seen a rapid rise in the use of nanomaterials, particularly in the development of advanced prosthetics, implants, and drug delivery systems. Nanolubricants can also be used to reduce friction in joint replacements, like knee or hip prostheses. The reduction in friction promotes the longevity and performance of the implants, leading to smoother movement and less wear during the engagement. Functional agents can be released from drug coatings, as seen in applications such as drug-eluting stents and drug-eluting intraocular lenses. Nanocoatings are also used to improve the biocompatibility of implants. These coatings also reduce the likelihood of infection and body rejection in addition to enhancing wear resistance. Another application of nanomaterials that scientists are studying is drug delivery systems, where nanoparticles are utilized to deliver drugs in a more efficient way to specific sites in the body, helping to decrease unintended effects and enhance therapy effects. Figure 12.6(a) provides a schematic representation of the biomedical applications covered in this section. Figure 12.6(b) illustrates a schematic of the photocatalytic process for TiO2, which involves three steps: excitation, diffusion, and surface transfer. Initially, the nanoparticles absorb photons from a light source during the excitation phase. In targeted drug delivery, magnetic nanoparticles are utilized to transport drugs to specific locations. These nanoparticles are typically coated with a biocompatible layer, such as gold or polymers, to functionalize them. This coating allows the anticancer drug to be either conjugated to the surface or encapsulated within the nanoparticle, as illustrated in Figure 12.6(c).

TABLE 12.1
Importance of Various Nanoparticles and Their Effect in Transportation (Mathew, Joy, and George 2019).

	Applications	Nanoparticles	Property Imparted	Advantages	Disadvantages
Automobiles	Paint coating	SiO_2	Anticorrosion, scratch resistant	Reduce the toxicity of paint and promote self-healing	Uniform dispersion is difficult to attain
		LDH	Flame retardant	Improve clear formation and absorb heat	Stability issues
		Titania, ZnO	UV resistant	Absorb harmful UV rays	Hardening problems
	Lightweight body parts	CNT	Substitute for steel parts	Much stronger and less weight compared to steel	Costly
		LFRT nanocomposites	Substitute for plastics	High-strength and high-impact resistance	Manufacturing is difficult
	Engines	NFs of CuO, Al_2O_3, CNT, Silica, TiO_2	Heat transfer	Improve heat transfer coefficient	Chances of agglomeration
		CuO_2, TiO_2, MoS_2, nanosheets	Lubrication	COF can be reduced to 50%	Short lifetime
	Tires	Carbon black, soot, silica	Better rolling resistance and reduced inner friction	Prolonged durability and higher fuel efficiency	Uniform dispersion is difficult
	Interior	Ag, Au, TiO_2, ZnO, Cu, CNT	Antimicrobial and antistatic	Non-toxic and high efficiency	Comparatively costly
Aerospace	Lightweight	CNT	Enhanced thermal, mechanical, and electrical properties	High strength and light weight	Costly
	Engines	ZrO, nanoclay	Heat resistant	High heat resistance even when added with small amount	Loose its property after some period of time
	Communication	SnO_2, graphene, CNT	High sensitivity	Accurate measure of pressure and temperature	High manufacturing precision is required
Marine	Coating	Nanostructured zirconia coating	Scratch and abrasion resistance	Provides good scratch and abrasion resistance for a longer period of time	Coating is difficult
	Body parts (hull)	Al	Light and tough	High strength and lightweight by cyromilling	Comparatively costly
		Nanoclay, Al	Anticorrosion, flame retardant	High heat resistance and impart self-healing to prevent corrosion	Uniform dispersion is difficult
		TiO_2, ZnO, MgO, Al_2O_3	Antimicrobial and UV resistant	High anti-biofouling activity	Costly

TABLE 12.2
Comparison Between Three Different Types of Multi-Step Methods for the Synthesis of Cutting NFs (Reverberi et al. 2019).

Types of Multi-Step Process	Strengths	Weakness
Chemical synthesis by plant/animal extracts	Sustainability Process safety	Low reaction yield Product contamination
Biochemical synthesis by living cells	Sustainability High reaction yield	Poor reaction control Product contamination
Chemical synthesis by cementation	High reaction yield High product purity	Kinetic sensitivity Cost of reactants

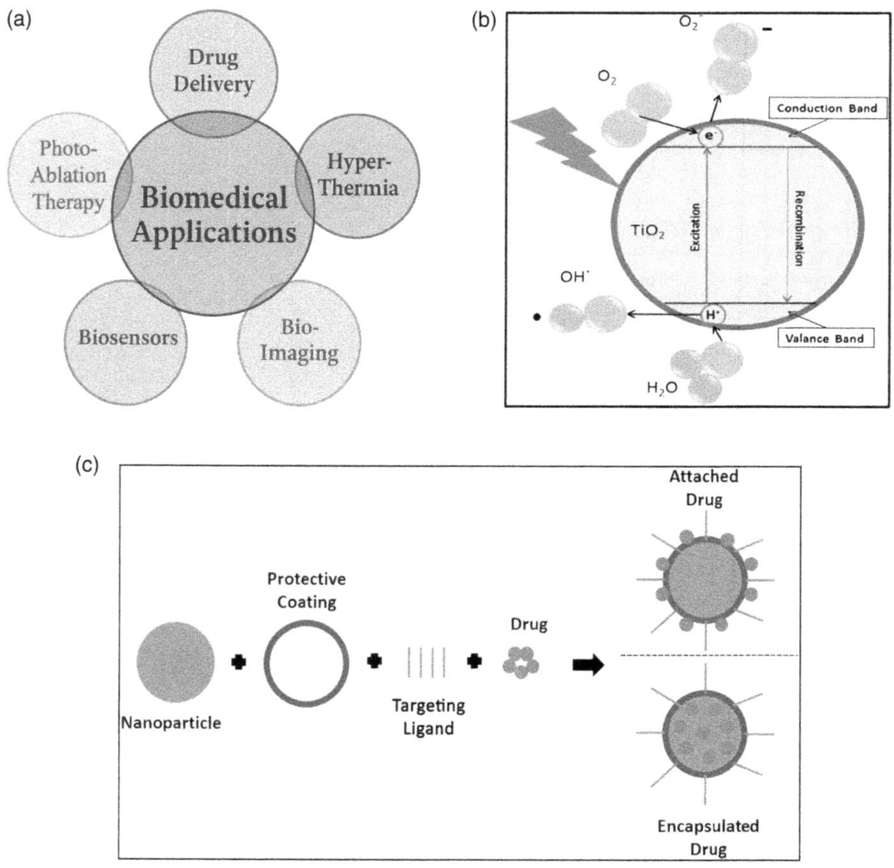

FIGURE 12.6 (a) Schematic showing some of the biomedical applications of nanoparticles. (b) Schematic showing the photocatalytic process of TiO_2 nanoparticles. (c) Schematic of drug loading options in targeted drug delivery (McNamara and Tofail 2017).

12.5 CASE STUDIES

12.5.1 Nanoparticles and the Sliding Wear of Short Fiber-Reinforced Polymer Composites (Chang and Friedrich 2010)

Short fiber-reinforced polymer composites have been used as advanced engineering materials because of their lightweight and high specific strength characteristics, accompanied by processing simplicity. The main applications are in the field of self-lubrication and cleanliness, in automotive and industrial machinery applications. These SFRPs have few significant limitations on exposure to extreme operating conditions, such as high pressure, velocity, and temperature, which may lead to matrix degradation and fiber detachment, alongside a rapid increase in wear rates, thus reducing their life. Researchers have investigated the use of nanoparticles as fillers for improving the tribological performance of SFRPs to get rid of these calamities. Nanoparticles, because of their extremities in area and specificity in mechanical properties, can modify wear mechanisms, improve load distribution, and stabilize the wear process under extreme conditions. This chapter focuses on experimental studies and evaluations concerning nanoparticle influence on the wear behavior of SFRPs. The primary ingredient in SFRPs has normally been a polymer matrix reinforced with short fibers and various fillers. Typical matrices for these composites have included epoxy (EP) or polyamide 66, being the most frequently used polymers for mechanical strength and thermoplasticity. The incorporation of nanoparticles in this composite structure would thus involve careful selection and preparation. Nano-TiO, which has approximately 300nm in diameter, is characterized as being reinforced by a volume average concentration of between 0.01% and 2% with use as an example of much interest to improve tribological performance. Thus, the volume of nanoparticles in these composites usually ranges very low—around 5%—for a uniform distribution while keeping the overall structure of the composite intact. The epoxy-based composites are made using dissolver mixing, while PA66 and other thermoplastics are made using twin-screw extrusion methods. Figure 12.7 shows contact mechanism in SFRP composite reinforced with additional nanoparticles.

***Wear test procedure*:** The wear behavior was determined using a pin-on-disk test, where a polymer composite pin was made to slide against a hardened steel disk with the control of relevant parameters to simulate dry sliding conditions within about room temperature for 20 h. Tests simulated conditions of wear for extended periods. The following key parameters were recorded: coefficient of friction, penetration depth, and surface temperature. Scanning electron microscopy and atomic force microscopy were employed for damaged surface analysis, while the presence of nanoparticle wear regions was confirmed through wavelength-dispersive X-ray spectrometry.

***Key findings*:**

***Wear resistance*:** Compared with normal SFRP composites, nanocomposites showed much lower values of weight loss and were thus found to be advantageous wear-resistant materials under high pressure–velocity (pv) conditions. In comparison to progressive wear characteristics observed in conventional composite materials, nanocomposites demonstrated stable wear rates.

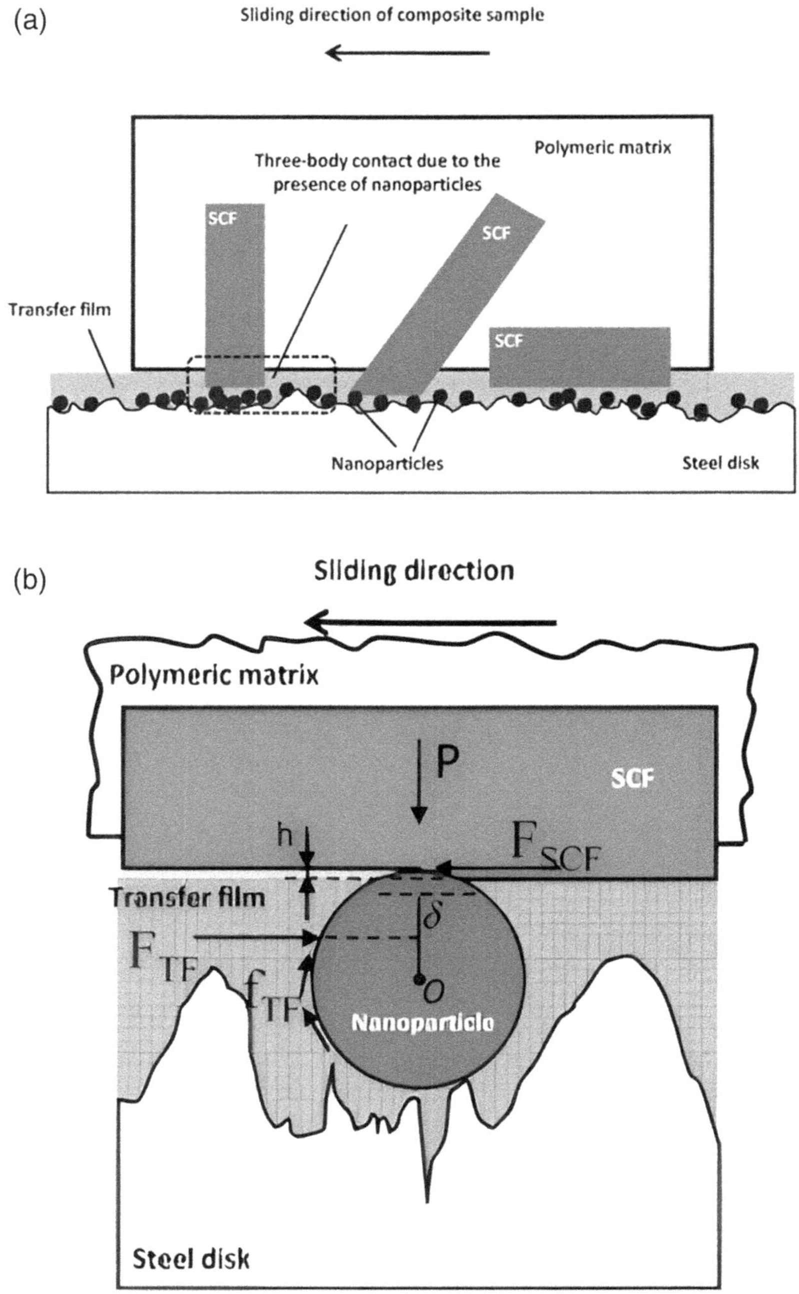

FIGURE 12.7 (a) Illustration of the contact mechanism in SFRP composites reinforced with additional nanoparticles. For clarity, the fiber diameter is depicted as relatively small compared to the surface roughness of the steel counterpart and the size of the nanoparticles. (b) Force analysis for an individual nanoparticle with a radius of R (Chang and Friedrich 2010).

Friction reduction: Introducing nanoparticles consistently decreased the coefficient of friction for all testing conditions, which could be attributed to modified adhesive and rolling action effects resulting from nanoparticles on contact mechanics.

Surface integrity: Nanocomposite wear suffered slightly even under extreme conditions. Nanoparticles acted as spacers besides shielding the polymer matrix from thermal-mechanical breakdown.

Nanoparticle abrasiveness provides a polishing effect on the worn surfaces in such a way that their average roughness is kept constant (close to ~300 nm), about the size of nanoparticles. This is causing up to very low stress concentration, which ultimately restores high durability for the composite.

Comparative analysis- With and Without Nanoparticles:

Epoxy-based composites: In ordinary epoxy composites, brittle failure occurs at higher pressures with the rapid detachment of the fibers. By contrast, nanocomposites show a smooth surface finish and gradual fiber removal that incur stable wear rates.

PA66-based composites: Ductility of PA66 further impeded any matrix breakage; however, high temperatures and softening caused fiber pull-out during extreme application conditions. Nanoparticles played a role in averting such scenarios by retaining the matrix and improving wear grooves.

Conclusion: The incorporation of nanoparticles into the SRFPs represents a phenomenal advancement in tribological materials. By minimizing friction, stabilizing wear processes, and protecting the matrix from thermal-mechanical failure, nanoparticles enhance the performance of composites under extreme conditions. These findings also offer the promise of developing next-generation, high-performance materials designed and built specifically for challenging environments.

12.5.2 Metallographic Study of SI_3N_4-Reinforced Al Matrix Composites (Kumar et al. 2022)

Metal matrix composites like aluminum-based AMMC materials play a significant role in industries as compared to others. They possess lightweight when measured against performance for mechanical strength and wear resistance. AMMCs are tailored to enhance their mechanical properties with a special focus on tribological improvements in application as aerospace and automotive; this relies on using ceramic nanoparticles that act as reinforcement. The wear resistance and hardness of high-strength aluminum alloy AA7068 are improved when reinforced with silicon nitride (Si_3N_4) nanoparticles. This chapter focuses entirely on the history of fabrication, characterization, and optimization of wear performance of these Si_3N_4-reinforced AA7068 composites.

1. ***Base material (high-strength AA7068 aluminum alloy)***: AA7068 is a zinc-rich aluminum alloy known for high strength, excellent resistance to corrosion, and good workability, as well as for its significant amount of zinc, magnesium, copper, and Aluminum as base material.

2. ***Reinforcement: silicon nitride (SI_3N_4) nanoparticles*****:** Owing to their outstanding hardness and thermal stability, coupled with a peculiar behavior of refining the grain structure of the matrix, Si_3N_4 nanoparticles have been selected as suitable additions for enhancing the wear resistance as well as the hardness of the AA7068 matrix.
3. ***Fabrication methodology*****:** The ultrasonic-assisted stir casting method as seen in Figure 12.8 has proven to be an economical and effective technique for producing nanocomposites with excellent and uniform nanoparticle distribution:
   i. ***Stir casting process*****:** Molten AA7068 was mechanically stirred at 500 rpm for 10 minutes to introduce the nanoparticles.
   ii. ***Ultrasonic treatment*****:** Ultrasonic waves at 2 kHz were applied to the molten slurry to break nanoparticle clusters and promote homogeneous distribution.
   iii. ***Bottom pouring*****:** The processed mix was poured into preheated steel molds to ensure the lowest possible porosity and defects.

Reinforcement levels of 0.5%, 1.0%, 1.5%, and 2.0% weights of Si_3N_4 (see Figure 12.8) were taken up to study the effects of nanoparticle concentration on the properties of composites.

1. ***Grain growth*****:** The addition of Si_3N_4 nanoparticles has turned the coarse grains of AA7068 into a fine equiaxed structure.
2. ***Microstructural analysis*****:** Figure 12.9 shows investigation into the refined microstructure, which was performed using FESEM and EDS:
   i. ***Uniform distribution*****:** Up to 1.5% of nanoparticle reinforcement, uniform distribution was found, resulting in a better mechanical performance of the matrix. Above 2.0%, clustering and porosity became apparent, damaging performance.

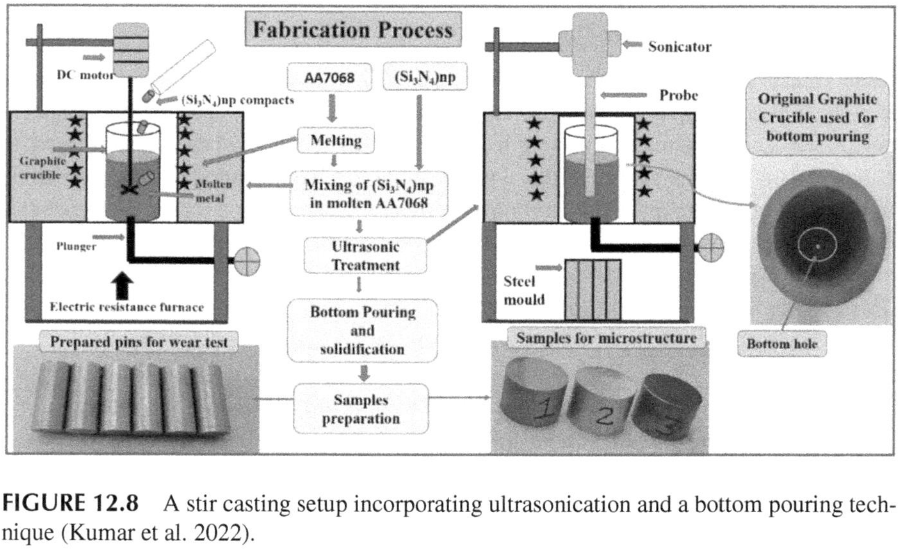

FIGURE 12.8 A stir casting setup incorporating ultrasonication and a bottom pouring technique (Kumar et al. 2022).

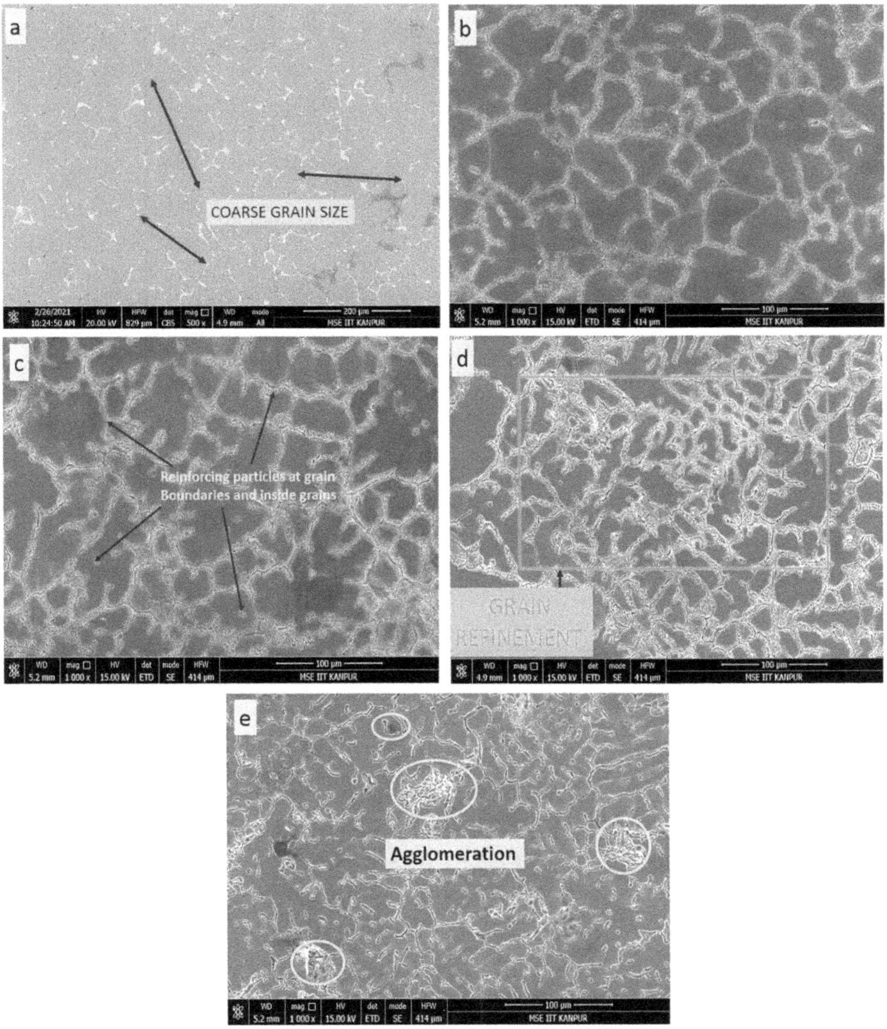

FIGURE 12.9 FESEM images of 7068 alloy and its nanocomposites: (a) pure alloy; (b) with 0.5 wt.% Si_3N_4; (c) with 1.0 wt.% Si_3N_4; (d) with 1.5 wt.% Si_3N_4; (e) with 2.0 wt.% Si_3N_4 (Kumar et al. 2022).

ii. ***Phase analysis*:** Results from X-ray diffraction showed that Si_3N_4 and $MgZn_2$ were phases present and were responsible for the enhanced hardness and wear resistance of the samples.

3. ***Hardness enhancement*:** Figure 12.10(a) and (b) shows substantial enhancement in hardness was added through the reinforcement in the composites: Maximum hardness improvement of 32.72% at 1.5% Si_3N_4 reinforcement. The possible reason for the loss of hardness at 2.0% reinforcement is nanoparticle clustering and increased porosity.

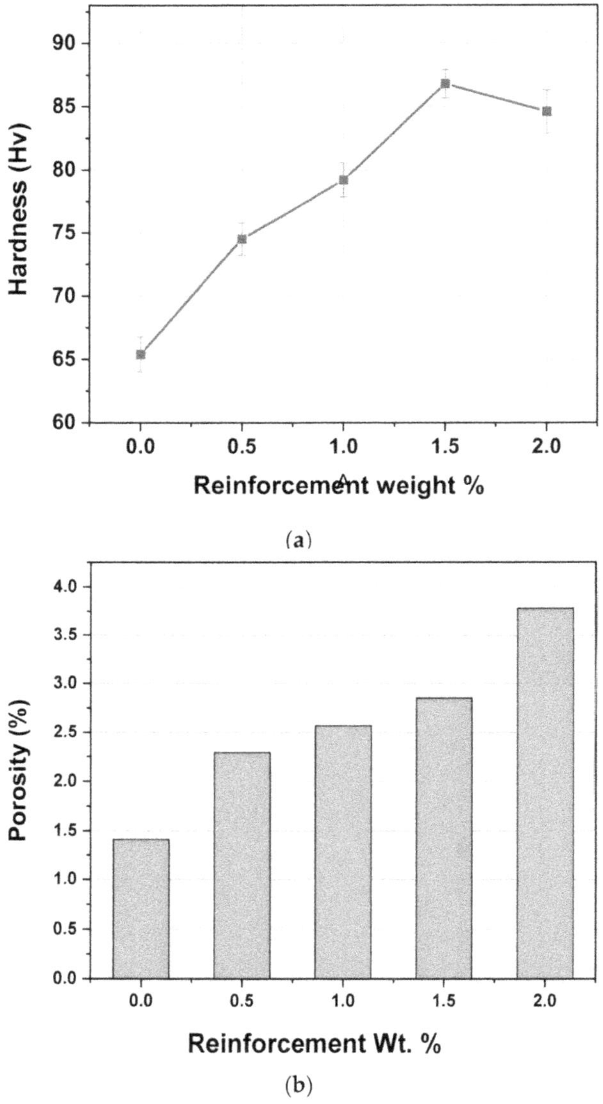

FIGURE 12.10 (a) Hardness of the alloy and nanocomposites; (b) porosity percentage (%) of the alloy and nanocomposites (Kumar et al. 2022).

***Porosity analysis*:** Porosity increased due to the air entrapping during mechanical stirring and clustering of nanoparticles. Ultrasound treatment reduces porosity, because it breaks clusters, but this effect diminishes at higher reinforcement levels.

***Wear testing, optimization, and setup*:** The parameters for optimization using the Taguchi method, as seen in Figure 12.11, include five applied loads ranging from 10 N to 50 N, sliding lengths between 200 m and

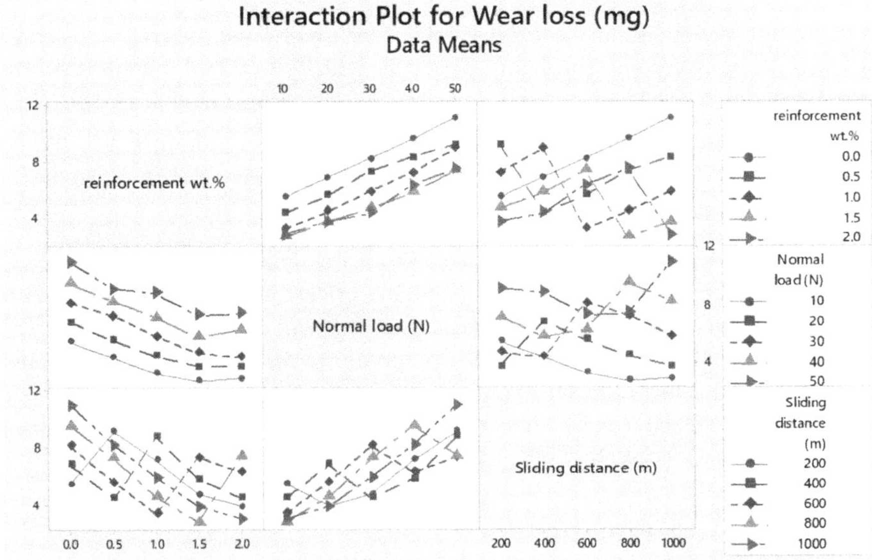

FIGURE 12.11 Interaction plot (Kumar et al. 2022).

1000 m, and reinforcement percentages of 0%, 0.5%, 1.0%, 1.5%, or 2.0% for dry sliding conditions, as per the L25 Orthogonal Array. The most favorable parameters include 1.5% reinforcement, 10 N load, and 400 µm sliding distance. Load was found to affect maximum 65.67% of wear loss, while wear loss was affected by 33.23% due to reinforcement weight and sliding distance by 0.75%.

The dominant wear mechanisms were identified as adhesive wear, two-body abrasion, and three-body abrasion. Minimal wear loss was observed under optimal conditions, along with improvements in hardness and load-bearing capacity. A regression model was developed to predict wear loss based on reinforcement weight percentage, load (N), and sliding distance (m). The model demonstrated high accuracy, with an adjusted R^2 value of 98.51%, validating its predictive capability. Severe plastic deformation of AA7068 alloy and delamination were observed under high loads, signifying bad wear resistance.

Nanocomposites up to 1.5% Si_3N_4: Smoother wear surface with lesser plastic deformation due to increased hardness and formulating a protective tribo-chemical layer. At 2.0% Si_3N_4: Increased wear due to clumping and uneven reinforcement distribution.

***Conclusion*:** This chapter shows how much impact Si_3N_4 nanoparticles can have on the tribological performance of AA7068 aluminum matrix composites. 1.5% Si_3N_4 increase in hardness and wear resistance is significant, but the marginal gains disappear at high reinforcement levels because clumping and porosity build up.

12.6 CHALLENGES AND LIMITATIONS

Even though nanomaterials and NFs hold a lot of promise, several challenges will have to be overcome for them to become widely used and commercialized. Such hurdles pertain to dispersion uniformity, economic practicality, and long-term stability.

12.6.1 Dispersion Uniformity

One of the severe problems associated with NFs is the proper dispersion of nanoparticles in a continuous medium. Due to small-size effects, nanoparticles are likely to agglomerate into large clusters or sediment over time without any spatial order. This inhomogeneous clustering diminishes the properties of the NF. Using advanced dispersion techniques like ultrasonic agitation, surfactant-assisted dispersion, or incorporation of stabilizers, it is ensured that the nanoparticles are uniformly dispersed, differing in the enhanced thermophysical properties of the NF.

12.6.2 Economic Feasibility

High costs associated with nanoparticle synthesis and the associated difficulties in incorporating nanoparticles into industrial systems are significant hurdles in the commercialization of nanomaterials and NFs. Nanoparticle manufacture is particularly energy-intensive and costly, particularly for rare or expensive materials. Furthermore, scaling a process to achieve the necessary industrial levels requires additional specialized equipment and process steps, therefore securing raised costs. Some possible actions include smart responses to such hurdles like cheaper and simpler means of nanoparticle synthesis. Finally, more rational manufacturing processes to allow seamless integration of nanomaterials into existing systems will significantly reduce costs and open possibilities for new technology applications.

12.6.3 Long-Term Stability

Long-term stability of NFs and nanocoatings is yet to be dealt with. With time, nanoparticles have the tendency to settle, agglomerate, or degrade, leading to a decline in the performance of the NFs or nanocoatings. Besides this, the extraneous agents and external factors, such as wide temperature fluctuations, environmental conditions, etc., affect the degradation process. In order to enhance the long-term stability of nanomaterials, the research is concentrating on the establishment of stable nanoparticles and coatings capable of withstanding adverse conditions. Stabilizing agents, surface modifications, and durable nanocoatings can help minimize these issues and ensure that nanomaterials keep their performance throughout longer periods.

12.7 FUTURE DIRECTIONS

Nanomaterials and NFs have several potentialities and the inability to realize them to their fullest must be explored in terms of opportunity and limitation. The next frontiers of innovation will be driven by advancements in manufacturing processes, integration with emerging technologies, and enhanced collaborations between industry and academia.

12.7.1 Advanced Manufacturing Techniques

Nanoparticle synthesis and coating technologies are anticipated to facilitate co-optimization of cost and quality of the nanomaterials. Advanced and promising coating technologies such as chemical vapor deposition (CVD), atomic layer deposition (ALD), etc., are available to produce high-quality and durable nanocoatings and nanocomposites. This reductive, high-yield method will help pave the way for the use of nanomaterials across multiple industrial processes.

12.7.2 Integration with Emerging Technologies

Nanomaterials have already played and will continue to play an important role in emerging technologies for renewable energy, energy storage, and advanced robotics. NFs are used for solar thermal systems to improve heat transfer efficiency, while nanocoatings enhance the durability of robotic components in extreme environmental conditions. Advancements are rapidly developing due to the implementation of nanomaterials in developing technologies, creating a step in the direction of more effective and greener solutions.

12.7.3 Collaboration Between Academia and Industry

Increased collaboration between industry and academia will blow the nascent field of nanotechnology wide open, speeding up the development and commercialization of nanotechnologies. The collaboration will help researchers to explore real-life problems and discover appropriate nanomaterial structures and designs for the specific application required. Through collaborative effort, we can shorten the distance between the lab and large-scale, real-world use and adoption of nanotechnology in real-world application.

12.8 CONCLUSION

Nanomaterials and NFs have revolutionized tribology and thermal management, offering unparalleled improvements in performance, efficiency, and durability. These advanced materials have demonstrated their potential in a wide range of applications, from automotive and aerospace to biomedical and manufacturing industries. While challenges such as dispersion uniformity, economic feasibility, and long-term stability remain, ongoing research and innovations in manufacturing techniques and collaboration between academia and industry will help unlock their full potential. As nanotechnology continues to evolve, it promises to drive the next wave of advancements, transforming industries and improving the quality of life for individuals worldwide.

REFERENCES

Afonso, Inês S., Glauco Nobrega, Rui Lima, José R. Gomes, and João E. Ribeiro. 2023. "Conventional and Recent Advances of Vegetable Oils as Metalworking Fluids (MWFs): A Review." *Lubricants* 11 (4): 160. https://doi.org/10.3390/lubricants11040160

Ahmed, Rashed, and Ahmed Nabhan. 2018. "Effects of TiO_2 and SiO_2 Nano Additive to Engine Lubricant Oils on Tribological Properties at Different Temperatures." In *Proceedings of the 20th International Conference on Aerospace, Mechanical, Automotive and Materials Engineering*, Rome, Italy, 30–31.

Alimuzzaman, Sk Md, Muhammad P. Jahan, C.S. Rakurty, Nithin Rangasamy, and Jianfeng Ma. 2023. "Cutting Fluids in Metal AM: A Review of Sustainability and Efficiency." *Journal of Manufacturing Processes* 106 (November):51–87. https://doi.org/10.1016/j.jmapro.2023.09.075

Asmatulu, Ramazan, P. Nguyen, and E. Asmatulu. 2013. "Nanotechnology Safety in the Automotive Industry." In *Nanotechnology Safety*, 57–72. Elsevier. https://doi.org/10.1016/B978-0-444-59438-9.00005-9

Chang, Li, and Klaus Friedrich. 2010. "Enhancement Effect of Nanoparticles on the Sliding Wear of Short Fiber-Reinforced Polymer Composites: A Critical Discussion of Wear Mechanisms." *Tribology International* 43 (12): 2355–64. https://doi.org/10.1016/j.triboint.2010.08.011

Devillez, A., F. Schneider, S. Dominiak, D. Dudzinski, and D. Larrouquere. 2007. "Cutting Forces and Wear in Dry Machining of Inconel 718 with Coated Carbide Tools." *Wear* 262 (7–8): 931–42. https://doi.org/10.1016/j.wear.2006.10.009

Erdi Korkmaz, Mehmet, and Munish Kumar Gupta. 2024. "Nano Lubricants in Machining and Tribology Applications: A State of the Art Review on Challenges and Future Trend." *Journal of Molecular Liquids* 407 (August):125261. https://doi.org/10.1016/j.molliq.2024.125261

Farfan-Cabrera, Leonardo I., Alejandro Rojo-Valerio, Juan de Dios Calderon-Najera, Karina G. Coronado-Apodaca, Hafiz M.N. Iqbal, Roberto Parra-Saldivar, Mariana Franco-Morgado, and Alex Elias-Zuñiga. 2023. "Microalgae Oil-Based Metal Working Fluids for Sustainable Minimum Quantity Lubrication (MQL) Operations—A Perspective." *Lubricants* 11 (5): 215. https://doi.org/10.3390/lubricants11050215

Ghodbane, Mokhtar, Zafar Said, Arun Kumar Tiwari, L. Syam Sundar, Changhe Li, and Boussad Boumeddane. 2022. "4E (Energy, Exergy, Economic and Environmental) Investigation of LFR Using MXene Based Silicone Oil Nanofluids." *Sustainable Energy Technologies and Assessments* 49 (February):101715. https://doi.org/10.1016/j.seta.2021.101715

Grzesik, Wit, and Joel Rech. 2019. "Influence of Machining Conditions on Friction in Metal Cutting Process – A Review." *Mechanik* 4 (April), 242–48. https://doi.org/10.17814/mechanik.2019.4.33

Kalsi, Sujata, Sunil Kumar, Anil Kumar, Tabish Alam, and Dan Dobrotă. 2023. "Thermophysical Properties of Nanofluids and Their Potential Applications in Heat Transfer Enhancement: A Review." *Arabian Journal of Chemistry* 16 (11): 105272. https://doi.org/10.1016/j.arabjc.2023.105272

Kilincarslan, Sena Kabave, and M. Huseyin Cetin. 2020. "Improvement of the Milling Process Performance by Using Cutting Fluids Prepared with Nano-Silver and Boric Acid." *Journal of Manufacturing Processes* 56 (August):707–17. https://doi.org/10.1016/j.jmapro.2020.05.027

Kumar, Ashish, Ravindra Singh Rana, Rajesh Purohit, Kuldeep K. Saxena, Jinyang Xu, and Vinayak Malik. 2022. "Metallographic Study and Sliding Wear Optimization of Nano Si_3N_4 Reinforced High-Strength Al Metal Matrix Composites." *Lubricants* 10 (9): 202. https://doi.org/10.3390/lubricants10090202

Kuntoğlu, Mustafa, Kübra Kaya, and Rüstem Binali. 2023. "Investigation of Surface Roughness Changes in The Machining of Carbon Steel Under Sustainable Conditions." *International Conference on Pioneer and Innovative Studies* 1 (June):163–67. https://doi.org/10.59287/icpis.823

Li, Yanhong, TianTian Liu, Yujuan Zhang, Pingyu Zhang, and Shengmao Zhang. 2018. "Study on the Tribological Behaviors of Copper Nanoparticles in Three Kinds of Commercially Available Lubricants." *Industrial Lubrication and Tribology* 70 (3): 519–26. https://doi.org/10.1108/ILT-05-2017-0143

Liñeira del Río José, M., Enriqueta R. López, and Josefa Fernández. 2021. "Tribological Properties of Graphene Nanoplatelets or Boron Nitride Nanoparticles as Additives of a Polyalphaolefin Base Oil." *Journal of Molecular Liquids* 333 (July):115911. https://doi.org/10.1016/j.molliq.2021.115911

Mathew, Jinu, Josny Joy, and Soney C. George. 2019. "Potential Applications of Nanotechnology in Transportation: A Review." *Journal of King Saud University - Science* 31 (4): 586–94. https://doi.org/10.1016/j.jksus.2018.03.015

McNamara, Karrina, and Syed A. M. Tofail. 2017. "Nanoparticles in Biomedical Applications." *Advances in Physics: X* 2 (1): 54–88. https://doi.org/10.1080/23746149.2016.1254570

Ogedengbe, T.S., A. P. Okediji, A. A. Yussouf, O. A. Aderoba, O. A. Abiola, I. O. Alabi, and O. I. Alonge. 2019. "The Effects of Heat Generation on Cutting Tool and Machined Workpiece." *Journal of Physics: Conference Series* 1378 (2): 022012. https://doi.org/10.1088/1742-6596/1378/2/022012

Perumal, Anand Babu, Lingxia Huang, Reshma B Nambiar, Yong He, Xiaoli Li, and Periyar Selvam Sellamuthu. 2022. "Application of Essential Oils in Packaging Films for the Preservation of Fruits and Vegetables: A Review." *Food Chemistry* 375 (May):131810. https://doi.org/10.1016/j.foodchem.2021.131810

Reverberi, A.P., D.M. D'Addona, A.A.G. Bruzzone, R. Teti, and B. Fabiano. 2019. "Nanotechnology in Machining Processes: Recent Advances." *Procedia CIRP* 79:3–8. https://doi.org/10.1016/j.procir.2019.02.002

Ross, K. Nimel Sworna, and G. Manimaran. 2020. "Machining Investigation of Nimonic-80A Superalloy Under Cryogenic CO2 as Coolant Using PVD-TiAlN/TiN Coated Tool at 45° Nozzle Angle." *Arabian Journal for Science and Engineering* 45 (11): 9267–81. https://doi.org/10.1007/s13369-020-04728-8

Ross, Nimel Sworna, Peter Madindwa Mashinini, Ritu Rai, and Munish Kumar Gupta. 2023. "Carbon Nano Dots Mixed Rice Bran Oil as a Cutting Fluid for Enhanced Lubrication/Cooling in Milling of Additively Manufactured 316 Stainless Steel." *Journal of Molecular Liquids* 391 (December):123200. https://doi.org/10.1016/j.molliq.2023.123200

Sen, Binayak, Shravan Kumar Yadav, Gaurav Kumar, Prithviraj Mukhopadhyay, and Sudarsan Ghosh. 2023. "Performance of Eco-Benign Lubricating/Cooling Mediums in Machining of Superalloys: A Comprehensive Review from the Perspective of Triple Bottom Line Theory." *Sustainable Materials and Technologies* 35 (April):e00578. https://doi.org/10.1016/j.susmat.2023.e00578

Şirin, Şenol. 2022. "Investigation of the Performance of Cermet Tools in the Turning of Haynes 25 Superalloy under Gaseous N2 and Hybrid Nanofluid Cutting Environments." *Journal of Manufacturing Processes* 76 (April):428–43. https://doi.org/10.1016/j.jmapro.2022.02.029

Sivalingam, Vinothkumar, Yanzhe Zhao, Ramkumar Thulasiram, Jie Sun, Guo Kai, and Thangapandian Nagamalai. 2021. "Machining Behaviour, Surface Integrity and Tool Wear Analysis in Environment Friendly Turning of Inconel 718 Alloy." *Measurement* 174 (April):109028. https://doi.org/10.1016/j.measurement.2021.109028

Tarafdar, Ayon, Ranjna Sirohi, Taru Negi, Shikhangi Singh, Prarabdh C. Badgujar, Navin Chandra Shahi, Sunil Kumar, Sang Jun Sim, and Ashok Pandey. 2021. "Nanofluid Research Advances: Preparation, Characteristics and Applications in Food Processing." *Food Research International* 150 (December):110751. https://doi.org/10.1016/j.foodres.2021.110751

Ukueje, Wisdom Etabiese, Fidelis Ibiang Abam, and Anthony Obi. 2022. "A Perspective Review on Thermal Conductivity of Hybrid Nanofluids and Their Application in Automobile Radiator Cooling." *Journal of Nanotechnology* 2022 (May):1–51. https://doi.org/10.1155/2022/2187932

Uysal, Cuneyt, and Mehmet Erdi Korkmaz. 2018. "Estimation of Entropy Generation for Ag-MgO/Water Hybrid Nanofluid Flow through Rectangular Minichannel by Using Artificial Neural Network." *Journal of Polytechnic*. https://doi.org/10.2339/politeknik.417756

Wang, Yubin, Yan Ren, Pei Yan, Siyu Li, Zhicheng Dai, Li Jiao, Bin Zhao, Siqin Pang, and Xibin Wang. 2023. "The Effect of Cutting Fluid on Machined Surface Integrity of Ultra-High-Strength Steel 45CrNiMoVA." *Materials* 16 (9): 3331. https://doi.org/10.3390/ma16093331

Waqas, Muhammad, Rehan Zahid, Muhammad Usman Bhutta, Zulfiqar Ahmad Khan, and Adil Saeed. 2021. "A Review of Friction Performance of Lubricants with Nano Additives." *Materials* 14 (21): 6310. https://doi.org/10.3390/ma14216310

Xu, Wenhao, Changhe Li, Yanbin Zhang, Hafiz Muhammad Ali, Shubham Sharma, Runze Li, Min Yang, et al. 2022. "Electrostatic Atomization Minimum Quantity Lubrication Machining: From Mechanism to Application." *International Journal of Extreme Manufacturing* 4 (4): 042003. https://doi.org/10.1088/2631-7990/ac9652

13 Exploring the Use of Nanoparticles to Improve the Properties of Denture Resins

Lakshay Sain, Pardeep Sharma, Abinet Gosaye Ayanie and Leliso Hobicho Dabelo

13.1 INTRODUCTION

Denture base resins have remained indispensable in the field of prosthodontics due to their ease of processing, cost-effectiveness, and acceptable aesthetics (Majeed et al., 2024; Gendreau & Loewy, 2011). Materials such as polymethyl methacrylate (PMMA) have been used for many years to produce complete and partial denture bases. Nevertheless, despite its pervasiveness, PMMA remains compromised mechanically in terms of impact resistance, fatigue resistance, and with regard to fracturing under the constant loads of mastication (Gad & Al-Thobity, 2021). These limitations frequently result in clinical failures requiring repair or replacement of the prosthesis, which in turn may negatively impact patient satisfaction and treatment duration (Aldegheishem et al., 2021). The necessity for prosthetic materials exhibiting a better mechanical adaptation has been increased due to a growing aging world population and the need for minimally invasive and long-term dental strategies (Naguib et al., 2023). This has motivated the investigation in the area of advanced material modifications that are beyond the classical fillers or fiber-reinforced composites. Nanotechnology in dental materials is a new and interesting field in recent years (Tijana et al., 2021). Nanoparticles (NPs) (particles characterized by at least one dimension in the range of 1–100 nm) possess unique physicochemical properties as a result of their very large surface area-to-volume ratio and quantum size effect (Abd El-Aziz et al., 2024). By the addition of dental polymers, these nanofillers have demonstrated great promise for enhancing strength, stiffness, toughness, and wear resistance. The study of nanoparticle types, such as silver and zirconium oxide (ZrO2), has revealed various material developments in terms of compatibility, strength, and durability. For instance, significant increases in the tensile and impact strength of heat and microwave-cured acrylic resin were observed after adding 3% ZrO2 nanoparticles, with little interference with the chemical aspects of the material (Bacali et al., 2020). The versatility of other nanoparticles to reinforce denture base materials is also demonstrated in the case of 1 wt% aluminum trioxide nanoparticles,

DOI: 10.1201/9781003560234-16

which were reported to increase the flexural strength in heat-polymerized PMMA (Pourhajibagher & Bahador, 2022). The characteristic features of adding nanoparticles to denture resin are shown in Figure 13.1.

Furthermore, diphenylurea incorporated silver nanoparticles provided by nystatin inhibited the growth of *Streptococcus mutans*, a common oral pathogen, and *Candida albicans*, another frequent dentist's issue related to denture users and enhanced the mechanical properties and biocompatibility of the resin-based (Cevik & Yildirim-Bicer, 2018). ZrO2 nanoparticles are also supposed to reduce water sorption and porosity in denture resins, which are the significant factors contributing to the performance and service life of the materials, as well as provide mechanical and antibacterial function (ABdallah et al., 2015). Although there are several known benefits of using nanoparticles in denture resins, it is important to consider any disadvantages, such as their long-term stability, and the need for a standard protocol for their incorporation into dental materials (Habibzadeh et al., 2020). Due to the molecular-level insertion of the nanoparticles within the polymer matrix, more efficient stress transfer and crack deviation are possible, which resists the crack propagation. Some nanoparticles also give multifunctional properties, such as antimicrobial ability as well as esthetic improvement, allowing them to be used on oral implants as well. The incorporation of nanomaterials into denture base resins is promising for next-generation prosthetic systems that satisfy both mechanical and biological demands (Ashour Ahmed et al., 2016).

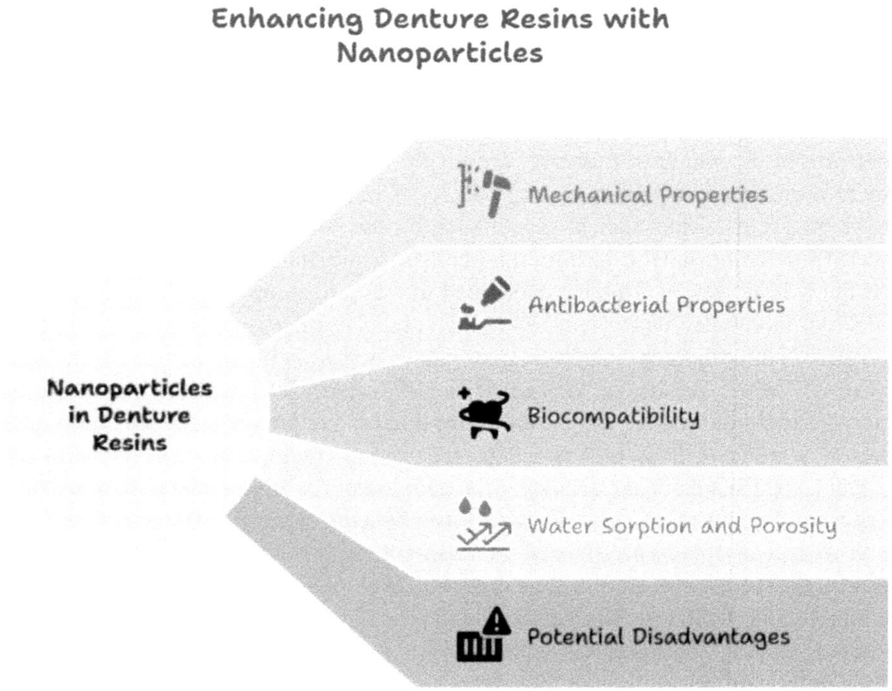

FIGURE 13.1 Features of adding nanoparticles to the denture resins.

This chapter is a comprehensive review of the feasibility of nanoparticles to improve the mechanical characteristics of denture resins. This book covers the proliferation of benefits provided by the development of nanotechnology and smart composites in orthopedic applications and examines new trends in material science and their durability, biodegradability, wear resistance, and the performance of smart implants in regenerative therapy. By summarizing the latest experimental research and shedding light on existing drawbacks, this chapter aims to pave the way for future research and clinical applicability in the fast-growing area of nanotechnology-modified dental applications.

13.2 OVERVIEW OF DENTURE BASE RESIN

Denture base materials are a structural basis for a removable dental prosthesis and under normal functional conditions support the teeth and distribute masticatory forces to the underlying oral tissues (Gad et al., 2018). The material demands from these resins are multifaceted, including sufficient strength, dimensional stability, biocompatibility, aesthetics, and ease of processing. Vulcanized rubber, metals, and thermoplastics, including PMMA, have been investigated in history, and PMMA has become the most used denture base resin (Abd-Ali, 2022) since it was introduced to the market in the 1930s. This is mostly because it has favorable characteristics such as good esthetics, easy handling, modern cost, and reparability (Aldegheishem et al., 2021). Denture base acrylic resins are typically categorized according to their polymerization process: heat-cured, self-curing (auto-polymerizing), light-curing, and microwave-curing. CAD/CAM milled and 3D-printed resins have been recently known for their uniformity and digital workflow properties (AATI et al., 2022). Although considerable progress has been made in the fabrication technology, the base materials used are still almost the same, and the mechanical limits are still obviously presented. The mechanical properties to be considered in designing the denture resin is shown in Table 13.1.

Even though it is extensively used, PMMA is inherently brittle and has an increased risk of fracture in its midline (of maxilla) due to tensile stresses (Abaza et al., 2021). Such a failure is frequently responsible for poor fatigue resistance, inadequate impact strength, and stress concentration resulting from microcracks in these regions. Several methods have been suggested to address this problem involving the addition of metal wires, glass or polyethylene fibers, and traditional filler particles in the micron size range. These enhancements, however, have provided only marginal enhancements to this end, usually with a sacrifice of esthetics or an increase in the difficulty of handling the materials (Krishna Alla et al., 2015). From the evolutionary point of view, a new breed of reinforcements marked by nanotechnology has emerged in materials science. Owing to their small size and special surface characteristics, nanoparticles provide a better dispersion in the resin matrix and the opportunity to improve mechanical properties while maintaining esthetics and ease of handling (Andreescu et al., 2018). Therefore, the incorporation of nanoparticles into denture base resins is a new wave in prosthodontic material fabrication with which high-performance prostheses possessing superior clinical performance can be developed (Bacali et al., 2020).

TABLE 13.1
Various Mechanical Properties and Their Relevance in Denture Resin

Property	Typical Values for PMMA	Clinical Significance	Remarks with Nanoparticles
Flexural strength	65–100 MPa	Ensures resistance to fracture under masticatory forces; crucial for midline fracture prevention.	Nanoparticles like ZrO_2, Al_2O_3, and SiO_2 significantly improve flexural strength via matrix reinforcement.
Impact strength	5–10 kJ/m^2	Critical for accidental falls or unexpected loading; fractures during handling or insertion often due to low impact strength.	Toughening achieved using rubber-modified nanoparticles or hybrid fillers (e.g., ZrO_2-SiO_2).
Hardness	15–25 (VHN)	Affects scratch resistance and long-term surface integrity; influences polishability and esthetics.	TiO_2, SiO_2, and graphene oxide (GO) nanoparticles enhance surface hardness by promoting denser cross-linking.
Fatigue resistance	~10^5–10^6 cycles before failure	Essential for long-term function under repetitive chewing forces; low fatigue resistance leads to microcrack propagation.	Nanoparticles deflect cracks, improve load transfer, and delay crack initiation—key for fatigue enhancement.
Fracture toughness	1.0–2.0 MPa·m½	High toughness needed to resist crack propagation; low toughness results in brittle behavior and sudden failure.	Carbon nanotubes (CNTs), nanodiamonds (NDs), and ZrO_2 improve fracture toughness via energy dissipation and crack bridging mechanisms.
Elastic modulus	1.8–2.4 GPa	Affects the load-bearing capacity and resistance to deformation; too high can result in rigidity, too low in flexibility.	Controlled enhancement achievable with ceramic nanoparticles while maintaining balance with flexibility.
Water sorption and solubility	Sorption: <32 μg/mm^3; Solubility: <1.6 μg/mm^3	Affects dimensional stability, color retention, and long-term integrity in oral fluids.	Some nanoparticles (e.g., hydrophobic SiO_2, TiO_2) reduce water sorption and solubility due to tighter matrix packing.
Dimensional stability	Minimal change (<0.5%)	Vital for accurate fit of dentures over time; warpage can lead to discomfort and poor function.	Nanoparticles help reduce polymer shrinkage and improve stability through polymer chain interaction.
Wear resistance	Moderate (varies by test)	Determines lifespan of occlusal surfaces and polish retention.	Reinforcement with hard nanoparticles like ZrO_2 and graphene improves resistance to surface degradation.

13.3 NANOPARTICLES: CLASSIFICATION AND CHARACTERISTICS

Nanoparticles are the ultra-micro materials widely spread between 1 and 100 nm and exhibit novel physicochemical properties due to their small size compared to their bulk forms (Bai et al., 2020). Their high surface area-to-volume ratio, quantum confinement effect, and adjustable surface chemistry have rendered them irreplaceable in broad scientific disciplines such as materials science, electronics, medicine, and dentistry (Barbin et al., 2020). Application of nanoparticles in prosthodontics has attracted attention as a class of materials with potential for improving properties (mechanical, biological, esthetic) of dental materials such as denture base resins, especially PMMA (Chung et al., 2018). The details and effects of some nanoparticles in denture resin are shown in Table 13.2.

TABLE 13.2
Various Nanoparticles Utilized in Denture Resin

Nanoparticle Type	Chemical Composition/ Structure	Enhanced Properties	Remarks
Zirconium dioxide (ZrO_2)	Inorganic ceramic oxide	Flexural strength, fracture toughness, fatigue resistance	Biocompatible, widely studied, opaque at high loadings
Silicon dioxide (SiO_2)	Inorganic silica (nanosilica)	Surface hardness, dimensional stability, flexural strength	Transparent, excellent dispersion, requires silanization
Titanium dioxide (TiO_2)	Metal oxide semiconductor	Hardness, wear resistance, antimicrobial effect	White pigment, UV resistance, enhances esthetics
Aluminum oxide (Al_2O_3)	Inorganic ceramic oxide	Wear resistance, hardness, stiffness	High thermal stability, cost-effective
Silver (Ag)	Metal nanoparticle	Antimicrobial properties	Can discolor resins, limited mechanical enhancement
Zinc oxide (ZnO)	Metal oxide	Antimicrobial activity, moderate hardness	Low toxicity, UV shielding effect
GO	Carbon-based layered material	Fracture toughness, flexural strength, antimicrobial	High aspect ratio, functionalizable, dark color
CNTs	Cylindrical carbon nanostructures	Flexural strength, modulus, fracture resistance	Difficult to disperse, requires surface treatment
NDs	Spherical carbon nanoparticles	Fracture toughness, impact resistance	High hardness, good biocompatibility
Polymeric nanoparticles	PMMA, polystyrene, polycaprolactone	Toughness, flexibility	Used for thermal or drug-delivery customization
Hybrid nanoparticles	For example, ZrO_2-GO, SiO_2-TiO_2	Multifunctional: strength + antimicrobial + esthetics	Customizable blends, emerging research area

13.3.1 Classification of Nanoparticles in Denture Resins

Nanoparticles effective in denture base resins reinforcement can be categorized into inorganic metal nanoparticles, ceramic nanoparticles, carbon-based nanomaterials, polymeric nanoparticles, and hybrid systems according to the type of the nanoparticle within its composition (Dawadi et al., 2021). Both classes have distinct benefits in enhancing the mechanical and functional properties of denture resins, especially PMMA.

Inorganic metal nanoparticles, including silver (Ag), zinc oxide (ZnO), titanium dioxide (TiO_2), and gold (Au), are reviewed in various research in terms of both mechanical- and antibacterial-associated features. Silver nanoparticles are particularly recognized for their powerful antibacterial activity against a wide range of oral microorganisms such as *Candida albicans* and *Streptococcus mutans* (Abd Elhamid Abo Salh et al., 2025). Their addition to denture base materials may decrease oral microbial adhesion and biofilm development, and consequently prevent denture stomatitis. Yet they have the drawback of bleaching resins and changing their optical properties. ZnO nanoparticles contribute with a medium reinforcement and with the antimicrobial and UV-protecting properties. There are several reasons using TiO_2-nanoparticles due to their increase of surface hardness, wear resistance, and hydrophilicity, and it can be used in patients susceptible to xerostomia or plaque accumulation (AlGhamdi et al., 2025). Gold nanoparticles, which are not as prevalent in denture bases, because of their high cost and reinforcement effect not being as good as the other nanoparticles, are also biocompatible and chemically stable.

Inorganic ceramic nanoparticles such as zirconium dioxide (ZrO_2), aluminum oxide (Al_2O_3), silicon dioxide (SiO_2), as well as silica fume (SF) are the most effective fillers for mechanical reinforcement. For instance, ZrO_2 NPs especially have been widely reported to improve in flexural strength, fracture toughness, and fatigue resistance (Cristache et al., 2020). They have inherent hardness and biocompatibility that are favorable for the long-term intraoral use. The increased wear resistance and stiffness caused by the Al_2O_3 nanoparticles may be particularly beneficial for high load-bearing sites in partial dentures. The SiO_2 nanoparticles (nanosilica) offer superior surface area, clarity, and ability to chemically bond to PMMA via silane coupling agents, thus providing both strength and esthetic bonding (Khattar et al., 2023).

Carbon nanoparticles are one of the most promising nanomaterials for use as a reinforcement as they exhibit high strength-to-weight ratios and have easily modified surfaces. CNTs, however, have high aspect ratio (length/diameter), tensile strength, and modulus, but are likewise hydrophobic and tend to agglomerate in the polymer matrix (Clarissa et al., 2022). Functionalized CNTs can get over this problem, and substantial improvements in stiffness and fracture toughness have been reported. Graphene and its derivatives such as GO deliver high strength, high barrier capability, and mild antibacterial activity. They can build up as a strong network in the resin to resist both crack initiation and crack propagation. NDs are another group of reinforcing agents having a similar spherical shape and chemically active surface, as well as they are intrinsically biocompatible and compatible with PMMA matrices.

Polymeric nanoparticles prepared from materials such as polystyrene, PMMA, or polycaprolactone are seldom used for mechanical strengthening, yet are also capable of being tailored to enhance flexibility, toughness, or drug loading of denture bases. These may be added in different amounts to alter thermal and/or rheological properties, without substantially detracting from esthetics or imparting rigidity (Goodacre et al., 2016). Mixed nanoparticles containing metal, ceramic, or carbon have surfaced as multitasking adjuncts that jointly contribute to the reinforcement, desired antibacterial features, and esthetics simultaneously. For instance, SiO_2-TiO_2 or ZrO_2-GO hybrid combines the advantages of the two components and has the advantage of overcoming several limitations of the traditional denture resins at a time. These are potential subjects of smart and bioactive prosthodontic material in the future (Salgado et al., 2022).

13.3.2 Characteristics of Nanoparticles for Denture Applications

The effectiveness of nanoparticles to strengthen denture base resins is related to a number of overlapping factors, which include particle size, shape, surface functionality, surface coverage, dispersion level, biocompatibility, and concentration requirement (Unkovskiy et al., 2018). The size and shape of the particle crucially dictate how well a nanoparticle can interface with the polymer matrix. Nanoparticles are generally between 1 and 100 nm, and smaller particles will generally have a greater surface area available for interaction with the matrix. However, too small a particle (<10nm) is easy to operate van der Waals forces to agglomerate, reducing their type of reinforcing effect. Spherical NPs are often preferred for easier expersion and lower surface energy, but other types of rod-like or plate-like particles such as CNT or GO nano-sheet have demonstrated better mechanical reinforcement due to their high aspect ratio, allowing for better load transfer at the resin matrix (Yodmongkol et al., 2014). The key research focus areas in nanoparticle-reinforced denture resin development are shown in Figure 13.2.

Surface chemistry and modification are key for nanoparticle-matrix adhesion. Nanoparticles were frequently treated with coupling agents to achieve a homogenous distribution and to make them compatible with the organic matrix (Municoy et al., 2023). Silanization is the most popular surface treatment method for inorganic oxides including SiO_2, ZrO_2, and Al_2O_3 due to its ability to improve adhesion to the PMMA matrix and reduce particle agglomeration (Sánchez-Salcedo et al., 2023). Regarding carbon nanoparticles, treatment with plasma, oxidation with acids, or grafting with polymers may be used to incorporate functional groups, which contribute to increase the hydrophilicity and dispersion. Such surface modulation is necessary to optimize interfacial adhesion and the functioning of high strength or the stressed situation.

Another important aspect is the stability of dispersion. Even if high-performance nanoparticles are used, none of them can play a role as a reinforcing agent unless dispersed homogeneously in the polymer matrix (Alrahlah et al., 2018). Agglomerated clusters become stress risers and may cause early failure of the material. It is often necessary to disperse the clusters and to achieve good molecular dispersion through advanced mixing techniques such as ultrasonication,

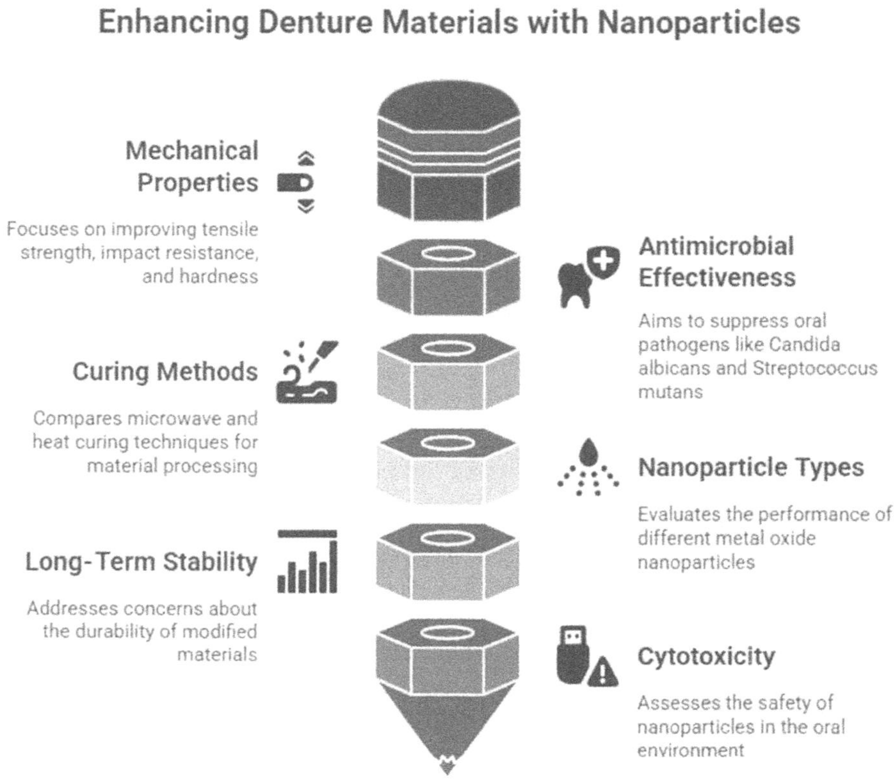

FIGURE 13.2 Key research focus areas in nanoparticle-reinforced denture resin development.

high-speed mechanical homogenization, and ball milling. The formation of a homogenous and uniform suspension without stratification of graphite in the suspension is the most important part of the control process as inconsistent material properties are a result of a poor stable state of suspension before polymerization.

Biocompatibility is an essential requirement for any additive material used in oral applications. Although the great majority of ceramic and carbon-based nanoparticles are biocompatible and inert, it has also become evident that metallic nanoparticles (especially silver, copper) in general tend to release cytotoxic agents (ions) in case of high concentrations. Hence, the use of nanomaterials should be restricted to problem-free dosages, and their long-term leach-into and breakdown should be controlled to a minimum (Kaurani et al., 2023). Their safety profiles need to be assessed by standardized cytotoxicity assays, such as ISO 10993. Establishing nanoparticle concentration also is critically important in regards to reinforcing efficiency. Optimal enhancement in mechanical properties has been reported at low to moderate concentrations (usually 0.5–2.5 wt. %). Beyond this point, the agglomeration of particles in the material may be more likely, the stress distribution may be less uniform, and denture esthetics and process ability may deteriorate (Khattar et al., 2023). The identification and preservation of a proper loading range of every type of nanoparticle is thus crucial, depending on the specific mechanical or functional aim to be achieved.

13.4 MECHANISM OF REINFORCEMENT IN DENTURE RESINS BY NANOPARTICLES

The addition of nanoparticles to denture base resin improves the mechanical properties through various inter-relation mechanisms in material science. Load transfer is one of the key mechanisms. Nanoparticles embedded within the polymer matrix behave as microfillers and share in the external load application while chewing or function purpose (Ragheb & Borg, 2021). This decreases the stress concentration in PMMA matrix and enhances resistance for deformation and fracture resistance.

Crack deflection and bridging is another important toughening mechanism. As cracks initiate and advance in the resin, such as in fatigue or impact, well-dispersed nanoparticles impede the advance of these cracks. This deflection needs some added energy for the crack tip to proceed, and in doing so delays or prevents failure. Some of the nanoparticles can also serve as bridges between the crack faces, absorbing energy and enhancing toughness, which is particularly useful for enhancing the fracture resistance (Fareed Al-Sammraaie & A Fatalla, 2023). The adhesion at the interface of the nanoparticle with the polymer matrix is also of great importance. Functionalized NPs, particularly silyl and plasma-treated, establish a chemical bond within the resin. This interfacial bonding ensures good stress transfer and reduces the appearance of weak zones, which are responsible for early failure (Siddiqui et al., 2015). High interfacial adhesion is necessary to increase flexural and impact strength, particularly under cyclic or critical loading conditions. The critical factors influencing the performance of nanoparticle-reinforced denture resins are shown in Figure 13.3.

The micromolecular structure of the polymer is also affected by nanoparticles. The polymer matrices, during curing, can be restricted in their mobility due to the presence of nanocharges, resulting in denser and more evenly distributed crosslinked network. This decreases internal voids and micropores that are the origin of generation of crack. Sometimes, nanoparticles may also serve as nucleation sites, producing a more crystalline polymer like-structure resulting in improved hardness and wear

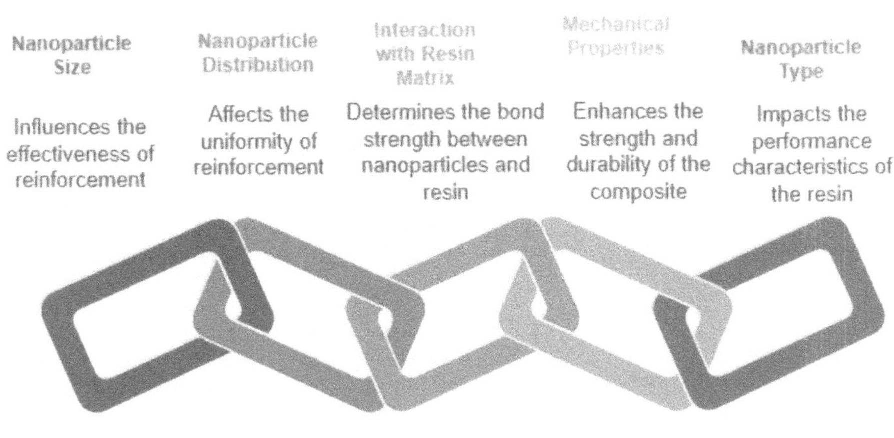

FIGURE 13.3 Critical factors influencing the performance of nanoparticle-reinforced denture resins.

resistance (Gowri et al., 2014). Finally, synergistic effects may be obtained by employing hybrid nanoparticles. For example, multifunctional prosthesis materials that would have more clinical applications could be obtained using ZrO_2 for mechanical resistance and Ag as an antimicrobial factor. Knowledge of these mechanisms and further optimization of these is essential for the design of nanocomposite denture bases with improved mechanical and functional properties.

13.5 INFLUENCE OF NANOPARTICLES ON MECHANICAL PROPERTIES

Many studies have shown that when nanoparticles are added, the mechanical properties of the denture base resin are significantly enhanced in vitro, and when used, the type, concentration, surface treatment, and dispersion quality of the nanoparticles will affect the specific mechanical properties of denture base resins (Elshereksi et al., 2017). Figure 13.4 depicts the key properties which are enhanced by adding the nanoparticles to dentures. Flexural strength, one of the important indicators of resistance to bending load of a material, becomes one of the most widely enhanced properties. For example, introduction of 1–2 wt% ZrO_2, SiO_2 or GO has been demonstrated to lead to the flexural strength of 15–35% as compared to unmodified PMMA.

These improvements are ascribed to the good interfacial adhesion and the high modulus of nanoparticles, which is protruded by the MWNTs for mechanical rearrangement at the microscale (Billings et al., 2021). The impact strength, or the capability of the material to absorb energy when suddenly loaded (e.g., impact loading), has also improved with the addition of the nanoparticles. Nanoparticles, such as rubber-modified SiO2 or NDs, contribute to plastic deformation due to their capability to retard the crack growth. But when the concentration is increased (>2.5 wt%), agglomeration increase the brittleness and impact may decrease. Hence, concentration and distribution are key demons of the package. Nanoparticles such as TiO_2, Al_2O_3 and GO, ceramic and carbon-based, greatly increase its surface hardness and wear resistance. The surface resistance to indentation and wear is improved by these nanoparticles, which is crucial for the longevity of dentures, especially in bruxer

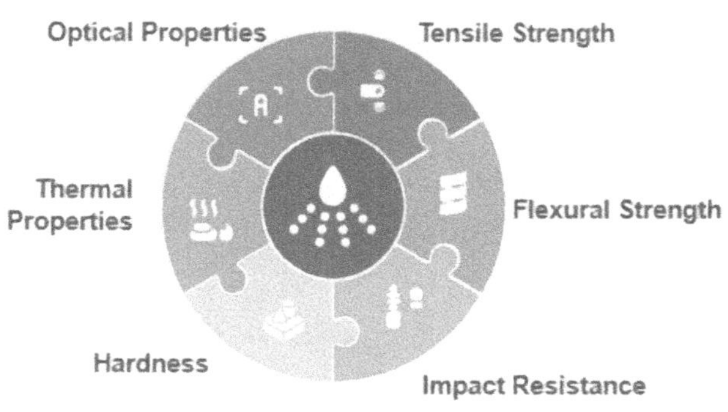

FIGURE 13.4 Key property enhancements achieved through nanoparticle reinforcement in denture resins.

patients or those with high occlusal forces. An enhanced hardness in addition provides for continued surface polish and cosmetic maintenance (Wang et al., 2017).

Fatigue properties, which dictate how well the material can withstand multiple chewing cycles, are improved via mechanisms such as crack bridging and an improved distribution of energy. Test results indicated that ZrO_2 and CNTs could increase the fatigue life by suppressing microcrack initiation and stress concentration (Saini et al., 2024). Fracture toughness (the material resistance to catastrophic failure formed a crack) also increases with the use of high aspect ratio or stiffness nanoparticles. In addition, G-O, CNTs, and ND show enhanced energy absorption and stress distribution near the crack tip. That in their sum return a more durable and clinically resistant denture base material, as long as its formulation is further properly adjusted in concentration, dispersion, and biocompatibility.

13.6 PROCESSING TECHNIQUES AND DISPERSION STRATEGIES

Nanoparticle addition to denture base resins is largely dependent on nanoparticle type as well as the processing method which incorporating into the resin matrix. The displacement has also an influence on the mechanical improvement that must be performed properly (Emam et al., 2024). Causal to this can be particle agglomeration leading to a stress-rising-point due to defective dispersion of the nanoparticles, rather than a reinforcement. Mechanical stirring is the easiest and most convenient method for mixing the NPs into the resin monomer at high speed. This process is cheap but frequently presents the problem of inhomogeneous dispersion, in particular for hydrophobic particles, as CNT or GO. Agglomerates are decomposed by ultrasonic energy with the nanoparticles being finely homogenized in the monomer. It works better on ceramic and carbon-based nanoparticles, but long sonication may decompose the polymer chains or even heat the mixture (Abar et al., 2024).

Higher level techniques are ball milling (which is OK for dry nano powder mixing before polymerization) and sol-gel processing (for making in situ nanoparticles in the network). Such methods also provide better interfacial interaction between filler and matrix, especially when surface-modified particles are employed. In situ polymerization is also a good candidate method because the nanoparticles are incorporated directly into the monomer prior to the addition of initiators. This approach causes the polymer chains to grow around the nanoparticles and form a connected structure. However, strict control of the reaction conditions is necessary to prevent premature gelation and non-homogeneous curing (Sivakumar et al., 2014).

Surface modification methods such as silanization of ceramic nanoparticles have been used to enhance compatibility with PMMA. This is not only favorable for dispersion but also for covalent bonding with the matrix; the second effect will further enhance mechanical properties. Modification of the carbon-based fillers with plasma treatment and chemical grafting has been employed to enhance the engagement in the entire preparation system (Ogle & Byles, 2014). In the end, the selection of the method of processing is a function of the particle type, its application, and the degree of trade-off between mechanical performance and process ability that is being sought. Protocolization of these methods for clinical denture fabrication is essential for the transformation of their laboratory advantages into clinical practice.

13.7 BIOLOGICAL AND CLINICAL CONSIDERATIONS

The mechanical properties of nanoparticles in the denture base resins have been proven; however, their clinical applicability depends on their biocompatibility, safety, and long-term stability in the oral cavity. Denture base materials are in contact with oral mucosa for extended periods of time, and are hence perceived to be cytotoxic and tissue-compatible materials. The majority of ceramic NPs, for example, ZrO_2, SiO_2, and TiO_2, are highly inert and exhibited remarkable in vitro and in vivo biocompatibility (Kassaee et al., 2008). And the nanoparticle does not leach ions or degrade quickly in saliva, which means it is safe for oral use over extended periods. Carbonated fillers such as GO and ND have also shown good compatibility, but more long-term researches are needed to clarify their response to the complicated oral microflora.

Metallic nanoparticles, such as silver (Ag) and zinc oxide (ZnO), which possess both antimicrobial and mechanical reinforcement properties, are used. The impact of Ag nanoparticles to reduce *Candida albicans* colonization on denture surfaces is of particular interest in elderly patients who are susceptible to denture stomatitis. But the fact that they have the potential to release ions gradually and cause cytotoxicity in high doses calls for careful management of dosage and surface area. Also relevant is the resistance to biofouling. Surface-modified nanoparticles can decrease bacterial adherence and plaque biofilm formation, resulting in better hygiene and less infection. This is of particular advantage in full dentures, which are frequently worn for continuous periods and particularly in elderly or dependent subjects (Garcia et al., 2022).

The addition of the nanoparticles also affects the color stability and esthetics. Although ceramic and silica nanoparticles are colorless or white, metal nanoparticles could have a grayish or yellowish appearance that may alter the appearance of the denture base. Therefore, mechanical and biological performance must be weighed against aesthetic demands. The stability of the nanoparticles in the oral environment over long periods of time, including pH, temperature, mechanical loading, and enzymatic action, should be measured as well. Certain nanoparticles can clump or shed over time, which can degrade performance and safety. Hence, durability testing and in vitro aging simulations are essential for their clinical use. Regulatory complexity comes from the use of nanoscale materials and the fact that they can be regulated in different categories based on how they are formulated and behave. One of the highlights for future studies to facilitate clinical translation is the requirement of evaluating the biocompatibility and toxicity, in accordance with ISO and FDA recommendations, of this method.

13.8 RECENT ADVANCES AND SMART NANOCOMPOSITES

In recent years, there has been a considerable progress in the development of smart nanocomposite denture base materials with the goal of not only compensating for the mechanical deficiencies in traditional PMMA-based resins but also with functionality tailored for future prosthodontics. This technological evolution has been possible due to a multi-disciplinary combination of nanotechnology, polymer science, biomedical engineering, and surface chemistry, creating denture materials with improved mechanical properties, durability, antimicrobial activity, self-healing capacity, and even sensory potential (Alluhaidan et al., 2024). The utilization of hybrid nanocomposites, i.e., a combination of two or more types of nanoparticles, has gained significant interest and has great potential for exploiting the synergistic effects to achieve desired properties. For example, the combination of ZrO_2 for mechanical reinforcement with Ag nanoparticles for antibiotic function part has demonstrated that a simultaneous reinforcing, impact strength, and antibacterial effect could be achieved. Secondly, composites of GO and TiO_2 can also offer superior wear resistance as well as UV-induced self-cleaning characteristics. Their double nature makes these hybrids particularly desirable for geriatric and compromised patients who demand inexpensive but robust prosthetic materials with low maintenance (Vasiliu et al., 2021).

A further exciting development is the addition in to denture base resins of stimuli-responsive or so-called "smart" nanoparticles. If stimuli-responsive, these nanoparticles can be made to respond to the surrounding environment in terms of (e.g.) pH, temperature, and mechanical strain. For instance, the embedded thermoresponsive nano gels within the PMMA network are capable of swelling or deswelling by oral temperature, facilitating dynamic stress release against the PMMA-based material leading to reduced breakage. A pH-responsive nanoparticle carrying an antibiotic can release the drug upon exposure to acidic presence, which corresponds to the bacteria growth and can counteract denture stomatitis in situ. Self-healing nanocomposites have also been investigated recently by incorporating microcapsules or nanocontainers containing healing agents, i.e., monomers or initiators. Mechanical loading or crack growth under stress causes the capsules to break and to spill their content, which polymerizes and closes the crack autonomously. This innovation can greatly improve the useful life of denture bases and make them less likely to require repair or replacement (Szerszeń et al., 2022). 3D printing and AM have offered novel strategies to incorporate NPs with a high degree of accuracy and handling within a space. Thanks to nanoparticle-loaded resins, the 3D printing techniques such as SLA (stereolithography) and DLP (digital light processing) can be used to fabricate customized denture into which several functions are embossed. Graphene and silica nanoparticles have also been successfully printed in complex geometries with mechanically and aesthetically uniform properties (SONI & SAHU, 2022).

Another frontier is the investigation of biomimetic nanocomposite materials, which use biomimetic inspiration for material design to improve properties. For example, nacre-mimetic layered materials with nanoclay or graphene nanoplatelets have been demonstrated to dissipate energy with a good level of toughness. These designs are derived from nature and are further optimized by computational modeling and machine learning, which can predict optimal formulations and patterns. Moreover, machine learning and artificial intelligence (AI) are also employed for designing and predicting the properties of nanocomposite. By parsing extensive data sets on nanoparticle properties, resin attributes, and environmental conditions, AI can assist in designing optimal pairings, cutting down on trial-and-error during laboratory synthesis. For commercialization and translational applications toward the clinic, there are a few companies that have been developing nanoparticle-reinforced denture resins in the market. Nanofilled PMMA disks for CAD/CAM milling, for example, already available with better wear and aesthetic properties. But more widespread clinical use of smart nanocomposites will have to be approved by regulators and backed up by longer clinical data.

These developments represent a departure from inert prosthetic materials to active adaptive denture base materials that provide not only form and function but also intelligent interaction with the oral environment. Wireless biosensors, personalized nanocomposite formulations based on the patient's biomechanics, and private and public acceptance are the next steps. In the end, the combination of intelligent material with digital dentistry presents a great potential to further the quality of removable prosthodontics treatment.

13.9 CHALLENGES AND LIMITATIONS

Despite the encouraging potential and fast-paced changes in nanoparticle-reinforced denture base resins, there are several limitations and factors that prevent their widespread application in the clinic. These issues include material compatibility, processing conditions, complexity of clinical validation, cost, regulation, and long-term performance and demand cautious and scientifically sound advancement. The first of these issues is the nanoparticle dispersion. It is necessary to achieve a homogeneous distribution of nanoparticles in the resin matrix to materialize the designed enhancement of mechanical properties. Bad dispersion is easy to agglomerate, between the particles together by Van der Waals force or surface energy. These "agglomerates" are defects in the matrix which not only negate mechanical reinforcement but also act as stress raisers which result in premature failure. This is especially so in high surface area fillers, such as CNTs or graphene. All of the above methods are complex and expensive processes due to utilizing various dispersion methods like ultrasonication, magnetic stirring, and surface functionalization.

A major restriction is the ideal concentration of nanoparticles. It is occasionally prophesied that low concentrations (< 2 wt.%) can generally reinforce properties, while high amounts could cause the performance to decrease due to a saturation of the matrix, poor polymerization kinetics, or an increase in viscosity. In addition, it is an empirical problem to find out an optimal compromise between mechanics and processability, and often such optimization needs to be conducted on a case-by-case basis.

The compatibility of the nanoparticles with the polymer matrix is also a key factor. Inert nanoparticles such as SiO_2 or Al_2O_3 can exhibit weak interfacial adhesion with PMMA if not properly surface-modified. Adhesion-promoting agents, such as silanization or grafting agents, are required, but they add extra steps, chemicals, and verifications of processes that may make the manufacturing pipeline more cumbersome. In terms of biocompatibility, not all nanoparticles are benign. Ceramics and silicates are more compatible materials although metallic nanoparticles such as Ag, CuO, or ZnO could release ions in time, especially in contact with saliva or enzymatic degradation. These ions can induce cytotoxicity, allergic reactions, or even systemic toxicities if not sufficiently retained in the matrix. Thus, there is a strong requirement of full ISO- and FDA-approved toxicologic studies prior to clinical use.

The color and aesthetics are also brought out by the nanoparticles. Discoloration or clouding of the denture base occurs with certain fillers and can render the denture base unattractive. For instance, adding silver/titanium oxide may cause the material to appear gray in color, which generally is unsuitable for use in anterior restorations where translucency is desired. As far as processing and scalability are concerned, the co-entrapment of NPs typically requires cleanroom conditions, custom mixing machines, and specific polymerization recipes. These demands are not fulfilled by conventional denture laboratories based on standard PMMA processing. Therefore, the majority of nanocomposites are still limited to academic/industrial R&D when it comes to everyday prosthodontic use. It is costly at the moment in that the nanoparticle and processing equipment are not adjusted for it. High-purity nanomaterials, particularly if functionalization or hybridization is included, can be costly to fabricate, handle, and store. This expense is passed on to the final product, and this cost (and many other associated costs) may not necessarily be justifiable for all patients or even healthcare systems. In addition, long-term clinical data are missing for nanoparticle-reinforced denture bases. To date, the majority of studies are in vitro testing or short in vivo tests. Factors impacting the performance and safety of these materials in the real world, including patient oral hygiene, dietary behavior, and wear cycles, may potentially alter these materials in unpredictable ways. Hence, randomized control trials and post-market surveillance are required to demonstrate credibility.

Another big obstacle is regulatory restriction. Nanomaterials are subjected to a more stringent regulation compared to their bulk in conventional policies at the same time that the classification of these materials remains not clear under the guidelines of medical devices. To satisfy regulatory requirements, developers are required to submit detailed information on the composition, behavior, toxicity, and degradation of their NPPs as well as on the spread of the NPPs in the environment; this can slow down product registration and lead to higher development costs. Lastly, there is a relative lack of awareness among clinicians and technicians. The advantages and disadvantages of nanocomposites are not well realized in the prosthodontic literature, thus professional education and continuing education programs are essential. In conclusion, although the full potential of nanoparticle-reinforced denture resins remains uncertain, the application of these materials in the clinic is currently limited by challenges in distribution and compatibility, cost, regulatory issues, and the scarcity of long-term data. These barriers can be overcome through multi-disciplinary approaches, thorough testing, and translational initiatives that bring laboratory innovation to the bedside.

13.10 CONCLUSIONS AND FUTURE SCOPE

The integration of nanoparticles into denture base resins has shown significant potential in improving mechanical properties such as flexural strength, impact resistance, wear resistance, and fracture toughness. Compared with these laboratory successes, bringing promising laboratory results to be acceptable in clinical standards needs rigorous research and development. For the future of this area, an optimal nanoparticle concentration, type, and dispersion approach will need to be defined in order to find the optimal balance between mechanical improvement, cosmetic acceptability, and biocompatibility. Decomposable smart nanocomposites: there is also a significant field for moving forward for intelligent nanocomposites, which not only enhance their strength, but load more capabilities such as antibacterial resistance, self-healing, and pH sensitivity. These could provide substantial advantages for prosthetic users by avoiding infections and extending the lifetime of the appliance. Novel nanomaterials such as bioactive glass, layered double hydroxides, or stimuli-responsive polymers hold great potential in this respect.

At the same time, long-term stability and safety are essential parameters. Longer-term studies that simulate the oral environment, with all its complexities of saliva, enzymes, pH, and temperature variations, are required to determine the aging behavior of nanocomposites. All such issues, including nanoparticle leaching, tissue accumulation, and the immunological response, that these materials elicit must be adequately addressed through proper biocompatibility testing in compliance with ISO and FDA regs. The standardization of the processing procedures is also critical for clinical translation. Although current studies employ a broad range of incorporation techniques, resin formulations, and nanoparticle treatments, in future efforts, the objective should be to develop reproducible and clinically pertinent protocols. Computational modeling and AI predictive tools can be used to optimize formulations and predict mechanical performance under various conditions.

On the whole, nanotechnology in denture base resins appeared to be a promising interdisciplinary concept for improving the disadvantages of traditional PMMA-based materials. It has been proved in recent years that nanoparticles such as ZrO_2, SiO_2, TiO_2, CNTs, which can dramatically increase specific mechanical properties, could be used to prepare the future stronger, more fatigue resistant, even multifunctional (example: antimicrobial or self-cleaning) denture base materials. Nevertheless, further studies on the long-term implications, biocompatibility, scale-up, standardization, and economically viable production methods are necessary in order for them to be extensively used in a clinical setting. With further innovation and cooperation among materials science, dentistry, and biomedical engineering, NPI-reinforced denture resins will be expected to be a widespread clinical practice in the future, greatly enhancing patient satisfaction and prosthetic life.

REFERENCES

Aati, S., Aneja, S., Kassara, M., Leung, R., Nguyen, A., Tran, S., Shrestha, B., & Fawzy, A. (2022). Silver-loaded mesoporous silica nanoparticles enhanced the mechanical and antimicrobial properties of 3D printed denture base resin. *SSRN Electronic Journal*. https://doi.org/10.2139/ssrn.4166940

Abar, E. S., Vandghanooni, S., Memar, M. Y., Eskandani, M., & Torab, A. (2024). Enhancing antifungal and antibacterial properties of denture resins with nystatin-coated silver nanoparticles. *Scientific Reports*, 14(1). https://doi.org/10.1038/s41598-024-74465-7

Abaza, S. F., Elbialy, N. S., & Mohamed, N. (2021). Incorporating silver nanoshell-coated mesoporous silica nanoparticles improves physicochemical and antimicrobial properties of chitosan films. *International Journal of Biological Macromolecules*, 189, 792–801. https://doi.org/10.1016/j.ijbiomac.2021.08.161

Abd-Ali, N. K. (2022). the effect of adding nanoparticles on the mech-anical properties of acrylic removable denture. *Journal of Engineering Science and Technology*, 17(4), 2983–2996.

Abd El-Aziz, A. B. E. D., El-Mokadem, M., Abo-Ghalia, H. H., Mattar, Z. A., & Sallam, A. I. (2024). Biosynthesis of a novel composite resin incorporating gamma radiation synthesized pomegranate extract–coated zinc oxide nanoparticles and in vitro assessment against streptococcus mutans causing dental caries. *BioNanoScience*, 14(5), 5017–5034. https://doi.org/10.1007/s12668-024-01410-3

Abd Elhamid Abo Salh, I. B., Add El-Salam, E. B., Ezzat, A. A., Aboshama, M., & Elhagali, A. F. (2025). Antimicrobial effect of three different nanoparticles-modified 3D-printed denture resin: An in vitro study. *Journal of International Society of Preventive and Community Dentistry*, 15(1), 42–49. https://doi.org/10.4103/jispcd.jispcd_53_24

ABdallah, R. M., Emera, R. M. K., & Gebreil, A. S. (2015). Antimicrobial activity of silver nanoparticles and their effect on acrylic resin mechanical properties. *Egystian Dental Journal*, 61(1), 1039–1052.

Aldegheishem, A., Aldeeb, M., Al-Ahdal, K., Helmi, M., & Alsagob, E. I. (2021). Influence of reinforcing agents on the mechanical properties of denture base resin: A systematic review. *Polymers*, 13(18). https://doi.org/10.3390/polym13183083

AlGhamdi, M. A., Fouda, S. M., Al-Dulaijan, Y. A., Khan, S. Q., El Zayat, M., Al Munif, R., Albazroun, Z., Amer, F. H., Al Ammary, A. T., Mahrous, A. A., & Gad, M. M. (2025). Nanocomposite 3D printed resins containing titanium dioxide (TiO_2) nanoparticles: An in vitro analysis of color, hardness, and surface roughness properties. *Frontiers in Dental Medicine*, 6. https://doi.org/10.3389/fdmed.2025.1581461

Alluhaidan, T., Qaw, M., Garcia, I. M., Montoya, C., Orrego, S., & Melo, M. A. (2024). Seeking endurance: Designing smart dental composites for tooth restoration. *Designs*, 8(5). https://doi.org/10.3390/designs8050092

Alrahlah, A., Fouad, H., Hashem, M., Niazy, A. A., & AlBadah, A. (2018). Titanium Oxide (TiO2)/polymethylmethacrylate (PMMA) denture base nanocomposites: Mechanical, viscoelastic and antibacterial behavior. *Materials*, 11(7). https://doi.org/10.3390/ma11071096

Andreescu, C. F., Ghergic, D. L., Botoaca, O., Hancu, V., Banateanu, A. M., & Patroi, D. N. (2018). Evaluation of different materials used for fabrication of complete digital denture. *Materiale Plastice*, 55(1), 124–128. https://doi.org/10.37358/mp.18.1.4977

Ashour Ahmed, M., El-Shennawy, M., Althomali, Y., & Omar, A. A. (2016). Effect of titanium dioxide nano particles incorporation on mechanical and physical properties on two different types of acrylic resin denture base. *World Journal of Nano Science and Engineering*, 06(03), 111–119. https://doi.org/10.4236/wjnse.2016.63011

Bacali, C., Baldea, I., Moldovan, M., Carpa, R., Olteanu, D. E., Filip, G. A., Nastase, V., Lascu, L., Badea, M., Constantiniuc, M., & Badea, F. (2020). Flexural strength, biocompatibility, and antimicrobial activity of a polymethyl methacrylate denture resin enhanced with graphene and silver nanoparticles. *Clinical Oral Investigations*, 24(8), 2713–2725. https://doi.org/10.1007/s00784-019-03133-2

Bai, X., Lin, C., Wang, Y., Ma, J., Wang, X., Yao, X., & Tang, B. (2020). Preparation of Zn doped mesoporous silica nanoparticles (Zn-MSNs) for the improvement of mechanical and antibacterial properties of dental resin composites. *Dental Materials*, 36(6), 794–807. https://doi.org/10.1016/j.dental.2020.03.026

Barbin, T., Velôso, D. V., Del Rio Silva, L., Borges, G. A., Presotto, A. G. C., Barão, V. A. R., & Mesquita, M. F. (2020). 3D metal printing in dentistry: An in vitro biomechanical comparative study of two additive manufacturing technologies for full-arch implant-supported prostheses. *Journal of the Mechanical Behavior of Biomedical Materials*, 108. https://doi.org/10.1016/j.jmbbm.2020.103821

Billings, C., Cai, C., & Liu, Y. (2021). Utilization of antibacterial nanoparticles in photocurable additive manufacturing of advanced composites for improved public health. *Polymers*, 13(16). https://doi.org/10.3390/polym13162616

Cevik, P., & Yildirim-Bicer, A. Z. (2018). The effect of silica and prepolymer nanoparticles on the mechanical properties of denture base acrylic resin. *Journal of Prosthodontics*, 27(8), 763–770. https://doi.org/10.1111/jopr.12573

Chung, Y. J., Park, J. M., Kim, T. H., Ahn, J. S., Cha, H. S., & Lee, J. H. (2018). 3D printing of resin material for denture artificial teeth: Chipping and indirect tensile fracture resistance. *Materials*, 11(10). https://doi.org/10.3390/ma11101798

Clarissa, W. H. Y., Chia, C. H., Zakaria, S., & Evyan, Y. C. Y. (2022). Recent advancement in 3-D printing: Nanocomposites with added functionality. *Progress in Additive Manufacturing*, 7(2), 325–350. https://doi.org/10.1007/s40964-021-00232-z

Cristache, C. M., Totu, E. E., Iorgulescu, G., Pantazi, A., Dorobantu, D., Nechifor, A. C., Isildak, I., Burlibasa, M., Nechifor, G., & Enachescu, M. (2020). Eighteen months follow-up with patient-centered outcomes assessment of complete dentures manufactured using a hybrid nanocomposite and additive CAD/CAM protocol. Journal of *Clinical Medicine*, 9(2). https://doi.org/10.3390/jcm9020324

Dawadi, S., Katuwal, S., Gupta, A., Lamichhane, U., Thapa, R., Jaisi, S., Lamichhane, G., Bhattarai, D. P., & Parajuli, N. (2021). Current research on silver nanoparticles: Synthesis, characterization, and applications. *Journal of Nanomaterials*, 2021. https://doi.org/10.1155/2021/6687290

Elshereksi, N. W., Ghazali, M. J., Muchtar, A., & Azhari, C. H. (2017). Studies on the effects of titanate and silane coupling agents on the performance of poly (methyl methacrylate)/barium titanate denture base nanocomposites. *Journal of Dentistry*, 56, 121–132. https://doi.org/10.1016/j.jdent.2016.11.012

Emam, A.-N. M., Ahmed, K.-E. I., Shaaban, A. M., Alqhtani, M. A., Gad, M. M., & Helal, M. A. (2024). Color and flexural properties of nanoparticles-modified denture base resin: An in vitro comparative study. *International Journal of Health Sciences*, 18(3), 23–29. http://www.ncbi.nlm.nih.gov/pubmed/38721136%0Ahttp://www.pubmedcentral.nih.gov/articlerender.fcgi?artid=PMC11075442

Fareed Al-Sammraaie, M., & Fatalla, A. (2023). The effect of ZrO_2 nanoparticles addition on candida adherence and tensile strength of 3D printed denture base resin. *Journal of Nanostructures*, 13(2), 544–552.

Gad, M. M., Abualsaud, R., Rahoma, A., Al-Thobity, A. M., Al-Abidi, K. S., & Akhtar, S. (2018). Effect of zirconium oxide nanoparticles addition on the optical and tensile properties of polymethyl methacrylate denture base material. *International Journal of Nanomedicine*, 13, 283–292. https://doi.org/10.2147/IJN.S152571

Gad, M. M., & Al-Thobity, A. M. (2021). The impact of nanoparticles-modified repair resin on denture repairs: A systematic review. *Japanese Dental Science Review*, 57, 46–53. https://doi.org/10.1016/j.jdsr.2020.12.004

Garcia, A. A. M. N., Sugio, C. Y. C., de Azevedo-Silva, L. J., Gomes, A. C. G., Batista, A. U. D., Porto, V. C., Soares, S., & Neppelenbroek, K. H. (2022). Nanoparticle-modified PMMA to prevent denture stomatitis: A systematic review. *Archives of Microbiology*, 204(1). https://doi.org/10.1007/s00203-021-02653-4

Goodacre, B. J., Goodacre, C. J., Baba, N. Z., & Kattadiyil, M. T. (2016). Comparison of denture base adaptation between CAD-CAM and conventional fabrication techniques. *Journal of Prosthetic Dentistry*, 116(2), 249–256. https://doi.org/10.1016/j.prosdent.2016.02.017

Gowri, S., Rajiv Gandhi, R., & Sundrarajan, M. (2014). Structural, optical, antibacterial and antifungal properties of zirconia nanoparticles by biobased protocol. *Journal of Materials Science and Technology*, 30(8), 782–790. https://doi.org/10.1016/j.jmst.2014.03.002

Habibzadeh, S., Omidvaran, A., Eskandarion, S., & Shamshiri, A. R. (2020). Effect of incorporation of silver nanoparticles on the tensile bond strength of a long term soft denture liner. *European Journal of Dentistry*, 14(2), 268–273. https://doi.org/10.1055/s-0040-1709923

Kassaee, M. Z., Akhavan, A., Sheikh, N., & Sodagar, A. (2008). Antibacterial effects of a new dental acrylic resin containing silver nanoparticles. *Journal of Applied Polymer Science*, 110(3), 1699–1703. https://doi.org/10.1002/app.28762

Kaurani, P., Hindocha, A. D., Jayasinghe, R. M., Pai, U. Y., Batra, K., & Price, C. (2023). Effect of addition of titanium dioxide nanoparticles on the antimicrobial properties, surface roughness and surface hardness of polymethyl methacrylate: A systematic review. *F1000Research*, 12. https://doi.org/10.12688/f1000research.130028.1

Khattar, A., Alghafli, J. A., Muheef, M. A., Alsalem, A. M., Al-Dubays, M. A., AlHussain, H. M., AlShoalah, H. M., Khan, S. Q., AlEraky, D. M., & Gad, M. M. (2023). Antibiofilm activity of 3D-printed nanocomposite resin: Impact of ZrO_2 nanoparticles. *Nanomaterials*, 13(3). https://doi.org/10.3390/nano13030591

Krishna Alla, R., Raghavendra Swamy, K., Krishna Alla Scholar, R., Swamy Professor, R. K., Vyas, R., Konakanchi, A., Rama Krishna Alla Scholar, C., & Swamy, R. K. (2015). Conventional and contemporary polymers for the fabrication of denture prosthesis: Part I-overview, composition and properties. *International Journal of Applied Dental Sciences*, 1(4), 82–89. www.oraljournal.com

Gendreau, L., & Loewy, Z. G. (2011). Epidemiology and etiology of denture stomatitis. *Journal of Prosthodontics*, 20(4), 251–260. http://ovidsp.ovid.com/ovidweb.cgi?T=JS&PAGE=reference&D=emed10&NEWS=N&AN=21463383

Majeed, H. F., Hamad, T. I., & Bairam, L. R. (2024). Enhancing 3D-printed denture base resins: A review of material innovations. *Science Progress*, 107(3). https://doi.org/10.1177/00368504241263484

Municoy, S., Antezana, P. E., Bellino, M. G., & Desimone, M. F. (2023). Development of 3D-printed collagen scaffolds with in-situ synthesis of silver nanoparticles. *Antibiotics*, 12(1). https://doi.org/10.3390/antibiotics12010016

Naguib, G., Maghrabi, A. A., Mira, A. I., Mously, H. A., Hajjaj, M., & Hamed, M. T. (2023). Influence of inorganic nanoparticles on dental materials' mechanical properties. A narrative review. *BMC Oral Health*, 23(1). https://doi.org/10.1186/s12903-023-03652-1

Ogle, O. E., & Byles, N. (2014). Nanotechnology in dentistry today. *West Indian Medical Journal*, 63(4), 344–348. https://doi.org/10.7727/wimj.2013.178

Pourhajibagher, M., & Bahador, A. (2022). Effects of incorporation of nanoparticles into dental acrylic resins on antimicrobial and physico-mechanical properties: A meta-analysis of in vitro studies. *Journal of Oral Biology and Craniofacial Research*, 12(5), 557–568. https://doi.org/10.1016/j.jobcr.2022.07.010

Ragheb, N. A., & Borg, H. S. (2021). Antimicrobial effect of titanium oxide (Tio_2) nano particles in completely edentulous patients. a randomized clinical trial. *Advanced Dental Journal*, 3(4), 173–184. https://doi.org/10.21608/adjc.2021.91184.1105

Saini, R. S., Bavabeedu, S. S., Quadri, S. A., Gurumurthy, V., Kanji, M. A., Okshah, A., Binduhayyim, R. I. H., Alarcón-Sánchez, M. A., Mosaddad, S. A., & Heboyan, A. (2024). Mapping the research landscape of nanoparticles and their use in denture base resins: A bibliometric analysis. *Discover Nano*, 19(1). https://doi.org/10.1186/s11671-024-04037-1

Salgado, H., Gomes, A. T. P. C., Duarte, A. S., Ferreira, J. M. F., Fernandes, C., Figueiral, M. H., & Mesquita, P. (2022). Antimicrobial activity of a 3D-printed polymethylmethacrylate dental resin enhanced with graphene. *Biomedicines*, 10(10). https://doi.org/10.3390/biomedicines10102607

Sánchez-Salcedo, S., García, A., González-Jiménez, A., & Vallet-Regí, M. (2023). Antibacterial effect of 3D printed mesoporous bioactive glass scaffolds doped with metallic silver nanoparticles. *Acta Biomaterialia*, 155, 654–666. https://doi.org/10.1016/j.actbio.2022.10.045

Siddiqui, M. N., Redhwi, H. H., Vakalopoulou, E., Tsagkalias, I., Ioannidou, M. D., & Achilias, D. S. (2015). Synthesis, characterization and reaction kinetics of PMMA/silver nanocomposites prepared via in situ radical polymerization. *European Polymer Journal*, 72, 256–269. https://doi.org/10.1016/j.eurpolymj.2015.09.019

Sivakumar, I., Arunachalam, K. S., Sajjan, S., Ramaraju, A. V., Rao, B., & Kamaraj, B. (2014). Incorporation of antimicrobial macromolecules in acrylic denture base resins: A research composition and update. *Journal of Prosthodontics*, 23(4), 284–290. https://doi.org/10.1111/jopr.12105

Soni, S., & Sahu, S. (2022). Dental application of properties of smart materials. https://www.google.com/books?hl=id&lr=&id=SNVuEAAAQBAJ&oi=fnd&pg=PT8&dq=%22toothbrushing%22+AND+(method+OR+technique)+AND+%22plaque%22+-AND+(mental+OR+blind+OR+deaf+OR+speech+OR+intellectual+OR+hearing+OR+visual)+AND+%22children%22+-therapy+-diagnostic+&ots=5xpLCF9ej3&sig=GwIdm1hE_cPjUO9ZQkrJaU231WE

Szerszeń, M., Cierech, M., Wojnarowicz, J., Górski, B., & Mierzwińska-Nastalska, E. (2022). Color stability of zinc oxide poly(methyl methacrylate) nanocomposite—A new biomaterial for denture bases. *Polymers*, 14(22). https://doi.org/10.3390/polym14224982

Tijana, A., Valentina, V., Nataša, T., Miloš, H. M., Suzana, G. A., Milica, B., Yoshiyuki, H., Hironori, S., Ivanič, A., & Rebeka, R. (2021). Mechanical properties of new denture base material modified with gold nanoparticles. *Journal of Prosthodontic Research*, 65(2), 155–161. https://doi.org/10.2186/JPR.JPOR_2019_581

Unkovskiy, A., Bui, P. H. B., Schille, C., Geis-Gerstorfer, J., Huettig, F., & Spintzyk, S. (2018). Objects build orientation, positioning, and curing influence dimensional accuracy and flexural properties of stereolithographically printed resin. *Dental Materials*, 34(12), e324–e333. https://doi.org/10.1016/j.dental.2018.09.011

Vasiliu, S., Racovita, S., Gugoasa, I. A., Lungan, M. A., Popa, M., & Desbrieres, J. (2021). The benefits of smart nanoparticles in dental applications. *International Journal of Molecular Sciences*, 22(5), 1–24. https://doi.org/10.3390/ijms22052585

Wang, L., Hu, C., & Shao, L. (2017). The antimicrobial activity of nanoparticles: Present situation and prospects for the future. *International Journal of Nanomedicine*, 12, 1227–1249. https://doi.org/10.2147/IJN.S121956

Yodmongkol, S., Chantarachindawong, R., Thaweboon, S., Thaweboon, B., Amornsakchai, T., & Srikhirin, T. (2014). The effects of silane-SiO_2 nanocomposite films on Candida albicans adhesion and the surface and physical properties of acrylic resin denture base material. *Journal of Prosthetic Dentistry*, 112(6), 1530–1538. https://doi.org/10.1016/j.prosdent.2014.06.019

Part 4

Specialized Applications and Case Studies

14 Case Study
Nonlinear Dynamic Behavior of Boron Nitride Nanotube-Based Mass Sensors in Smart Manufacturing

Dinesh Deshwal, Anil Kumar Narwal, Monika and Vivek Srivastava

14.1 INTRODUCTION

The field of nanotechnology has recently adopted boron nitride nanotubes (BNNTs) due to the combination of their unique mechanical properties with thermal and chemical features. The nanotubes share a similar structure with carbon nanotubes (CNTs) but use boron and nitrogen atoms in place of carbon atoms to provide suitable traits for smart manufacturing sensor applications (Yin et al. 2016; Bae, Lee, and Kim 2024). The wide 5.5 eV bandgap in BNNTs makes them great electrical insulators that function by blocking electrical signals under harsh conditions. The ability of BNNTs to resist temperature changes along with chemical attack makes them resilient against damaging conditions that could compromise alternative nanomaterials. The combined properties of BNNTs enable their suitability for sensitive sensor applications which demand extended durability (Shtansky et al. 2022; Clegg 2018).

Combining automation and real-time data analytics facilitates smart manufacturing in optimizing production processes, improving product quality, and reducing waste generation. In this ecosystem, mass sensors function as essential components by delivering substance properties data, pollutant detection, and validating product compliance. Traditional sensor materials fail to reach the advanced sensitivity needed to detect minimal mass changes that these sensors need to detect. A solution emerges from BNNTs because their mechanical strength, thermal conductivity, and chemical inertness make them suitable for maintaining accurate operation across diverse challenging environments. This foundation on using BNNTs in mass sensing introduces the necessary knowledge to study their nonlinear behaviors while demonstrating

DOI: 10.1201/9781003560234-18

their severe value in sensor development for manufacturing applications (Sohrabi et al. 2023; Wong et al. 2023).

The unique characteristics of BNNTs emerge directly from their physical tube-shaped organization. BNNTs consist of connected boron and nitrogen atoms within a hexagonal lattice which forms their strong tubular structure. The hexagonal arrangement of BNNTs produces super-strong tensile behavior and fabulous Young's modulus which reaches 1 TPa while ensuring complete resistance under stressful conditions. The combination of boron and nitrogen atoms gives BNNTs exceptional chemical stability together with oxidation resistance starting from elevated temperatures, thus extending their usefulness in corrosive environments. Heat dissipation capabilities of BNNTs serve to preserve measurement precision through high thermal conductivity above many metal and CNT levels in varying temperature environments (Hong, Kang, and Fan 2024).

The electronic properties of BNNTs create advantages for smart manufacturing because they provide important characteristics that support the field. The intrinsic electronic characteristics of BNNTs differ from those of CNTs because they always remain insulating due to their large bandgap feature. Manufacturing sensor performance remains reliable thanks to the electrical insulation characteristics BNNTs provide in these environments. The electrically insulating properties of BNNTs create prospects to develop applications that demand a safe, non-conducting material for biological systems. BNNTs present particular value to biomedical manufacturing because their biocompatibility helps achieve essential purity and safety requirements (Zhang et al. 2020; Mane, Shaishta, and Manjunatha 2024; Zhi et al. 2010).

The use of BNNTs demonstrates one of the most interesting characteristics of smart manufacturing because they function as sensitive mass detectors. The detection of minimal mass variations depends heavily on mass sensors because these variations reveal defects, contamination, or manufacturing problems. The mechanical capability of BNNTs allows them to track minimal mass changes through shifts in their resonant frequency detection. The precise operation of BNNTs in harsh environments depends on their structural quality which establishes enhanced sensitivity that stands out compared to conventional materials (Mohammed et al. 2024; Golberg et al. 2007).

BNNT detects changes in resonance frequencies that arise from modifications in applied mass. BNNT-based sensors detect small modifications in mass because they possess exceptional sensitivity to minuscule changes and work well for quality control within smart manufacturing environments. Mass sensors based on BNNTs succeed in detecting small pollutants which threaten product quality within pharmaceutical and electronics among other production sectors. Smart manufacturing demands require precise monitoring, and the BNNT-based sensors serve this objective because they can detect exact mass changes (Roy et al. 2021).

The occurrence of nonlinear dynamic response in BNNTs results from natural structural abnormalities that introduce surface waviness. The synthesis methods, along with processing techniques of BNNTs, typically result in imperfect straightness because they cause length irregularities throughout the nanotube structures. The combination of structural flaws creates nonlinear sensor characteristics in BNNT-based devices when these devices experience mass loading situations. Nonlinear characteristics of dynamic responses in BNNT-based sensors enhance their ability to detect very small mass changes because of amplified sensitivity. The structural waviness in BNNTs creates second- and third-order nonlinearities that modify the resonant frequency that directly depends on the BNNT-loaded mass. The sensing mechanism of BNNT-based sensors remains sensitive because their nonlinear aspects compensate for fluctuations in the operational conditions (Sharma et al. 2024; Deshwal and Narwal 2023b).

Research into BNNT nonlinear dynamics provides important information about these sensors while operating in real environments. The sensor frequency response develops enhanced sensitivity when the waviness factor of a BNNT grows more substantial. The enhanced sensitivity of this technique turns out to be advantageous for precise mass detection in industries depending on quality control and environmental monitoring. The physical control of BNNT dimensions, together with the waviness factor, allows manufacturers to optimize sensor performance specifications depending on specific manufacturing needs (Deshwal and Narwal 2021; 2023a).

The mechanical, thermal, and chemical characteristics of BNNTs create a unique proposition that makes them suitable for advanced sensing solutions. BNNTs fit perfectly into smart manufacturing mass sensing applications because they use their resistant structure together with their strong mass response-ability and distinctive nonlinear properties to boost their measurement results. The integration of Internet of Things (IoT) with BNNT sensors expands their use for automated continuous monitoring of manufacturing environments. The advancement of BNNT technology will make these sensors necessary components of future manufacturing systems where they will improve accuracy levels, efficiency rates, and sustainability benchmarks of production activities

14.2 BACKGROUND AND RATIONALE

Scientists currently consider BNNTs a promising nanomaterial for superior sensor applications because of their exceptional resistance combined with sensitivity and stability features. BNNTs stand out from the rest of nanomaterials with their exceptional thermal resistance in addition to high tensile durability, excellent electrical isolation, and strong chemical resistance. CNTs have achieved substantial use through their mechanical and electrical properties, yet their sensitivity to oxidation and chemical degradation makes them unusable in highly demanding environments. Thermally resistant BNNTs demonstrate better chemical stability than other nanomaterials which makes them an attractive solution for sensing applications requiring durability under demanding conditions.

The atomic configuration of BNNTs underpins their exceptional characteristics. BNNTs consist of boron and nitrogen atoms organized in a hexagonal lattice, creating tubular formations akin to CNTs. The atomic arrangement confers a bandgap of roughly 5.5 eV to BNNTs, rendering them stable electrical insulators. The insulating properties of BNNTs remain largely invariant despite variations in the tube's structure, including chirality. The stability and predictability of BNNT-based sensors confer a significant advantage in applications requiring minimal electrical interference, thereby improving the accuracy and reliability of mass sensing in precision manufacturing and environmental monitoring (Ali 2022).

14.2.1 Mechanical and Chemical Properties of BNNTs

BNNTs exhibit remarkable mechanical strength, possessing a Young's modulus that exceeds 1 TPa. Their strength, coupled with a lightweight structure, allows BNNTs to endure significant mechanical stresses, essential for preserving sensor performance under dynamic loading conditions. The robust covalent bonds between boron and nitrogen atoms provide significant structural integrity, enabling BNNTs to maintain their mechanical properties in high-stress or high-temperature conditions. The mechanical properties of BNNTs render them an optimal material for nanoscale devices that require consistent structural stability, such as mass sensors utilized in quality control and environmental monitoring within manufacturing processes (Jing 2018).

The stability properties of BNNTs separate them from CNTs since CNTs prove to be more susceptible to environmental chemicals and less durable when exposed to corrosive substances. BNNTs demonstrate chemical resistance against acids, bases, and other chemical agents, making them appropriate for applications exposed to corrosive substances. This characteristic provides exceptional benefits to environmental monitoring sensors because they must analyze pollutants, contaminants, and reactive chemicals. BNNTs maintain their reliable service life in harsh chemical environments thanks to their resistance which prevents sensitivity loss from chemical deterioration (Capek 2009; Turhan et al. 2022).

14.2.2 The Role of Nonlinear Dynamics in BNNT-Based Sensors

The research field dedicated to BNNT-based sensors focuses on understanding how nonlinear dynamics affect their sensitivity together with their total operational effectiveness. The introduction of nonlinearity in BNNTs occurs through structural irregularities such as surface waviness that emerge from the synthesis process. The dynamic behavior of wavy BNNTs differs compared to straight nanotubes since their structural geometry causes nonlinear response patterns during mass detection applications. The frequency response of BNNTs shows improved mass detection capability because surface waviness generates quadratic and cubic nonlinearities. Applications benefiting from ultra-sensitive mass detection have better performance through nonlinear behavior since it increases the sensor's responsiveness to minimal mass fluctuations (Aligayev et al., 2022).

Related to structural waviness, there are unique effects which modify BNNT frequency output characteristics. High levels of quartz nonlinearity emerge because of surface deviations but decrease after reaching an optimal factor point. The increase in waviness reduces the effects of Cubic nonlinearity which manifests through midplane stretching behavior. The nonlinearity enables BNNT-based sensors to deliver accurate and sensitive results throughout various operating conditions that manufacturers require for precise mass measurement (Vaccarini et al.).

14.2.3 Rationale for Focusing on BNNT-Based Mass Sensors

The research focus on BNNT-based mass sensors happens because they meet the requirements of advanced smart manufacturing alongside environmental monitoring systems. Modern manufacturing needs highly sensitive sensors because its automation processes and data-driven operations have intensified. Smart manufacturing relies on time-sensitive quality monitoring systems, so sensor reliability must be maintained at high levels. The combination of BNNTs' resistance to damage and their special sensitivity to weight alterations fulfills essential characteristics needed for industrial sensor development.

The analysis of nonlinear dynamic properties of BNNTs introduces major benefits for sensor performance enhancement. Research shows that instead of being a disadvantage, the structural irregularities in BNNTs can improve mass sensor sensitivity. The design of BNNT-based sensors for specific applications becomes more effective through researchers' understanding of surface waviness, frequency response behavior, and sensitivity characteristics (Vaccarini et al. 2000; Ertekin 2024).

BNNTs serve as an alternative solution for CNT-based applications when high sensitivity needs to be combined with stability in demanding extreme conditions. The sensor technology field benefits from BNNTs because of their exceptional combination of mechanical durability, chemical resistance, and nonlinear dynamic responses. The investigation analyzes the nonlinear behavior of BNNT-based mass sensors because researchers aim to enhance sensor performance in precision manufacturing and environmental application scenarios (Vaccarini et al. 2000; Yang et al. 2023).

14.3 EXPERIMENTAL SETUP AND MODELING APPROACH

A finite element analysis (FEA) serves as the foundation to understand BNNT-based mass sensor dynamic behavior characteristics. The analysis investigates how surface waviness and mass loading affect BNNT resonator performance by studying both their frequency response and sensitivity levels. The FEA model received parameters and specifications from the thesis while considering nonlinear effects that come from surface disturbances and mid-strain elongation effects. This section details the entire model structure alongside mathematical equations and simulation parameters that represent real BNNT sensor performance.

14.3.1 Mathematical Model of BNNT-Based Mass Sensors

A mathematical model for the wavy BNNT-based mass sensor operates using continuum mechanical principles to develop equations of motion that evaluate surface deviation (waviness factor) phenomena together with nonlinear dynamic reactions. A sensor based on a doubly clamped BNNT with an attached mass positioned at the midpoint operates through dynamic loading conditions observed in Figure 14.1.

The waviness factor w of the BNNT is defined by equation (14.1):

$$w = \frac{\mathrm{e}}{L} \tag{14.1}$$

where e is the eccentricity (surface deviation) and L is the length of the BNNT.

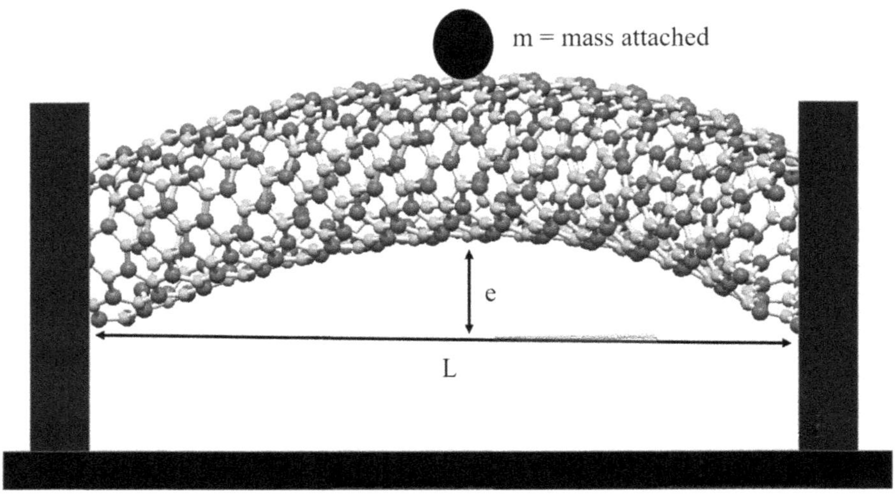

FIGURE 14.1 Doubly clamped single-walled BNNT with attached mass at the center.

14.3.2 Kinetic and Potential Energy Equations

To obtain the dynamic behavior, the kinetic energy (E_{KE}) and potential energy (E_{PE}) of the BNNT with the attached mass are derived in equation (14.2) and (14.3). The kinetic energy E_{KE} for the system is given by:

$$E_{KE} = \frac{1}{2}\left(\rho_{BNNT} \times A \times \int_0^L \left(\frac{\partial u}{\partial t}\right)^2 dx\right) + \frac{1}{2} m\_att \times \left(\frac{\partial u}{\partial t}\right)^2 \tag{14.2}$$

where:

ρ_{BNNT} is the density of the BNNT,
A is the cross-sectional area,
m_att is the mass attached at the midpoint of the BNNT, and
u represents the transverse displacement.

The potential energy E_{PE} for the system is represented as:

$$E_{PE} = \frac{1}{2}\left(E_{BNNT} \times A \int_0^L \left(\frac{\partial^2 u}{\partial x^2}\right)^2 dx\right) \tag{14.3}$$

where E_{BNNT} is the modulus of elasticity (Young's modulus) of the BNNT.

14.3.3 Equation of Motion and Nonlinearities

Using the Galerkin method, the equation of motion for the wavy BNNT, including the nonlinear terms due to waviness and stretching, is derived in equation (4):

$$(1+M)\ddot{\varnothing}+\alpha\varnothing+\beta\varnothing^2+\gamma\varnothing^3=0.816\left(\frac{F\cos(\omega t)}{L^2}\right) \tag{14.4}$$

where:

$M = m_{att}/(2L)$ is the mass ratio of the attached mass to the BNNT length,

$\alpha = 5.33\left(E_{BNNT}\times I_{BNNT}\times\frac{\pi^4}{L^4}\right)\times\left(E_{BNNT}\times A\times\pi^2\times\frac{e^2}{L^4}\right)$ represents the square of the first fundamental frequency,

$\beta = 4.352\times\left(E_{BNNT}\times A\times\pi^3\times\frac{e}{L^4}\right)$ is the coefficient of quadratic nonlinearity due to surface waviness, and

$\gamma = 0.89\left(E_{BNNT}\times A\times\frac{\pi^4}{L^4}\right)$ is the coefficient of cubic nonlinearity from mid-plane stretching.

14.4 SIMULATION PARAMETERS, MATERIAL PROPERTIES, AND KEY FINDINGS

The FEA model uses specific geometrical and material parameters for the BNNT, as tabulated in Table 14.1:

These parameters allow for a realistic simulation of the BNNT's response under mass loading conditions. The model was further enhanced by adding damping as depicted in equation (14.5), represented by the coefficient μ, in the equation of motion to simulate energy dissipation in real-world applications:

$$(1+M)\ddot{\varnothing}+\mu\dot{\varnothing}+\alpha\varnothing+\beta\varnothing^2+\gamma\varnothing^3=0.816\left(\frac{F\cos(\omega t)}{L^2}\right) \tag{14.5}$$

14.4.1 Frequency Response Analysis

A fast Fourier transform (FFT) was applied to analyze the BNNT's frequency response under different lengths and waviness factors. The dynamic behavior for BNNT lengths of 20 nm, 40 nm, and 60 nm, with a waviness factor of 0.05, was examined as shown in Figures 14.2, 14.3 and 14.4, respectively. It was found that wavy BNNTs exhibit an enhanced resonant frequency compared to straight BNNTs,

TABLE 14.1
Parameters of BNNT

Parameters	Value
Density (ρ_BNNT)	2180 kg/m^3
Young's modulus (E_BNNT)	1180 GPa
Outer diameter	1.5 nm
Inner diameter	1.34 nm
Attached mass	10^{-24} kg

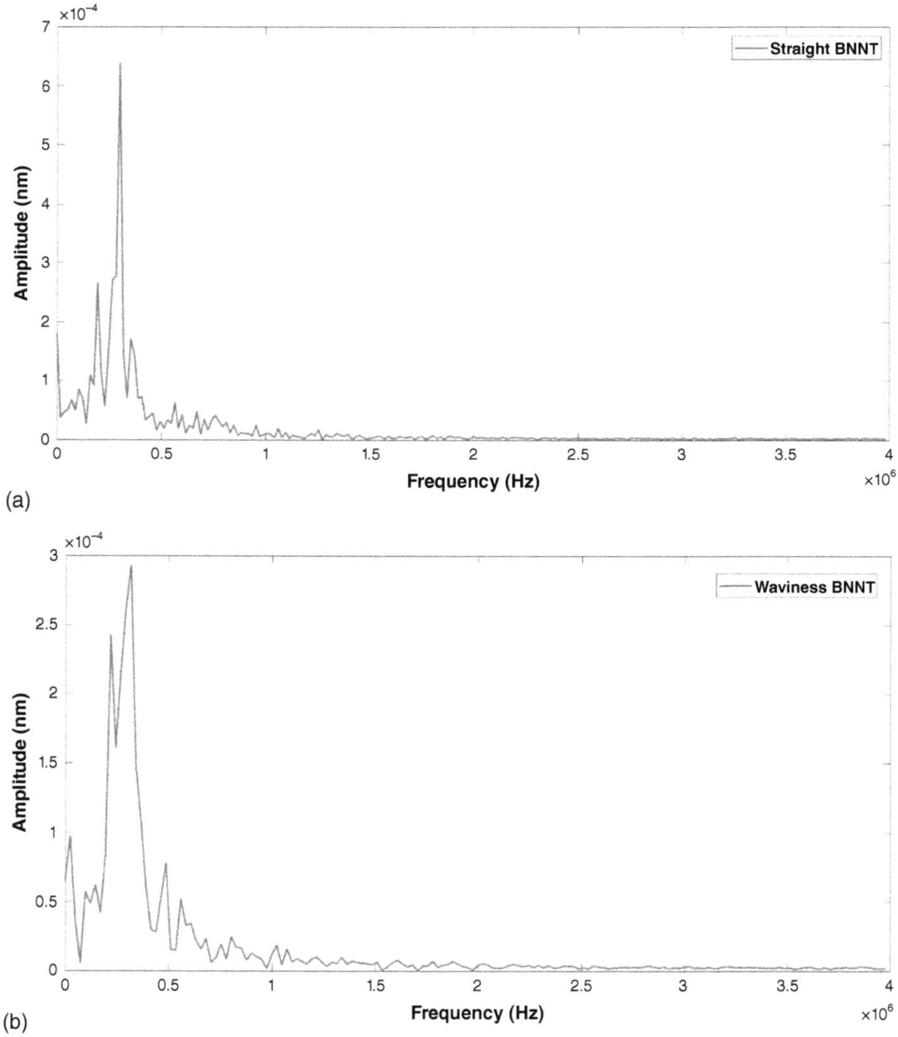

FIGURE 14.2 Dynamic response: FFT for L = 20 nm. (a) Straight BNNT and (b) eccentric (waviness = 0.05) BNNT.

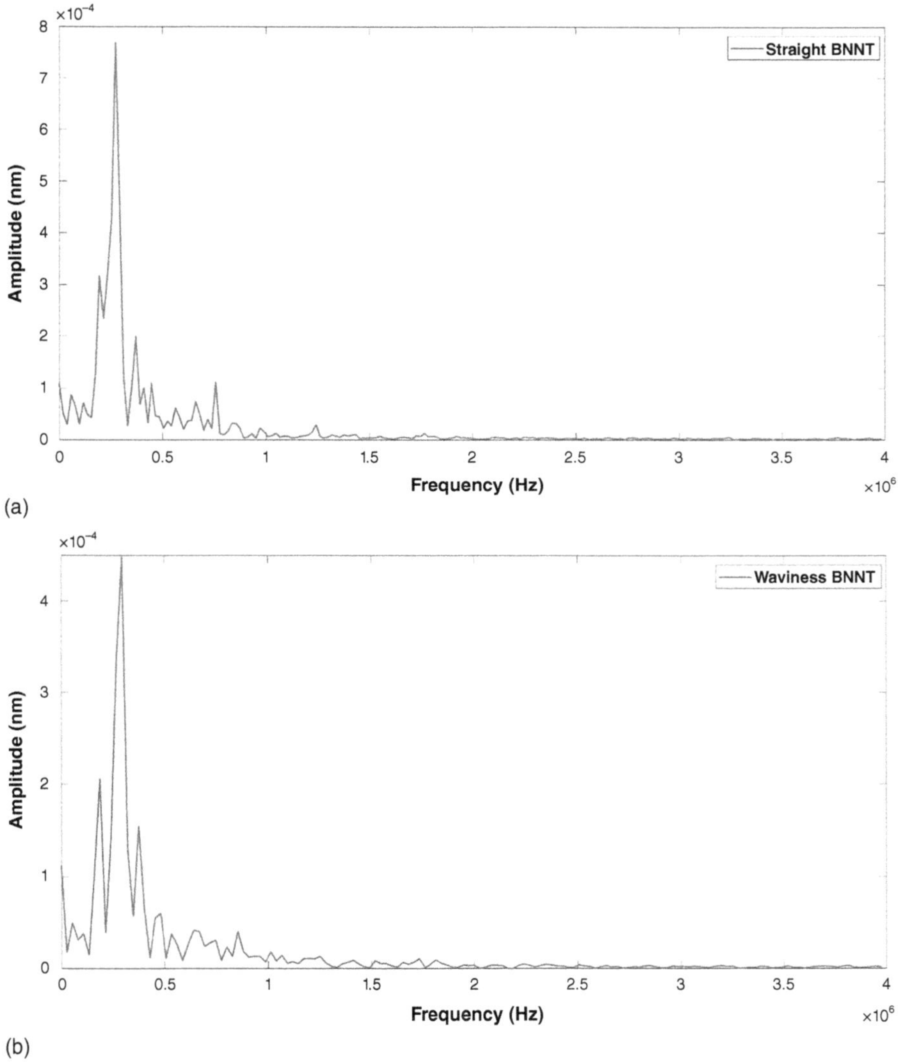

FIGURE 14.3 Dynamic response: FFT for L = 40 nm. (a) Straight BNNT and (b) eccentric (waviness = 0.05) BNNT.

and the response becomes chaotic or quasi-periodic depending on the level of waviness. This resonant behavior indicates the sensitivity of BNNT-based sensors, as even small changes in mass produce measurable shifts in frequency.

The finite element modeling (FEM) approach simulates the impact of surface waviness and nonlinear dynamics on the efficacy of BNNT-based mass sensors. The model enhances BNNT applications for precise sensing by simulating real-world conditions by incorporating waviness factors, nonlinear coefficients, and damping parameters. Altering the dimensions of BNNT alongside its waviness attributes enhances sensor sensitivity, enabling the detection of minimal mass fluctuations essential for environmental monitoring and quality evaluation in industrial production.

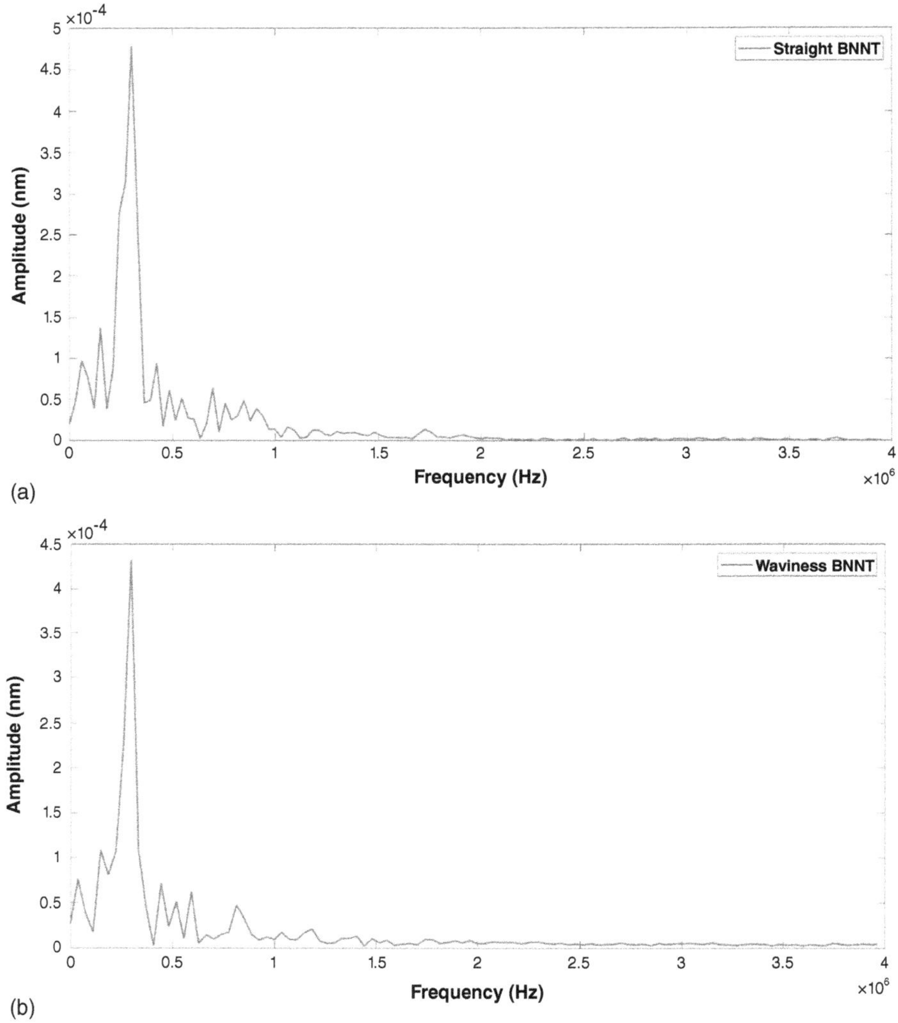

FIGURE 14.4 Dynamic response: FFT for L = 60 nm. (a) Straight BNNT and (b) eccentric (waviness = 0.05) BNNT.

14.4.2 Impact of Waviness on Frequency Response

The analysis showed that larger surface waviness factors among BNNTs generated significant frequency changes. The research examined patterned BNNTs, which were 20 nm and 60 nm long, for identifying frequency changes from 0.05 to 0.1 waviness factors. Wavy BNNTs demonstrated higher resonant frequencies than straight BNNTs, and larger waviness factors caused additional changes to these resonant frequencies. The data presented in Table 14.2, along with accompanying figures, demonstrates the greater sensitivity achieved by raising the waviness factor of BNNTs across various lengths.

TABLE 14.2
Resonant Frequencies of Different Length BNNT for Different Waviness Factors

Waviness Factor (h/L)	Resonant Frequency (L = 20 nm)	Resonant Frequency (L = 60 nm)
0	3.71E+05 Hz	1.24E+05 Hz
0.05	4.32E+05 Hz	2.54E+05 Hz
0.075	4.99E+05 Hz	3.55E+05 Hz
0.1	5.78E+05 Hz	4.60E+05 Hz

Table 14.2 and the following Figures 14.5 and 14.6 show the relationship between the waviness factor and resonant frequency for BNNTs of varying lengths. Figures demonstrate how increased waviness enhances the resonant frequency, particularly in shorter BNNTs.

14.4.3 Sensitivity Enhancement Through Nonlinear Dynamics

The study investigated the effects of quadratic and cubic nonlinearities induced by surface waviness which is tabulated in Table 14.3. These nonlinearities, which result from geometric deviations and mid-plane stretching, significantly impact the sensitivity of BNNT-based mass sensors. For example, quadratic nonlinearity was found to increase with waviness factor, reaching an optimal point before stabilizing. Cubic nonlinearity, however, showed an initial increase with waviness before decreasing, suggesting a point of diminishing returns for sensitivity enhancement.

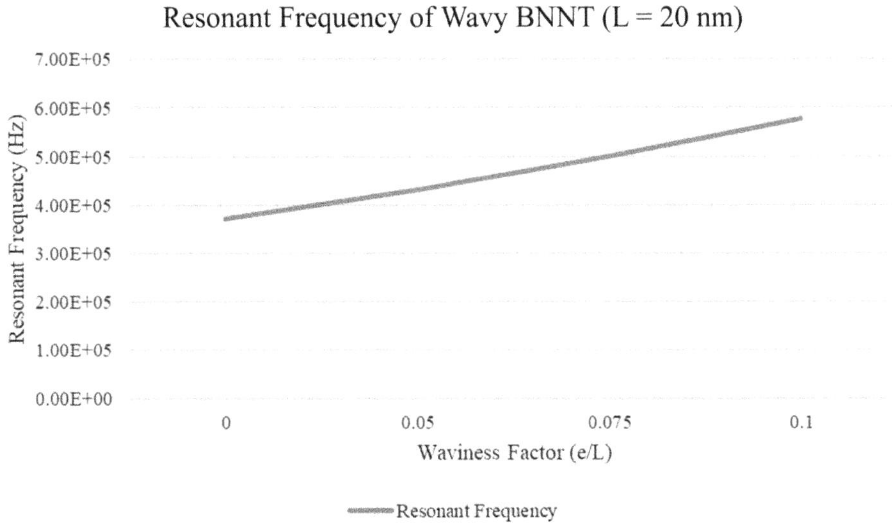

FIGURE 14.5 Shift in resonant frequency of wavy BNNT (L = 20 nm).

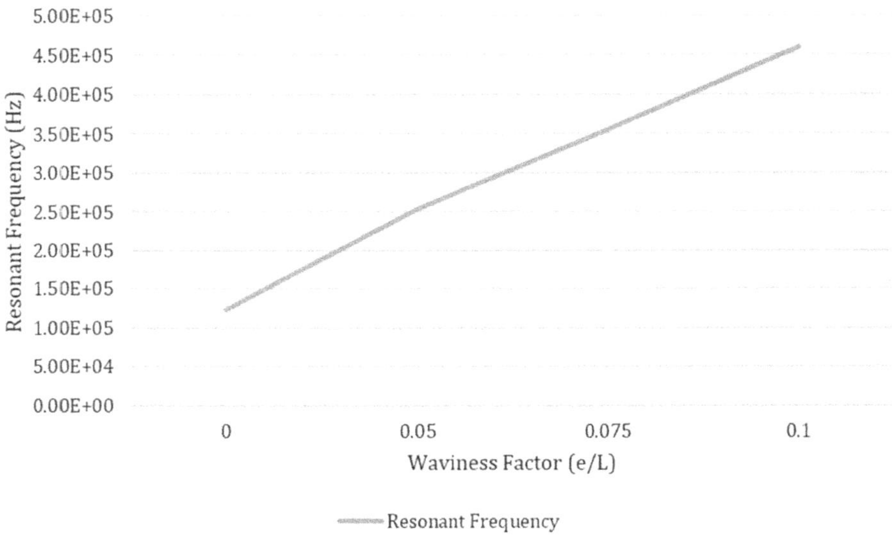

FIGURE 14.6 Shift in resonant frequency of wavy BNNT (L = 60 nm).

TABLE 14.3
Relationship Between Waviness and Quadratic and Cubic Nonlinearity

Waviness Factor (h/L)	Quadratic Nonlinearity	Cubic Nonlinearity
0.05	Increases and stabilizes	Increases slightly
0.075	Peaks	Starts to decline
0.1	Declines	Further decline

14.5 APPLICATIONS IN SMART MANUFACTURING

Smart manufacturing applications benefit greatly from BNNT-based mass sensors because these sensors offer exceptional sensitivity, chemical stability, and extreme condition resistance. Reliable real-time sensing technologies, which operate continuously while providing high-precision measurements, have become priority components for manufacturing processes that gain complexity from data-driven operations. The specifications required by smart manufacturing applications are perfectly matched by BNNT-based sensors, which makes them ideal solutions for multiple critical usage areas. Through their exceptional properties, BNNTs enable quality control management as well as environmental safety checks and the integration of IoT networks for advanced automation.

14.5.1 Mass Sensing for Quality Control

Quality control plays a vital role in manufacturing operations because it upholds product uniformity and binds business standards alongside guaranteeing safe customer products. The precise measurement capabilities of BNNT-based sensors enable their use in spotting minute changes in mass, hence identifying small elements that endanger product quality or operational ability. The application finds its most applicable use in industries governed by rigorous standards, especially pharmaceuticals and both semiconductors and food processing, that must adhere to strict regulations for product safety and effectiveness (Vatanpour et al. 2021; Kumar et al. 2022).

Microscopic contamination detection can be done through BNNTs because these nanotubes respond noticeably to variations in mass. The pharmaceutical industry uses BNNT sensors to assess raw materials as well as finished products, thus preventing dangerous contaminants from entering the supply chain. BNNT sensors in the electronics business protect semiconductor wafers from performance-damaging particles. Manufacturers can utilize BNNT-based mass sensors for ongoing process oversight, which reveals product defects immediately, thus decreasing operational losses while improving the reliability of their products.

14.5.2 Environmental Monitoring in Manufacturing

Current manufacturing operations focus heavily on environmental safety and compliance since regulatory emission and pollutant controls become more complex every year. The mass sensors using BNNT operate most effectively for environmental surveys in manufacturing plants because they effectively track pollutants along with dangerous chemicals and small airborne objects. The functionality of these sensors proves essential for manufacturing facilities which produce emissions in addition to handling volatile chemicals including automotive workshops and chemical manufacturing facilities as well as heavy industry sectors.

BNNTs remain operational under corrosive or reactive conditions thanks to their chemical inertness combined with their stability properties. Once incorporated into monitoring systems, BNNT sensors operate continuously to detect harmful gas concentrations and particle amounts; thus, operators gain early warning about impending dangers which prevent health problems and regulatory noncompliance. Production lines in the automotive industry would benefit from automotive BNNT sensors which protect air quality standards by monitoring particulate emissions. These sensors function optimally in metalworking and chemical processing plants because their high temperature tolerance makes them effective in hot environments (Li et al. 2022; Tiano et al. 2014).

Use of BNNT sensors in environmental monitoring helps manufacturers implement eco-friendly production systems. The sensors provide instant information about pollutants which helps manufacturers enhance their operations to decrease environmental consequences and save energy while minimizing waste. The need to track environmental impact alongside consumers and governments increases the value of these applications during sustainable production initiatives.

14.5.3 Potential in IoT-Enabled Smart Manufacturing

The adoption of IoT technology within manufacturing systems forms a main architectural element of the Industry 4.0 movement defined by automated production spaces that use real-time data for interconnection. The application of BNNT-based sensors reflects their suitability for IoT-enabled operations because of their adaptability to real-time monitoring as well as their stability and robust sensing capabilities (Panicker, Rajeev, and Thomas 2023).

A system based on IoT would let manufacturers utilize BNNT-based sensors to observe material quality metrics and machine performance levels in addition to environmental conditions. The sensors transmit operational data to centralized control systems for cloud-based platforms that allow users to analyze the information to detect anomalies and predict maintenance requirements. The sensors based on BNNTs serve as device-wide monitors by tracking dust buildup and other impurities on essential machine components to notify maintenance teams before operational failure occurs (Schulz, Kelkar, and Sundaresan 2005).

Moreover, BNNT-based sensors offer significant advantages in IoT networks owing to their durability and minimal maintenance needs. Their exceptional chemical stability and thermal resistance enable reliable operation in demanding environments, minimizing the necessity for frequent replacement or recalibration. The enduring reliability of BNNT sensors renders them cost-effective for manufacturers aiming to establish comprehensive sensor networks for condition monitoring and predictive maintenance throughout their facilities.

The incorporation of BNNT-based sensors into IoT systems facilitates remote monitoring, advantageous for multi-site manufacturing operations. Remote access to real-time data from BNNT sensors enables companies to oversee conditions at various locations from a centralized hub, facilitating a coordinated and efficient strategy for quality control, environmental monitoring, and process optimization. In this context, BNNT-based sensors facilitate smart manufacturing by promoting responsive, data-driven decision-making that increases efficiency, minimizes downtime, and enhances product quality (Jakubinek et al. 2019).

14.5.4 Advanced Process Control and Automation

As manufacturing advances in automation, the demand for sensors capable of delivering accurate, continuous feedback increases. BNNT-based mass sensors provide a benefit in automated process control through detecting critical variables and transmitting real-time feedback to automated systems that modify process parameters accordingly. This application is especially advantageous in high-precision sectors like aerospace and medical device manufacturing, where automated production lines must uphold stringent control over product specifications (Jalili 2010).

In process control, BNNT sensors can identify alterations in material composition or assess the thickness and weight of products as they traverse the production line. In the aerospace sector, BNNT-based sensors can assess the uniformity of composite materials, essential for structural integrity. BNNT sensors facilitate automated systems to make precise adjustments to material formulations or machine settings by supplying real-time data, thereby ensuring that each component adheres to exact specifications without the need for manual intervention.

Furthermore, BNNT-based sensors facilitate adaptive manufacturing processes, enabling production systems to dynamically adjust to fluctuations in material properties or environmental conditions. This adaptability is essential for maintaining product quality in scenarios where minor variations in raw material quality or temperature may affect the final product. Through the continuous monitoring of these variables,

BNNT-based sensors facilitate real-time responses in manufacturing systems, thereby improving efficiency and minimizing the probability of defects (Singh and Melnik 2022).

14.5.5 Health and Safety Monitoring

Prioritizing the health and safety of workers is essential in manufacturing, particularly in environments where employees might encounter hazardous substances or severe conditions. BNNT-based sensors, capable of detecting trace levels of hazardous substances, can significantly contribute to health and safety monitoring. BNNT sensors can be utilized to assess air quality in manufacturing facilities, identifying toxic gases, dust, or other deleterious particles (Barhoum et al. 2022).

The robustness of BNNTs against chemical and thermal stress makes these sensors suitable for use in inherently challenging environments, such as chemical processing facilities or metal fabrication workshops. In these environments, real-time data from BNNT sensors can notify safety personnel of increasing concentrations of hazardous substances, facilitating prompt interventions to safeguard workers. This capability is becoming increasingly vital in industries that emphasize occupational health and safety to adhere to regulatory standards (Lou, Chen, and Fan 2021).

BNNT-based sensors enhance safety monitoring in real time, thereby promoting a safer and more responsive work environment. This proactive strategy for worker safety corresponds with the objectives of smart manufacturing, wherein technology is employed not solely to enhance processes but also to establish safer and more sustainable work environments.

14.6 CHALLENGES AND FUTURE SCOPE

BNNTs possess significant potential for smart manufacturing applications; however, numerous challenges must be resolved to facilitate their widespread implementation. The main challenges consist of intricacies in BNNT synthesis, obstacles in large-scale production, constraints in functionalization, and integration issues with current manufacturing systems. Resolving these issues will be essential for the advancement of BNNT-based sensors and the actualization of their complete potential in industrial applications.

14.6.1 Challenges in BNNT Synthesis and Large-Scale Production

The synthesis of BNNTs is a complex and resource-demanding procedure necessitating meticulous regulation of reaction conditions, selection of catalysts, and operational parameters. Presently, techniques like chemical vapor deposition (CVD), arc discharge, and laser ablation are employed to synthesize BNNTs; however, each method exhibits constraints regarding scalability, cost, and product quality. Attaining elevated yields with uniform nanotube quality is challenging, as the synthesis process frequently produces BNNTs with disparate lengths, diameters, and surface properties. These discrepancies can impact the efficacy of BNNT-based sensors, especially in applications necessitating elevated sensitivity and precision (Radhakrishnan et al. 2024).

The large-scale production of BNNTs necessitates the resolution of issues pertaining to purity and defect management. Impurities introduced during synthesis can affect the material's stability and sensitivity, particularly in sensor applications where purity is essential. Present synthesis techniques face challenges in reliably generating high-purity BNNTs with few defects, thereby constraining their efficacy in precision applications. Establishing dependable and scalable techniques for the industrial production of defect-free, high-purity BNNTs will be crucial for their incorporation into conventional manufacturing.

14.6.1.1 Functionalization Challenges

Functionalization refers to the modification of BNNTs to improve their solubility, biocompatibility, or reactivity with particular substances. The inherent chemical stability of BNNTs is beneficial for durability, yet it complicates their functionalization. Functionalization is essential to customize BNNTs for particular sensing applications, including the attachment of receptors for chemical detection or the modification of surfaces to enhance sensitivity to environmental factors. Nonetheless, the functionalization of BNNTs frequently necessitates intricate procedures that may compromise their structural integrity, and achieving uniform functionalization across extensive batches of BNNTs continues to pose a challenge (Xu et al. 2022).

To improve the utility of BNNTs in sensing, it is crucial to develop functionalization techniques that can be applied consistently and reliably at scale. Moreover, these techniques must preserve the structural and functional characteristics of BNNTs, guaranteeing that functionalized BNNTs maintain their elevated sensitivity and stability. Future research should concentrate on determining effective chemical or physical modifications that improve BNNT reactivity and compatibility with various sensing targets while preserving their distinctive properties (Liu et al. 2021).

14.6.1.2 Integration into Manufacturing Systems

The implementation of BNNT-based sensors into current manufacturing systems presents additional challenges. Manufacturing systems are generally optimized for compatibility with existing sensing technologies, necessitating modifications in system design and protocols to integrate a new sensor material such as BNNT. Ensuring effective interfacing of BNNT-based sensors with data collection, processing, and feedback systems in an automated environment necessitates specialized calibration, software adaptation, and compatibility testing (Zhao et al. 2025).

Also, BNNTs must be designed to endure the mechanical, thermal, and chemical stresses encountered in various manufacturing settings. In numerous instances, this necessitates the creation of protective coatings or encapsulation techniques to maintain the functionality of BNNT sensors while ensuring their sensitivity remains intact. Adapting BNNT-based sensors for operation in extreme conditions while preserving accuracy and durability presents a challenge that must be resolved to guarantee seamless integration into industrial environments.

14.6.2 Future Scope

With these challenges, BNNTs present significant opportunities for advancement and innovation in sensor technology. Ongoing research aims to surmount these challenges, and advancements in BNNT synthesis, functionalization, and integration techniques will facilitate their broader utilization in smart manufacturing. The prospective applications for BNNTs encompass various promising domains:

14.6.2.1 Optimizing BNNT Synthesis for Scalability

Enhancing synthesis techniques to facilitate large-scale production while maintaining quality is essential for the adoption of BNNT. Advancements in catalyst selection, reaction management, and purification techniques are anticipated to render BNNT production more cost-effective and scalable. Advancing synthesis techniques to yield BNNTs with regulated length, diameter, and purity will facilitate wider industrial applications.

14.6.2.2 Developing Targeted Functionalization Techniques

Future research should concentrate on developing functionalization techniques customized for particular applications, including chemical detection and environmental monitoring. These methods are intended to augment the sensitivity of BNNTs to specific analytes, enabling BNNT-based sensors to identify a diverse array of substances beyond mass, including temperature, pressure, and humidity. By enhancing the sensing capabilities of BNNTs, functionalized BNNTs can evolve into multi-functional sensors capable of concurrently monitoring diverse environmental factors.

14.6.2.3 Multi-Functional Sensing Applications

The applicability of BNNTs transcends mass sensing, as their distinctive characteristics render them appropriate for detecting additional parameters, including temperature, pressure, and humidity. Utilizing the chemical and physical adaptability of BNNTs, future sensors may incorporate multiple functionalities, resulting in versatile devices for extensive environmental monitoring. A solitary BNNT sensor could yield concurrent measurements of temperature, humidity, and mass, proving indispensable in sectors such as aerospace, automotive, and environmental science.

14.6.2.4 Enhancing Durability and Resilience

Further research is required to enhance the durability and resilience of BNNT-based sensors as they are integrated into challenging environments. Advancements in protective coatings, structural improvements, and thermal regulation will enable BNNT sensors to preserve functionality and sensitivity under extreme conditions. These enhancements will facilitate the implementation of BNNT sensors in sectors such as energy, healthcare, and defense, where dependable performance in adverse conditions is critical.

14.6.2.5 Exploring Advanced IoT and Data Analytics Integration

The advancement of IoT and data analytics in manufacturing presents an opportunity to create BNNT-based sensors that integrate seamlessly into IoT networks. This integration would facilitate remote, automated monitoring and real-time data analysis, thereby enabling predictive maintenance, quality control, and environmental management. The exceptional sensitivity of BNNT sensors, coupled with IoT technology, has the potential to transform monitoring and decision-making in smart manufacturing by delivering manufacturers accurate, actionable insights.

14.7 CONCLUSION

This investigation into BNNT-based mass sensors proved the outstanding application potential of BNNTs for smart manufacturing technologies. BNNTs emerge as a promising substitute to other nanomaterials including CNTs because they exhibit thermal stability and tensile strength together with insulating properties and chemical resistance. FEA and simulation results from the study showed that the sensor precision and sensitivity increase due to surface waviness combinations with natural quadratic and cubic effects in BNNT-based sensors. The research demonstrates BNNTs provide exceptional performance for detecting small mass alterations which constitutes a vital feature for quality control systems, environmental surveillance, and automated manufacturing process feedback systems.

The implementation of BNNT-based mass sensors proves beneficial in smart manufacturing because it helps improve manufacturing processes through enhanced efficiency and precision along with adaptability capabilities. These devices reveal product contaminations and structural changes that defend product quality standards and monitor environmental pollutants effectively. Their durability and minimal maintenance needs together mark them as perfect candidates for IoT network integration which brings Industry 4.0-level continuous remote automatic monitoring systems.

Large-scale synthesis challenges along with functionalization methods and integration systems between BNNT sensors and current manufacturing approaches limit their widespread use. Upcoming research should concentrate on enhancing synthesis strategies for scaling operations, creating specific functionalization protocols, and expanding sensor functionality to measure both mass and additional factors like environmental conditions such as temperature, pressure, and humidity. Future development of BNNT sensors requires addressing manufacturing hurdles which will enable them to serve as foundational technology in the upcoming age of smart production.

Mass sensors developed from BNNTs show great potential to improve manufacturing smart systems' measurement accuracy and system reliability together with their operational capabilities. BNNTs demonstrate nonlinear dynamics, which makes them strategically valuable for detecting subtle masses, thus establishing them as crucial components for advanced manufacturing. Progress in both BNNT manufacturing techniques and utilization will probably discover new capabilities that will help manufacturing obtain greater accuracy alongside adaptation and sustainability.

REFERENCES

Ali, Fayaz. 2022. "Boron materials for energy applications." En *Fundamentals and Applications of Boron Chemistry*, 203–89. Elsevier. https://doi.org/10.1016/B978-0-12-822127-3.00004-1

Aligayev, Amil, Fazal Raziq, Ulkar Jabbarli, Nurlan Rzayev, and Liang Qiao. 2022. "Morphology and topography of nanotubes." *En Graphene, Nanotubes and Quantum Dots-Based Nanotechnology*, 355–420. Elsevier. https://doi.org/10.1016/B978-0-323-85457-3.00019-0

Bae, Dongsu, Kun-Hong Lee, and Myung Jong Kim. 2024. "Growth methodologies of boron nitride nanotubes and their neutron shielding applications: A review." *Nanoscale* 16 (8): 3817–37. https://doi.org/10.1039/D3NR06070E

Barhoum, Ahmed, María Luisa García-Betancourt, Jaison Jeevanandam, Eman A. Hussien, Sara A. Mekkawy, Menna Mostafa, Mohamed M. Omran, Mohga S. Abdalla, and Mikhael Bechelany. 2022. "Review on natural, incidental, bioinspired, and engineered nanomaterials: History, definitions, classifications, synthesis, properties, market, toxicities, risks, and regulations." *Nanomaterials* 12 (2): 177. https://doi.org/10.3390/nano12020177

Capek, Ignác. 2009. "Dispersions, novel nanomaterial sensors and nanoconjugates based on carbon nanotubes." *Advances in Colloid and Interface Science* 150 (2): 63–89. https://doi.org/10.1016/j.cis.2009.05.006

Clegg, Brian. 2018. *The Graphene Revolution The Weird Science of the Ultra-thin*. Icon Books.

Deshwal, Dinesh, and Anil Kumar Narwal. 2021. "An extensive review of nanotubes-based mass sensors." *Journal of Micro and Nano-Manufacturing* 9 (2). https://doi.org/10.1115/1.4051261

Deshwal, Dinesh, and Anil Kumar Narwal. 2023a. "Resonant frequency analysis of wavy boron nitride nanotube based bio-mass sensor." *International Journal on Interactive Design and Manufacturing (IJIDeM)*. https://doi.org/10.1007/s12008-023-01300-0

Deshwal, Dinesh, and Anil Kumar Narwal. 2023b. "Analysis of surface deviation impact on bio-mass sensing application of Boron Nitride Nanotubes." *Results in Engineering* 19 (setembre): 101282. https://doi.org/10.1016/j.rineng.2023.101282

Ertekin, Nilüfer. 2024. "Nano-electro-mechanical conduct of boron nitride nanotube as piezo-electric nanogenerators and nanoswitches." *Smart Materials and Structures* 33 (2): 025037. https://doi.org/10.1088/1361-665X/ad1fa8

Golberg, D., Y. Bando, C. C. Tang, and C. Y. Zhi. 2007. "Boron nitride nanotubes." *Advanced Materials* 19 (18): 2413–32. https://doi.org/10.1002/adma.200700179

Hong, Yang, Zhanxiao Kang, and Jintu Fan. 2024. "Revolutionizing textile: Advanced techniques for superior thermal conductivity." *Composites Part A: Applied Science and Manufacturing* 186 (novembre): 108380. https://doi.org/10.1016/j.compositesa.2024.108380

Jakubinek, Michael B., Behnam Ashrafi, Yadienka Martinez-Rubi, Jingwen Guan, Meysam Rahmat, Keun Su Kim, Stéphane Dénommée, Christopher T. Kingston, and Benoit Simard. 2019. "Boron nitride nanotube composites and applications." En *Nanotube Superfiber Materials*, 91–111. Elsevier. https://doi.org/10.1016/B978-0-12-812667-7.00005-7

Jalili, Nader. 2010. "Nanomaterial-based piezoelectric actuators and sensors." En *Piezoelectric-Based Vibration Control*, 419–61. Boston, MA: Springer US. https://doi.org/10.1007/978-1-4419-0070-8_12

Jing, Lin. 2018. *Novel composite architectures based on low dimensional C/BN with enhanced mechanical and thermal responses for sports applications*. Nanyang Technological University. https://doi.org/10.32657/10220/47389

Kumar, Sanjeev, Ritika Sharma, Bhawna, Akanksha Gupta, Prashant Singh, Susheel Kalia, Pankaj Thakur, and Vinod Kumar. 2022. "Prospects of biosensors based on functionalized and nanostructured solitary materials: Detection of viral infections and other risks." *ACS Omega* 7 (26): 22073–88. https://doi.org/10.1021/acsomega.2c01033

Li, Mengna, Gordon Huang, Xiujuan Chen, Jianan Yin, Peng Zhang, Yao Yao, Jian Shen, Yuwei Wu, and Jing Huang. 2022. "Perspectives on environmental applications of hexagonal boron nitride nanomaterials." *Nano Today* 44 (juny): 101486. https://doi.org/10.1016/j.nantod.2022.101486

Liu, Zhen, Alireza Dibaji, Da Li, Srikanth Mateti, Jingquan Liu, Fuhua Yan, Colin J. Barrow, Ying Chen, Katsuhiko Ariga, and Wenrong Yang. 2021. "Challenges and solutions in surface engineering and assembly of boron nitride nanosheets." *Materials Today* 44 (abril): 194–210. https://doi.org/10.1016/j.mattod.2020.11.020

Lou, Lun, Kaikai Chen, and Jintu Fan. 2021. "Advanced materials for personal thermal and moisture management of health care workers wearing PPE." *Materials Science and Engineering: R: Reports* 146 (octubre): 100639. https://doi.org/10.1016/j.mser.2021.100639

Mane, Sunil Kumar Baburao Naghma, Shaishta, and G. Manjunatha. 2024. "Biocompatibility, toxicity evaluations, environmental and health impact of hexagonal boron nitride." En *Hexagonal Boron Nitride*, 613–36. Elsevier. https://doi.org/10.1016/B978-0-443-18843-5.00002-1

Mohammed, Sohail, Pranjal Nautiyal, Tanaji Paul, Cheng Zhang, and Arvind Agarwal. 2024. "A critique on boron nitride nanotube reinforced metal matrix composites." *Critical Reviews in Solid State and Materials Sciences* 49 (3): 482–519. https://doi.org/10.1080/10408436.2023.2229867

Panicker, Swathy S., Sreenidhi Prabha Rajeev, and Vinoy Thomas. 2023. "Impact of PVDF and its copolymer-based nanocomposites for flexible and wearable energy harvesters." *Nano-Structures & Nano-Objects* 34 (abril): 100949. https://doi.org/10.1016/j.nanoso.2023.100949

Radhakrishnan, Sithara, Abhinandan Patra, G. Manasa, Mohammed Arkham Belgami, Sang Mun Jeong, and Chandra Sekhar Rout. 2024. "Borocarbonitride-based emerging materials for supercapacitor applications: Recent advances, challenges, and future perspectives." *Advanced Science* 11 (4). https://doi.org/10.1002/advs.202305325

Roy, Soumyabrata, Xiang Zhang, Anand B. Puthirath, Ashokkumar Meiyazhagan, Sohini Bhattacharyya, Muhammad M. Rahman, Ganguli Babu, et al. 2021. "Structure, properties and applications of two-dimensional hexagonal boron nitride." *Advanced Materials* 33 (44). https://doi.org/10.1002/adma.202101589

Schulz, Mark J., Ajit D. Kelkar, and Mannur J. Sundaresan, ed. 2005. *Nanoengineering of Structural, Functional and Smart Materials*. CRC Press. https://doi.org/10.1201/9780203491966

Sharma, Harsh, Sharad Shrivastava, Jitendra Singh Rathore, and Sandesh Trivedi. 2024. "Nonlinear vibration analysis of curvy single-walled boron nitride nanotube using mathematical modeling for dynamic responses." *International Journal of Modern Physics B* 38 (05). https://doi.org/10.1142/S0217979224500735

Shtansky, Dmitry V., Andrei T. Matveev, Elizaveta S. Permyakova, Denis V. Leybo, Anton S. Konopatsky, and Pavel B. Sorokin. 2022. "Recent progress in fabrication and application of BN nanostructures and BN-based nanohybrids." *Nanomaterials* 12 (16): 2810. https://doi.org/10.3390/nano12162810

Singh, Sundeep, and Roderick Melnik. 2022. "Coupled multiphysics modelling of sensors for chemical, biomedical, and environmental applications with focus on smart materials and low-dimensional nanostructures." *Chemosensors* 10 (5): 157. https://doi.org/10.3390/chemosensors10050157

Sohrabi, Hessamaddin, Omid Arbabzadeh, Mahdi Falaki, Vahid Vatanpour, Mir Reza Majidi, Nurbolat Kudaibergenov, Sang Woo Joo, and Alireza Khataee. 2023. "Advances in fabrication, physio-chemical properties, and sensing applications of non-metal boron nitride and boron carbon nitride-based nanomaterials." *Surfaces and Interfaces* 41 (octubre): 103152. https://doi.org/10.1016/j.surfin.2023.103152

Tiano, Amanda L., Cheol Park, Joseph W. Lee, Hoa H. Luong, Luke J. Gibbons, Sang-Hyon Chu, Samantha Applin, et al. 2014. "Boron nitride nanotube: Synthesis and applications." En , editat per Vijay K. Varadan, 906006. https://doi.org/10.1117/12.2045396

Turhan, Emine Ayşe, Ahmet Engin Pazarçeviren, Zafer Evis, and Ayşen Tezcaner. 2022. "Properties and applications of boron nitride nanotubes." *Nanotechnology* 33 (24): 242001. https://doi.org/10.1088/1361-6528/ac5839

Vaccarini, Laetitia, Christophe Goze, Luc Henrard, E. Hernández, Patrick Bernier, and Angel Rubio. 2000. "Mechanical and electronic properties of carbon and boron–nitride nanotubes." *Carbon* 38 (11–12): 1681–90. https://doi.org/10.1016/S0008-6223(99)00293-6

Vatanpour, Vahid, Seyed Ali Naziri Mehrabani, Basak Keskin, Negar Arabi, Bihter Zeytuncu, and Ismail Koyuncu. 2021. "A comprehensive review on the applications of boron nitride nanomaterials in membrane fabrication and modification." *Industrial & Engineering Chemistry Research* 60 (37): 13391–424. https://doi.org/10.1021/acs.iecr.1c02102

Wong, Tan Lo, Cristina Vallés, Adel Nasser, and Chamil Abeykoon. 2023. "Effects of boron-nitride-based nanomaterials on the thermal properties of composite organic phase change materials: A state-of-the-art review." *Renewable and Sustainable Energy Reviews* 187 (novembre): 113730. https://doi.org/10.1016/j.rser.2023.113730

Xu, Tao, Kai Zhang, Qiran Cai, Nanyang Wang, Liyun Wu, Qian He, Han Wang, et al. 2022. "Advances in synthesis and applications of boron nitride nanotubes: A review." *Chemical Engineering Journal* 431 (març): 134118. https://doi.org/10.1016/j.cej.2021.134118

Yang, Cui-Cui, Xu Duan, Li Li, Xue-Lian Zheng, Jiu Chen, Wei Quan Tian, Wei-Qi Li, and Ling Yang. 2023. "BN-doped carbon nanotubes and nanoribbons as nonlinear-optical functional materials for application in second-order nonlinear optics." *ACS Applied Nano Materials* 6 (3): 1549–61. https://doi.org/10.1021/acsanm.2c04068

Yin, Jun, Jidong Li, Yang Hang, Jin Yu, Guoan Tai, Xuemei Li, Zhuhua Zhang, and Wanlin Guo. 2016. "Boron nitride nanostructures: Fabrication, functionalization and applications." *Small* 12 (22): 2942–68. https://doi.org/10.1002/smll.201600053

Zhang, Jie, Shibo Ye, Honglei Liu, Xiaoliang Chen, Xiaoming Chen, Baotong Li, Wanhong Tang, et al. 2020. "3D printed piezoelectric BNNTs nanocomposites with tunable interface and microarchitectures for self-powered conformal sensors." *Nano Energy* 77 (novembre): 105300. https://doi.org/10.1016/j.nanoen.2020.105300

Zhao, Hongfa, Liguang Ruan, Zihan Wang, Mingrui Shu, Chuqiao Lyu, Bulathsinghala Rameesh Lakshan, Dharmasena Ishara, Changsheng Wu, and Wenbo Ding. 2025. "Flexible nanogenerators for intelligent robotics: Design, manufacturing, and applications." *International Journal of Extreme Manufacturing* 7 (2): 022012. https://doi.org/10.1088/2631-7990/ad94b8

Zhi, Chunyi, Yoshio Bando, Chengchun Tang, and Dmitri Golberg. 2010. "Boron nitride nanotubes." *Materials Science and Engineering: R: Reports* 70 (3–6): 92–111. https://doi.org/10.1016/j.mser.2010.06.004

15 Nanotube-Based Sensors and Actuators for Smart Manufacturing Systems

Monika, Boussehel Hamida and George Jacob

15.1 INTRODUCTION

The convergence of digital technologies, advanced materials, and intelligent systems is causing a paradigm shift in the manufacturing industry, which is aptly described as a revolution encompassed within the umbrella term Industry 4.0. Such a fourth industrial revolution focuses on the intelligent manufacturing environments where data sharing, execution of processes in real time, autonomous decision-making, and adaptive control are key. The critical importance of sensing and actuation technologies comes up as the global economy is becoming more dependent on the high precision, customized, and efficient processes. These elements are the building blocks of perception and control in cyber-physical systems (CPS) as the hallmarks of the new concept of smart factories.

History of industrial development can be categorized into the following four general revolutions. The former pioneered water- and steam-powered mechanized production, whereas the Second Industrial Revolution capitalized on the use of electricity and assembly line production. The third revolution starting in the late 20th century introduced computers and automation and programmable logic controllers (PLCs) into the factories. Nevertheless, the fourth and the most recent revolution, Industry 4.0, is a big step as it incorporates high-level information technology, artificial intelligence (AI), big data analysis, cloud computing, and interconnectivity in real time across all manufacturing layers.

The main customer target in this smart transformation revolves around the idea of CPS, in which digital systems strive to monitor and control the physical assets (e.g., machines, materials, products) by using sensors and actuators. The data backbone is the physical variables of the sensor which include force, pressure, temperature, position, and chemical composition. On the other hand, actuators convert digital inputs into physical motion, which allows robots movement, handling of materials, temperature control, and adjustments to the system.

Performance of an embedded sensing and actuation mechanism, therefore, has a direct proportion to the effectiveness of an innovative manufacturing system. It is common to find that traditional equipment cannot handle the requirements of contemporary factories, which are: complex (nonlinear process dynamics),

DOI: 10.1201/9781003560234-19

high-speed production cycles, miniaturized systems, microscale assembly, real-time reconfigurability, and harsh environments (high-temperature, corrosive conditions). With the rise in the number of data points and control actions, there is a rising demand for compact, multipurpose, and ultrasensitive sensing or actuation elements. This is where nanomaterials, more specifically, nanotubes, come into the limelight.

Linear actuators and sensors, which have traditionally been bulk material (e.g., piezoelectric stack) based and macro-electromechanical in their operating principle, have been the mainstay of automation over a long historical period. These are the strain gauge, thermocouple, piezoelectric ceramics, electromagnetic actuator, and hydraulic systems. But as industries shift to hyper-connected, responsive, and autonomous systems, their weaknesses become ever more evident. Some of the weaknesses of the conventional devices are—*Size and scaling down limitations*: The majority of the conventional elements cannot be simply miniaturized right down to a point at which sensitivity or mechanical integrity is not lost. This inhibits their incorporation into such compact systems as microelectromechanical systems (MEMS), soft robotics, or wearable industrial interfaces. *Restricted sensitivity and swiftness*: Very high resolutions or small strain, pressure, or temperature changes may be beyond the sensitivity of conventional material. Most of them need large signal amplification or conditioning circuits in order to work properly. *Material fatigue and degradation*: At cycling load, vibrations, or extreme temperature, conventional sensors tend to degrade with fatigue, causing signal DC drift, hysteresis, or actual failure. This impacts long-term reliability and consistency of automation. *Single functionality*: Majority of the traditional devices constitute single functionality, whether sensing or actuation. This constrains the flexibility and compactness needed in dynamic settings where there is a need to ensure that space and functionality are exploited. *Internal power supply deficiency*: Self-sustaining capability is also a limitation as far as power consumption is concerned, in wireless sensor networks and remote monitoring tasks. Conventional appliances tend to use an outside source of energy and are inexperienced with energy harvesting.

The combination of these limitations restricts the scalability, flexibility, and responsiveness of traditional devices in the contemporary manufacturing systems, which are data-intensive and decentralized. This therefore demands a timely approach in finding new materials and architectures that would fill such lacunas and at the same time allow new functionalities.

15.1.1 Motivation for Adopting Nanotube-Based Devices in CPS

Carbon nanotubes (CNTs) and boron nitride nanotubes (BNNTs) are one of the most promising nanomaterials to be used in sensing and actuation in intelligent manufacturing. Such 1D nanostructures, having both the diameter commonly being of 1–100 nanometers and lengths of several micrometers, have outstanding electrical, mechanical, and thermal properties. Their fine-grained atomistic design and programmable properties make them very convenient to be utilized in the next-generation CPS.

15.1.1.1 Exceptional Mechanical and Electrical Properties

CNTs have a very high tensile strength (~100 GPa), low density, and great electrical conductivity. These faculties enable them to be reinforcing in the structural arrangements and active electrical elements. The BNNTs are electrically insulating; however, they show good thermal stability and piezoelectricity and hence are very useful sensors in hostile environments. This also means they are very sensitive to deformation or any kind of environmental variations that can allow them to sense fine mechanical stimuli like micro-vibrations, any cracks on the surface, or wear (Deshwal and Narwal 2025).

15.1.1.2 Multifunctionality in a Single Material

As opposed to traditional elements, nanotubes are able to do all of the sensing, actuation, and in many cases, energy harvesting. As a representation, the CNTs have the capabilities of serving as strain sensors, field-effect transistors (FETs), and thermal actuators based on their alignment, functionalization, and the environment that they are subjected to. The BNNTs are also sensitive to vibration, temperature, and are impervious to oxidation and mechanical intactness (Yanar et al. 2020; Yao et al. 2021).

This multifunctionality is particularly advantageous in compact systems where space is limited such as in living spaces, workstations, collaborative robots (cobots), soft grippers, or real-time process monitoring tools making it essential to minimize the complexity of individual components.

15.1.1.3 Compatibility with Flexible and Wearable Systems

Stretchable, conformal, and flexible devices are needed as factory floors start to require using human–machine interfaces (HMI), wearable diagnostic, and skin-like tactile systems of the robotic arms. Polymers can be embedded with nanotube-based composites to design bendable sensors and actuator that will not lose performance when subjected to strain or curvature (Heng, Solomon, and Gao 2022).

15.1.1.4 High Responsiveness and Fast Signal Propagation

The low-delay and increased signal speed observed in nanotube-based devices are attributed to their nanoscale size and electron mobility, which reaches immense levels (in some cases, in CNTs). Such fast response is essential when used in real-time feedback control systems like vibration dampening, active alignment, or predictive fault detection (Abiri et al. 2020).

15.1.1.5 Integration with Internet of Things (IoT) and Edge Computing Systems

Multi-builder intelligence is expected to serve as the next step in smart manufacturing, where data will be collected, processed, and acted at the network edge. The sensors built on the nanotube are friendly when it comes to integration with the low-power IoT module. Other newer studies have investigated CNTs in wireless sensing, in energy harvest through triboelectric or piezoelectric processes, and even neuromorphic computing devices (Farzin, Naghib, and Rabiee 2024).

15.1.1.6 Potential for Self-Powered Operation

As the need for autonomous and wireless operation rises in manufacturing, the capability of gathering environmentally sourced energy is useful. The piezoelectric behavior of BNNTs allows them to generate usable electrical energy by converting mechanical stress (e.g., vibration found in machines), and they can be used to create self-powered sensing and communication modules (Shim et al. 2022).

15.1.1.7 Future-Proofing Through Scalability and AI Integration

With the growing popularity of AI-based control systems in manufacturing, high-frequency, high-resolution sensor data will be essential. Predictive maintenance and anomaly detection along with optimization of the systems can also be utilized by using nanotube-based sensors as a rich source of data used by machine learning (ML) algorithms. Moreover, the dimension of producing nanotube-incorporated flexible electronics at large scale is starting to become realistic as technologies in additive manufacturing and printing advance, and this progress is likely to enable scalable industrial applications (G. Chen and Tang 2024).

This has been triggered by the shortcomings of traditional sensor and actuator technologies when confronted with the challenges of Industry 4.0 that prompted the development of novel materials approaches as shown in Table 15.1. Boron nitride and CNTs have tremendous multifunctional potential and could be the way ahead. When added to CPS, they would allow automation, accuracy, and flexibility in manufacturing never before possible. Although the introduction of nanotubes into sensing and actuation applications faces some barriers (big fabrication, property consistency, and standardization included), the current research and collaboration in the industry indicate a bright future in the field.

This section will explore more with the working principles, mechanism, integration issues, and future directions of nanotube-based sensors and actuators, and how they will work in the framework of smart manufacturing ecosystems as we read on in this chapter.

TABLE 15.1
Summary of Motivation

Motivating Factor	Nanotube Advantage
Sensitivity and resolution	Atomic-scale deformation sensitivity (piezoresistive, FET-based)
Harsh environment operation	High thermal and chemical stability (especially BNNTs)
Flexibility and stretchability	CNT/BNNT-polymer composites enable flexible device fabrication
Multimodal functionality	Single structure can sense, actuate, and harvest energy
IoT and edge device integration	Low-power operation and small size
Real-time feedback and fast response	High-speed electron mobility and mechanical responsiveness

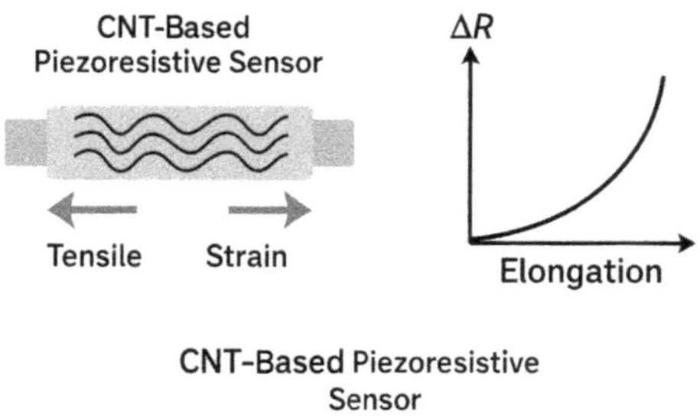

FIGURE 15.1 Schematic of CNT-based piezoresistive sensor showing resistance change under tensile strain (ΔR as a function of elongation).

15.2 SENSING MECHANISMS ENABLED BY NANOTUBES

15.2.1 Piezoresistive Sensing

Piezoresistive sensing is a very common transduction mechanism in the field of smart sensing (in structural health monitoring, tactile feedback, wearable electronics, or real-time mechanical diagnostics). CNTs, due to their exceptional electrical and mechanical characteristics, can be used as outstanding materials to realize the high-sensitivity piezoresistive sensors which have the lightweight, flexible, and possible integration into smart manufacturing setup.

Piezoresistive effect can be defined as how electrical resistance of a material depends on the mechanical strain imposed on it, i.e., tension strain, compression strain, or even bending strain. When traditional materials (silicon or metal foils) absorb the change, it is mostly because of the geometric changes (length and cross-sectional area). But in nanostructured material such as CNTs, the mechanism is more complex and involves:

a. ***Intrinsic band structure modulation***: In CNTs, strain induced by the application of an external stress causes a modulation of the interatomic π-bonding within the adjacent carbon atoms and can modify the band gap and carrier mobility (Qian et al. 2020).
b. ***Inter-tube tunneling*:** In networks or mats of CNTs, the effect of mechanical deformation on the distance and contact resistance between adjacent nanotubes occurs (Xia Wei et al. 2021).
c. ***Variation on contact resistance*:** In CNT-polymer composite or CNT film on a substrate, resistance varies with strain due to varying the quantity and characteristics of conductive path (Fazi et al. 2023).

The combination of these effects has the effect that even small strains have a large effect on the resistance of CNTs, so that changes in the resistance generally go many orders of magnitude compared to conventional materials.

In conventional metal strain gauges, the gauge factor (GF) is of the order of 2. Conversely, CNT-based devices can have GFs between 10 and 1000, depending on companionship and dispersion, and the particular arrangement of the CNT network

(Wagner et al. 2019). Such a drastic change in GF has served CNTs as the basis of ultrasensitive representation of strain detection. The GF of CNT is depicted in Table 15.2.

The GF in CNT-based sensors depends on various factors which include:

a. ***CNT type and quality:*** SWCNT breaks usually have greater GF since they exhibit more confinement of quantum and bandgap modulation (Xiaojun Wei et al. 2022).
b. ***Dispersion and percolation:*** The extent of dispersion, and the development of a percolating network, is what determines the number of current paths present which influence resistance response (Hunt 2001).
c. ***Alignment and orientation:*** Aligned CNTs give anisotropic behavior or more definite and sensitive in direction of orientation. Nonlinear or hysteretic behavior can be more easily fabricated and also the randomly oriented networks can have more relaxed orientations (Ponnamma et al. 2014).
d. **Substrate and composite matrix**: The elasticity and flexibility of the host material determine the need to transfer strain to the CNTs (Wen, Li, and Cheng 2016).

These sensors are strain-responsive and have high GF, which enables them to capture strain in a range of 0.001%, so they are appropriate for vibration sensing in rotating machinery, structural fatigue measurements, or force measurements in the robot's arms.

Wearable and bendable strain sensors are becoming highly popular in smart manufacturing particularly in HMI applications, extensive use of soft robotics, and integrations of condition monitoring measures. The CNT-based sensors have a number of strengths:

a. Flexibility and stretchability
   CNTs may also be incorporated into elastomeric materials like PDMS, Ecoflex, or TPU to create a composite that retains its conductivity despite being able to be elongated very far (typically, much more than 100%). This allows them to be applied to curved or moving surface without delamination effects or a decrease in performance (Faeghifard, Mehranpour, and Ghasemi 2025).

TABLE 15.2
Typical GF Values for CNT-Based Piezoresistive Sensors Under Different Configurations (Arana, Gamboa, and Avilés 2023).

CNT Configuration	GF	Notes
SWCNT film on polymer	50–200	High sensitivity, but may exhibit hysteresis
MWCNT in Polydimethylsiloxane (PDMS) composite	10–40	Flexible, lower GF due to interlayer shielding
Aligned CNT array	100–1000+	Directional sensing, complex fabrication

b. Conformal integration
   The CNT sensors are conformal, and due to their thinness and flexibility, they can be stuck onto tools, robotic joints, prosthetics, and even onto human skin so that the strain or motion can be monitored throughout. This plays a crucial role in manufacturing with robots where accuracy and feedback are key (Palumbo, Li, and Yang 2022).
c. Low power requirements
   The piezoresistive mechanism needs no active excitation (as in piezoelectric sensors); thus, these sensors can be easily used with low-powered electronics (such as battery-powered ITS nodes or edge computing devices) (Agar et al. 2016).
d. Robustness and environmental tolerance
   CNT-based sensors can be encapsulated, and with the appropriate encapsulation, they can provide a resistance in a harsh environment, whether it is the moisture content, vibrations, or thermal variations, and they have a long life compared to foil and/or silicon-based gauges (Yiqun Liu et al. 2023a).
e. High spatial resolution
   The dense arrays of CNT networks may be printed or patterned on flexible substrates to achieve a high-resolution strain mapping, which has applications including component quality check of the deformable components or tactile surfaces (Zhang et al. 2025a).

The piezoresistance of CNT-based strain sensors is significant in smartness manufacturing environments through the provision of real-time feedback which can be used in the following—Structural Health Monitoring: This can be done by acting as a crack initiation detector, residual strain test, fatigue test in the structure body, welds, or load-bearing element. Robotics Tactile Feedback: Sensing the grasp force and contact feedback in soft robots or collaborative robots (cobots). Human Posture Monitoring: Maintaining the safety and ergonomics of the worker by means of wearable CNT sensors, which are integrated with the garment or exoskeleton. Tool Condition monitoring: Strain sensing at the cutting regions or at the tool attachment areas to determine the wear or misalignment of the specific tool. Their compact sizes and wireless connectivity enable connecting them to distributed sensor networks, which serve as the most essential layer of the entire digital twin ecosystem, where real-world data are being synced with virtual simulations to create an optimal level of performance and enable anticipating possible failures.

CNT-based piezoresistive sensing represents an excellent next-generation strain measurement and mechanical feedback application. They are flexible and fit well into many active systems, have high sensitivity, and can be adapted to fit many types of intelligent systems due to fit into new advanced manufacturing processes (e.g., printing, coating, 3D patterning). CNT-based piezoresistive sensors will play a fundamental role in the smart manufacture infrastructures of the future whose market is expected to grow with the need to integrate cyber and physical systems, monitor real-time and predictive maintenance criteria (Sanati et al. 2018).

15.2.2 Piezoelectricity

Piezoelectricity is an electromechanical property that appears in noncentrosymmetric materials, in which a mechanical stress triggers a polarization of electric charges. To be effective, the material should not be a center-symmetrical crystal-only structure—that is, when deformed, it generates a net dipole moment.

BNNTs are one-dimensional analogues of hexagonal boron nitride (h-BN), with alternating atoms of boron (B) and nitrogen (N), which are in hexagonal lattices. As opposed to the centric arrangement of CNTs, which are built of the same carbon atoms and hence possess a centrosymmetric structure, a BNNT is built using an ionic lattice leading it to take the form of a non-centrosymmetric configuration. It is the polarization of their structure that is important in their inherent piezoelectric behavior.

- In BNNTs, piezoelectric effect can be attributed to the relative movement of charged atoms (B and N) with one another due to stretching, bending, or compressing the nanotube (Yang et al. 2025).
- Such mechanical deformation disrupts the symmetry of the lattice so that a dipole moment is generated, which translates into a measurable voltage between the length of a wall nanotube (Gartstein, Zakhidov, and Baughman 2003).

The piezoelectric coefficient (d33 or d11) of a piezoelectric material is usually used to characterize this effectiveness, a factor related to the electrical polarization produced as a consequence of an applied mechanical stress, divided by a factor that involves the magnitude of the mechanical stress. Although presently BNNTs do not compete with conventional piezoelectric ceramics in absolute d-values yet, there are several other advantages that they possess: High thermal stability (up to well above 900 °C), flexibility and nanoscale conformity, chemical resistance, and native compatibility into micro/nano systems.

Recent theoretical and experimental works have reported the piezoelectric coefficients of BNNTs as 10-30 pC/N, which is quite desirable in cases where it must have moderate signal strength, but high degrees of robustness and miniaturization. This voltage that is generated when a tube is deformed usually lies in the range of millivolts to volts, depending on the diameter of the tube, the thickness of the tube wall, the strain level, and the contact arrangement. Table 15.3 provides the piezoelectric coefficients and thermal stability of various materials.

Self-powered sensors are among the best prospects of piezoelectricity in BNNT; these are sensors that are not operated on batteries or external power. Rather, they scavenge energy available in the form of background mechanical stimuli (vibrations, impacts, strain) and turn it into usable electrical signals. The latter is very desirable to smart manufacturing where installation of wired sensors or a replacement of batteries may be logistically challenging, expensive, and disruptive.

Conventional piezoelectrics, including PZT, have proven to be quite successful at creating intense piezoelectric outputs but, as found in smart manufacturing, have a number of limitations—Brittle: During flexure or vibration, the ceramic sensors are fragile to fractures. Temperature restrictions: Most piezoceramics cannot withstand

TABLE 15.3
Comparative Piezoelectric Coefficients and Thermal Stability of Common Piezoelectric Materials

Material	d_{33} (pC/N)	Max Temp (°C)	Flexibility	Remarks
BNNTs (X. Chen et al. 2017)	10–30	>900	High	Excellent for harsh environments
Lead zirconate titanate (PZT) (ceramic) (Kulkarni, Saha, and Kumari 2023)	300–500	~300	Low	Brittle and non-biocompatible
Polyvinylidene fluoride (PVDF) (polymer) (Shalu et al. 2013)	20–30	~100	Very high	Limited thermal stability

temperatures in excess of 300 o C. Toxicity and biocompatibility: Materials like PZT contain elements such as lead that are hazardous to both the environment and human handling.

Piezoelectric sensors made of BNNT will change the industry of collecting, transmitting, and processing mechanics data about the production. Their sensing and energizing themselves at the same time eliminates much of the wiring in the systems, increases system reliability, and allows plug-and-play sensing deployment, even on rotating or sealed systems.

Key application areas include:

a. Predictive maintenance systems where BNNT sensor self-powered sensors keep a continuous track of machine health (L. Liu et al. 2021).
b. Smart tools, in which process control is done based on real-time stress data.
c. Dies, molding, and industrial robots were custom-fit with monitoring systems that enabled an awareness of condition presence in automation.
d. IoT-enabled platforms, where sensors report conditions without relying on power infrastructure in a centralized way.

BNNT-based piezoelectric sensing offers a novel category of durable, micro-minimized, and autonomous devices for smart manufacturing. They are particularly useful where conventional sensors cannot be used due to the harsh environment in which an algometric pressure sensor can directly convert mechanical stress into electrical energy even under extreme circumstances. With the development of techniques of material synthesis, alignment, and integration, the piezoelectric systems based on BNNTs are to become at the heart of developing intelligent, energy-autonomous, and maintenance-free sensing platforms in Industry 4.0 (Zhang et al. 2020b).

15.2.3 FET Sensing

The sensing platform based on the FET is one of the most sensitive and cross-purpose device platforms for analytical environmental, chemical, and biological stimulus detection. Semiconducting CNTs, in particular, are good candidate materials to fabricate nanostructured FET sensors, due to their great surface area, high mobility of carriers, and sensitivity to local electrostatic perturbation. Such properties allow building ultrasensitive, label-free, real-time monitoring devices, which are key elements of smart manufacturing, cleanroom monitoring, and bio-interfacing systems.

The FET consists of three terminals, which are represented as the source, the drain, and the gate. The electric field between the source and drain is commuted by the applied electric field between the gate terminal and the source. A typical FET has a channel made of a silicon-based semiconductor. This channel in CNT-FETs as shown in Fig. 15.2 is, however, replaced by the semiconducting CNT, and this allows the device to work at nanoscale size with a greater sensitivity (Zhang, Zhang, and Zhang 2025b).

The one-dimensional structure of CNTs exposes every atom to the environment, making them highly sensitive to perturbations and enabling the detection of extremely low analyte concentrations or subtle chemical shifts. Regulation of environmental safety, air purity, or process consistency are some of the things that also need real-time detection of chemical vapors, gases, or contaminants in smart manufacturing. CNT-FETs can act as chemiresistive sensors or chemicapacitive sensors, by showing high resistance or capacitance response, respectively, to a target species.

These CNT-FETs have shown good results in gas sensing as per the following gases: NH_3, NO_2, H_2S, CO, and volatile organic compounds (VOCs). When the gases are adsorbed on the CNT surface, they give away or take away electrons, altering the local carrier density and moving the threshold voltage or drain current.

- In the p-type CNT-FETs, holes will be decreased, lowering the current, when NH_3 (electron donor) is added (Z. Li et al. 2022).
- The hole concentration can be increased by the addition of NO_2 (electron acceptor) to create increased conductivity (Han et al. 2018).

Detection limits are capable of the parts-per-billion (ppb) range and are more sensitive than the traditional metal-oxide sensors.

Flexible substrates, an added advantage of CNT-FET chemical sensors, are that they can be manufactured on an inert flexible substrate which has the potential to form a selective membrane or be coated with a polymer film. This enables it to be used in high-humidity, temperature-variable, or chemically active environments that are likely to be found in a manufacturing facility, cleanroom, or chemical reactor.

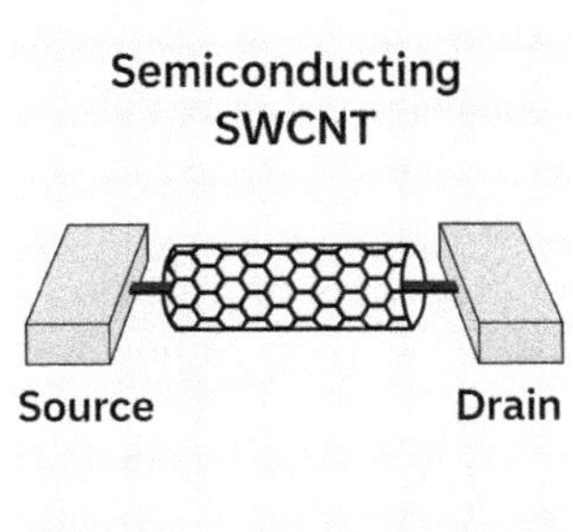

FIGURE 15.2 Schematic of a CNT-FET.

Another important aspect that promotes life-changing things by CNT-FETs is in biosensing. The conventional biosensors usually need labels, large transduction reagents, or multi-step procedures. CNTs-FETs allow extremely selective, real-time, and label-free detection of biomolecules by functionalizing CNT surfaces.

CNTs can be functionalized through covalent or non-covalent strategies to add biorecognition elements, e.g., antibodies, enzymes, DNA/RNA probes, aptamers. In FET, the local electrostatic or steric impact on the concentration of charge carriers in the channel is produced by the attachment of a target biomolecule to the functionalized CNT surface.

CNT-FET biosensors exhibit ultrasensitive responses, allowing detection in the femtomolar (fM) to attomolar (aM) range, making them highly suitable for early-stage diagnostics and preventive monitoring.

The CNT-FET chemical and biological sensors characteristically complement low-power, small-size electronic systems, and thus are well suited and fit to be used in distributed IoT sensor networks in smart factory environments. Key strengths include: rapid response times ranging from milliseconds to a few seconds; the ability to be miniaturized, with sensors fabricated at micrometer or submicrometer scales; low power requirements, allowing operation via batteries or energy harvesters; and reduced risk of skin irritation, as the sensors can be integrated onto flexible substrates and wearable materials such as fabrics.

The CNT-FET data streams can be applied along with the edge computing platforms to process data on-the-fly utilizing AI/ML algorithms to sense the tell-tales of wariness or programmatically respond to them, e.g., robotic cleaning. When combined with edge computing platforms, CNT-FET data streams can be analyzed in real time using AI/ML algorithms to detect anomalies, assess contamination risks, or trigger automated responses.

FET sensing application of semiconductor CNTs presents unprecedented sensitivity, speed, and miniaturization. CNT-FETs can be integrated into the smart manufacturing system since these components are useful in sensing chemical and biological weapons and can be used with the IoT and cyber-physical architectures. CNT-FET sensors will become a core feature of real-time, adaptive, and intelligent monitoring in Industry 4.0 with an increasingly mature fabrication technology and greater control of materials.

15.3 ACTUATION MECHANISMS IN NANOTUBE-BASED SYSTEMS

In the domain of smart manufacturing, systems possess the capability not only to sense environmental and mechanical stimuli but also to respond to them with precision through integrated mechanisms designed to perform specific actions within automation, robotics, and adaptive frameworks. Materials made in nanotubes, especially CNTs, have promising properties of actuation since they can achieve high levels of electrical and mechanical activity, responsiveness to thermal or molecular induction, and/or sensitivity.

Actuators based on nanotubes, in contrast to bulk actuators (motors, hydraulics, or piezoelectric ceramics), work at the micro- and nanometer scale, have a rapid response rate, and have the added advantage of embedded use in flexible, lightweight, and miniaturized systems. This allows them to be useful in MEMS/NEMS (Micro-/Nano-Electromechanical Systems), robotic microgrippers, adaptive optical systems, and smart textiles in manufacturing.

15.3.1 Electromechanical Actuation

The subject of electrical excitation causing mechanical deformation, i.e., bending, twisting, or axial strain of CNTs, is called electromechanical actuation in CNTs. In this process, the interplay of the electronic structure of CNTs and their atomic structure is exploited (Kong and Chen 2014).

In a CNT structure, motion can be induced by a number of different phenomena occurring simultaneously—Charge injection: This may either be charge introduced through injection or storage of electrons or holes in the CNT lattice, leading to repulsion or attraction of the carbon atoms, resulting in expansion or contraction. Electrostatic effects: In multi-component systems or CNT bundle, electric fields may be applied, and the Coulombic interaction may cause bending or deflection. Lorentz forces: The type of forces described by the Lorentz force law can deflect current-carrying CNTs in the presence of magnetic fields. Electrostrictive effects: It is a reversible deformation which arises because of the change of polarization when voltage is applied.

The displacement is proportional to the electric field applied and inversely proportional to nanotube or CNT array stiffness. In single-walled CNTs, the electrostatic forces on the scale of a piconewton will cause detectable deflections with nanometer scales (Ouakad 2016).

The features of electromechanical actuation on the performance—Fast response time: Since the electron mobility and mass of CNTs is so high, the response time is in the microsecond to millisecond regime. Low operating voltage: Operation may be commenced using as little as 110 V depending on configuration and the dielectric environment. Scalability: CNTs are suitable as single-nanotube actuators, or in macroscopic yarn or films to make motion on a larger scale. Reversibility and repeatability: Electromechanical deformation is extremely reversible, and is applicable to applications that need repeat, or cyclic, operation.

Electromechanical actuation in smart manufacturing—Microgrippers and robotic fingers: CNT vibrations/bundles or cantilevers can be actuated using electricity to achieve grip on or manipulation of small parts. Optical actuators: In adaptive optical systems, the change in position of a mirror or the length of an optical path can be effected by deflection of CNTs. MEMS/NEMS switches: Actuation of CNT beams for contact switching, enabling mechanical logic or signal routing in microdevices. Vibration dampers and tuners: CNT-based beams or films that dynamically adjust stiffness under electrical control (Gupta et al. 2021).

15.3.2 Thermal Actuation

Thermal actuation: Thermal actuation is mechanical deformation when CNT structures are exposed to temperature variations, especially by heating by Joule heating, that is, the local heating of a conductor by the passage of a current.

Electron–phonon interactions in CNT cause resistive heating when an electrical current flows through it (or a network of CNTs). The result is—Thermal expansion: CNTs straighten or warp because of differences in thermal expansion in the structure or with the substrate. Thermo-mechanical deformation: Thermal coefficients of CNTs and the matrix materials do not match in composite systems, resulting in flexing or bending that can be controlled (H. Li et al. 2011).

Performance proper of thermal actuation localized heating: Because of the large aspect ratio and the small heat capacity, heating is localized at specific spots, with minimal collateral thermal consequences. Good thermal conductivity: Thermal conductivity is reasonably well and performs in hot spots and life extenders significantly. Small size design: No moving mechanisms or motors are required; acting is inside the film or yarn. Deformability: Through custom shapes of electrodes, patterned areas of heating will be able to produce custom motion patterns.

Thermal actuation in smart manufacturing applications—Examples are microvalves and fluidic control: CNT structures with which it is possible to open/close microchannels using temperature as a stimulus (Sun and Qing 2011). Shape-shifting building blocks: Thermally activated CNT-polymer composites to be installed in adaptive molds, tool-fixtures, or robotic skin. Responsive seals and gaskets: These can contract or extend in such a way that they can fit to seal various temperature ranges. Actuated lenses or filters: Shape-tunable lenses or filters CNT elements in optical or photonic systems that are tuned by forced heating with current.

Thermal actuation can be especially useful in bodies of environment where already-existing temperature gradients are accessible (such as near a furnace, dies, or molding zones) to enable passively or semiactively driven actuation.

15.3.3 Chemically Responsive Actuation

Chemically responsive actuation is the mechanical effect that has come about as a result of exposure of CNT structures to particular chemical structures, solvents, and other agents in the environment. This reacts normally as swelling, contraction, or bending all of which can be very selective in the case of surface functionalization.

The CNTs react with the chemicals in the environment in a number of ways—Physical adsorption: The molecules adsorb on the CNT surface, which leads to charge redistribution or stress. Capillary force–driven swelling: CNT films or yarns can be soaked in a liquid or vapor, which can cause an actuation on swelling or expansion in volume or act anisotropically. Chemical bonding or reaction: Functionalized CNTs can reversibly bind with analytes, deforming them. pH or ion-sensitive coatings: CNTs coated or bound to stimuli-responsive hydrogels or polymers are susceptible to structural changes when exposed to a type of ion or pH (Tawfick et al. 2013).

The degree of deformation depends on the concentration, diffusion dynamics, and kind of chemical affinity of the analyte to the surface of CNTs (Schroeder et al. 2019).

CNTs are frequently integrated with chemical recognition agents to enhance sensor selectivity. Typical examples include polymer films such as Nafion and PVA that exhibit selective swelling toward solvents or humidity, enzyme coatings for biological analyte detection, and metal-oxide nanoparticles employed for gas sensing and redox reaction monitoring. These deposits enhance responses of the actuation and can be customizable to desired target chemicals.

In smart manufacturing, the applications of chemical-responsive actuation are—Smart valves and microfluidic controls: CNT actuators open or close channels based on chemical concentration (e.g., pH, solvent presence). Environmental triggers: Actuation as a passive safety device, e.g., opening of ventilation flaps in the event of VOC accumulation. Soft robotics and artificial muscles: CNT films bend in response to the humidity, or solvent vapor in the atmosphere, and allow biomimetic motion. Process monitoring: Devices or fittings that modulate with chemical composition of the environment to maintain the best level.

Chemically responsive actuation is especially appropriate when autonomous systems have to respond to the environment using only low-energy triggers and when only low levels of energy are desired (Zhao et al. 2025).

CNT actuation presents an extremely broad scale of applications of smart manufacturing systems, comprising fast nanoscale motions in MEMS, self-correction of environmental factors in large parts, to mention a few. Variability of the actuation mechanisms provided by each electromechanical, thermal, and chemical actuation allows solutions to be customized to the great variety of industrial requirements (Bayer 2022).

By its nature, these systems are compact, lightweight, energy-efficient, and can be designed to support flexible substrates, so it is natural that such systems will be important components of a range of next-generation intelligent systems, including robotic manipulation, self-adaptive surfaces, responsive materials, and cyber-physical feedback loops. It is in the future development of material science and fabrication technologies that CNT-based actuators will be used as the focal point in the shift to intelligent, autonomous, responsive manufacturing ecosystems.

15.4 INTEGRATION INTO SMART MANUFACTURING SYSTEMS

It is smart manufacturing, and the full promise of nanotube-enabled sensors and actuators should not only focus on the next level of manufacturing material properties but rather on how the underlying sensors and actuators can be flawlessly integrated into larger CPS as a sensible end goal. Smart manufacturing as conceived under the aegis of Industry 4.0 would require real-time data acquisition, autonomous decision-making, adaptive actuation, and interconnection at the system level. Such capabilities depend on the seamless combination of hardware-level sensing and actuation with such infrastructures as clouds, edges, and digital twins.

With their ability to miniaturize, speedy response, low power consumption, and multifunctionality, CNTs and BNNTs are the perfect applicants in terms of embedding them in distributed, smart industrial systems. In this section, the author discusses how these nanotube-based devices could be incorporated into the smart manufacturing with references to the five key aspects of integration: Embedding into physical industrial systems, interfacing with IoT platforms and digital twins, role in cyber-physical feedback and control loops, power management and signal processing, and wireless communication and network compatibility (Mittal, Sehrawat, and Kushwaha 2025).

15.4.1 Embedding Nanotube Devices into Industrial Systems

Physical embedding is the initial process involved in implementing nanotube-based components in a manufacturing context, and entails mechanical, electrical, and environmental usability with the infrastructural component, product, or equipment in which it is embedded.

The nanotube devices may have an implementation on industrial systems at different levels:

a. Metal or glass or polymer thin films or coating
b. Flexible sensors found in rotating shafts, man-made skins of a robot, or conveyor belts
c. MEMS with CNT cantilevers or FETs
d. Bulk composites reinforced with CNTs in the structure with load bearing

The concomitant versatility allows them to be used in structural health monitoring of machine frames, dynamic sensing on mobile robotic platforms, and on-tool vibration and wear diagnostics.

Nanotube devices have protective films (e.g., PDMS, epoxy, ceramic coatings) encapsulated and/or layered on them in order to survive industrial conditions and are impervious to dust and particulate contamination; high-temperature operation (in particular, BNNTs); vibration, shock, and mechanical fatigue; humidity and chemical exposure (Muresan 2023).

High-end integration methods have recently emerged in the form of laser direct writing, 3D printing of CNT inks, and inkjet patterned CNT inks which enable personalized sensor arrangements on functional items at the steam; certain examples of such items include molds, dies, and robot grippers (Abdolmaleki, Kidmose, and Agarwala 2021).

15.4.2 Interfacing with IoT Platforms and Digital Twins

The applications of nanotube-based devices in the IoT are natural since they can be deployed in a small form factor, powered in low-power consumption, and capable of multi-sensing applications. It is one of the nodes at the sensing and actuation layers of a hierarchical IoT implementation within a factory floor.

An industrial IoT (IIoT) usually has four layers of architecture:

a. Edge devices: Sensors and actuators operating at the asset level, such as CNT or BNNT units.
b. EDGE computing nodes: Appropriateness controllers (MCU) or gateways that filter data in real time.
c. Cloud platforms: Aggregation analytics visualization.
d. Digital twins: Virtual replicas that serve the aim of simulation, prediction, and optimization.

Nanotube-based sensors and devices can input high-resolution, high-frequency data into this hierarchy, making it possible to: Monitor processes in real time, perform predictive maintenance calculations, optimize energy consumption, and detect anomalies

Digital twins: A digital twin is a dynamic, virtual representation of a real, physical object and is being updated continuously based on real-world sensor data. Digital twins can be enabled using CNT/BNNT devices by enabling synchronized sensing—e.g., updating model angles of a robotic joint based on CNT strain sensors; Dynamic feedback—e.g., BNNT piezoelectric sensors adjust operating parameters in simulation; and Anomaly detection—monitoring the real-time activities of a process and flagging differences in model behavior (Elkington et al. 2021).

15.4.3 Role in Cyber-Physical Control Loops

The heart of smart manufacturing is the cyber-physical feedback loop, in which decisions are taking place based on real-time data, modeled response, and autonomous actuation. Nanotube-based systems seal this loop by doing the following: Sensing the environment (e.g., force, pressure, strain, gas concentration); transmitting data to control systems or edge nodes; and acting on commands via mechanical, thermal, or chemical actuations. Nanotube-based systems also provide feedback of results, to update decision algorithms.

Nanotube devices used in a typical control system may be placed as: Primary sensors which provide gauges of strain, vibration, temperature, and chemical exposure; secondary actuators which adjust flaps and guiding arms, tune components; or self-regulating elements acting on locally embodied intelligence, without external control.

CNT tactile sensors can be embedded in grippers to measure contact forces, while CNT thermal actuators can adjust grip tension based on the material type, enabling adaptive handling of objects, especially fragile items, so they can be firmly grasped without damage (Yuan et al. 2021).

Once ML models are coupled with the sensor measurements provided by nanotubes. The CPS is capable of learning the best operating regimes, failure predictions, energy optimizations, and manufacturing accuracy via adaptive control. The end result is self-optimizing systems.

15.4.4 Power Management and Signal Processing

Energy and power efficiency and signal integrity are two critical variables in mass deployment of sensors. Devices on nanotubes have natural flexibility to low-power operations and can be combined with low-power and advanced processing on board the device.

FETs and piezoresistive sensors based on CNTs work in a few μW to mW to keep the power intake minimum (Prakash, Mohana Sundaram, and Anto Bennet 2018). The BNNT piezoelectric sensors are capable of producing self-generated voltage under mechanical deformation and can be used as self-powered nodes. This enables the sensor to be deployed in locations where battery replacement is impractical, such as on moving components, embedded surfaces, or high-temperature areas.

The hybrid CNT devices can be paired with and the BNNTs can be incorporated in triboelectric nanogenerators (TENGs), thermoelectric modules (with the high thermal conductivity of CNTs), and micro-supercapacitors to store explosion energy. These devices have the ability to power data transmission modules only part of the time that would make the reliance on the fixed power infrastructure lesser (Kumar et al. 2022).

Edge processing circuits are introduced to alleviate the burden of communication as well as to boost autonomy in cooperation with nanotube sensors. The functions of these circuits include filtering raw sensor data, fast Fourier transforms (FFT) of the data to give in the frequency domain, and event classification (e.g., anomaly, normal operation, maintenance required). This facilitates local decision-making, reduces latency, and offloads computation to central systems.

15.4.5 Wireless Data Transfer Compatibility

To provide an easy entry to factory-wide networks, the nanotube sensors and actuators will have to be wireless communicative since in several of these machines that are in motion or rotate, it is impossible to wire the on-board machinery to provide wired communicative models.

Nanotube sensor nodes are usually used with low-power microcontrollers (e.g., ESP32, STM32, Arduino). These microcontrollers have the following sensors: Bluetooth Low Energy (BLE) short-range, high-efficiency communications, Zigbee or LoRa long-range low data rate, Wi-Fi high bandwidth local factory area network, and NFC or RFID passive reading or inventory management (Ahmed et al. 2024).

These nodes may be used in mesh, star, or hybrid networks and can scale up to provide solutions to small workshops and even large-scale manufacturing units. There are security measures (e.g., AES encryption, authentication handshakes), according to which it may be operated safely in the environments where the information integrity is of high value.

As the industry 4.0 standards, like OPC UA, MQTT, and Time-Sensitive Networking (TSN), evolve, nanotube nodes can interface with enterprise-level monitoring systems, SCADA dashboards, and even ERP systems with almost no custom interfacing required (Trifonov and Heffernan 2023).

Nanotube-based sensors and actuators are ready to be implemented into the smart manufacturing environments and are seen as one of the fundamental steps toward intelligent, adaptive, and energy-aware production environments. Its small size, reactivity, and versatility allow its installation on machines, materials, humans, and the infrastructure, making it possible to achieve a real cyber-physical convergence.

Whether CNT- and BNNT-based devices are involved in the real-time generation of data or autonomous actuation, or other aspects of predictive analytics elements in the smart manufacturing hierarchy, they have a role therein. With advances in integration technologies, AI frameworks, and digital twins, such components will continue to emerge as fundamental building blocks of responsive and resilient manufacturing ecosystems with the move toward Industry 4.0 and beyond becoming more attainable.

15.5 CHALLENGES IN INDUSTRIAL DEPLOYMENT

Although they demonstrate great potential, sensing and actuating based on nanotubes have not been widely applied on an industrial level beyond miniaturized or specialized applications. Although studies have been performed with success to show their mechanical strength, quick response, and the possibility of miniaturization, there are a number of steep challenges to overcome in order to bring all of these advantages to full fruition in the form of industrial products that are sound, repeatable, and scalable. Industrial ballrooms are usually noisy, complicated, and can be rough altogether, exposing the components to mechanical stress, thermal cycle, chemical, and electromagnetic interferences. To thrive in these environments and in others, nanotube-based devices will have to achieve high-performance robustness over repeated operations, both long-term reliability and packaging robustness, and also obey industrial protocols. In this section, the five most significant categories of problems with industrial integration of CNT/BNNT-based devices are identified and described and then a summary table of mitigation strategies is given.

15.5.1 Consistency in Nanotube Morphology and Chirality

One bottleneck of commercialization of the nanotube-based devices is morphological inconsistency, especially with CNTs. These are imprecisions in: diameter, length, wall number (single-walled vs. multi-walled), crystallinity, chirality (twist angle and direction) (Tao et al. 2024).

Of these, chirality is of extreme importance in specifying electronic properties. As an example, FET-based sensor necessitates semiconducting CNTs, but metallic CNTs are more appropriate in conductive films or in electromechanical actuators. Alarmingly, bulk production methods like chemical vapor deposition (CVD) produce a mixture of both types, resulting in unpredictable device performance. The result of discrepant morphology is non-reproducibility of GF and strain sensitivity in piezoresistive sensors, unstable threshold voltages in CNT-FETs, and heterogeneous or sporadic piezoelectric response in BNNTs because of non-uniform structure

Mitigation strategies: CNTs by chirality are isolated by use of post-synthesis sorting techniques to include density gradient ultracentrifugation (DGU), electrophoresis, or selective chemical functionalization. Template-guided synthesis and seeded growth may better control the structural parameters. The structural variability observed in ML models enables the model to predict how various devices will behave and alter the operation of the device (Janas 2018).

15.5.2 Signal Noise and Reliability Over Time

Any industrial-grade sensor or actuator must be able to respond with signal stability. Yet nanotube devices, in particular those based on surface interactions, do not perform well when exposed to (environmentally) long time spans because of environmentally introduced conditions, such as humidity, temperature, and chemical impurities. Charge trapping at the substrate–interface or within the functional layer. Oxidation or erosion of coating on the surface or degradation of the surface affects the aging of skin, leading to discoloration.

To give a specific example, the chemical CNT-FET sensors may potentially be unintentionally doped due to the exposure to polar molecules or humidity, which leads to the threshold voltage shift and change in sensor response (Deng et al. 2022).

In a dynamic task such as robotic touch sensing or vibration detection, the instability of signals can generate control error or false positive even with a small level of signal instability.

Mitigation strategies—Passivation: The surfaces of CNTs can be isolated by the application of passivation layers, e.g., PMMA or silicon nitride. Compensation of drift by self-calibrating circuits is made possible by reference sensors or incorporated diagnostics. Long-term reliability can be enhanced through encapsulation in inert atmospheres, or through the use of BNNTs of a chemically robust nature (Zorn et al. 2023).

Also, it is necessary to develop redundant sensing and periodic recalibration procedures to ensure reliability of sensor data even on long-termed, monthly, or year-long time scales.

15.5.3 Mechanical Fatigue and Adhesion to Substrates

Nanotube devices, in a number of industrial applications, are placed on flexible or rigid substrates, including metals, polymers, or ceramics. Delamination, cracking, or fatigue of the nanotube layer may take place after repeated cycles of mechanical deformation, especially in flexible systems. Challenges include weak van der Waals adhesion, making it to debond, crack under strain cycling, failing electrical continuity of conductive networks, and drift of performance as CNT networks rearrange themselves. In addition, BNNTs, despite their high thermal stability, may experience mechanical separation due to exposure to persistently vibrating or straining.

Mitigation strategies: Assessment of surface modification practices (e.g., plasma treatment or silane-coupling policies) to enhance adhesion between CNT and carries, embedding CNT in composite matrices (e.g., PDMS, polyurethane) to confer flexibility as well as structural stability, engineering mechanically decoupled layers of packaging that result in a reduced transfer of strain to the nanotubes units, and the embracing of strain island structures which confine mechanical deformation to regions that are not sensitive (Niranjana et al. 2025).

15.5.4 Packaging and Shielding in Noisy Industrial Environments

Industrial sites are electromagnetically polluted (EMI), electronically buzzed (E-transients), and mechanically vibrated. Nanotube devices and especially those running in microvolt or nanoampere scale are extremely vulnerable to EMI caused by motors, inverters, or power supplies in close proximity; electrostatic discharge (ESD); mechanical vibrational noise generated by nearby machines; and thermal noise due to ambient processes (Fernandes, Correia, and Mendes 2014).

Signals can be corrupted, and false actuation can occur, causing an accident in some cases. Moreover, the ingress of humidity, or particles, into the device package can result in damage to the integrity.

Mitigation strategies: Employing Faraday shielding (metal meshes or conductive coatings) to stop EMI; multilayer packaging with waterproofing, thermal insulation, and mechanical damping; application of differential sensing and signal processing algorithms (such as a low-pass filter or Kalman filter); and designing sensor enclosures with IP65/IP67 coded dust and moisture-proof cases (Othman et al. 2022).

Also, strong PCB-level design, appropriate grounding, and vibration isolation should be part and parcel of the final deployment strategy.

15.5.5 Lack of Standardization in Testing Protocols

The lack of standard characterization, environmental testing, electrical calibration, and long-term reliability measurements of nanotube devices, perhaps, constitutes the greatest systemic issue in transitioning nanotube devices to industry.

This is not a standardization, and therefore, benchmarking may not be consistent among the research groups and vendors, making it difficult to be evaluated and used by the industries.

Comparatively, classical sensors, strain gauges, thermocouples, and piezoelectrics are standardized via ASTM, ISO, and IEC and thus are interoperable and dependable (Hassani and Dackermann 2023).

Mitigation strategies: Development of test guidelines specific to industries by academic-industrial consortia (e.g., NIST, IEEE Nanotechnology Council), accelerated aging tests and validations in the real world of production within pilot lines, and acting in standardization groups to develop requirements of CNT/BNNT device testing and publicly posting open-source testing protocols and repositories to standardize sensors. Table 15.4 represents the challenges faced while deploying nanotube-based devices in Industry.

The future of smart manufacturing with embedded CNT- and BNNT-based technologies provides essentially limitless benefits, including anything from lightweight ultrahigh-resolution sensing to real-time self-powered actuation. Nevertheless, when it comes to the benefits associated with it, they can be achieved only in case the technical and deployment-related issues can be systematically addressed.

TABLE 15.4
Summary of Key Challenges in Industrial Deployment of Nanotube-Based Devices

Challenge	Impact on Device	Mitigation Strategies
Inconsistent morphology/chirality	Variability in performance (e.g., semiconducting vs. metallic behavior)	Post-synthesis sorting, guided synthesis, ML compensation
Signal drift and noise	Data inaccuracy, false readings	Passivation, calibration algorithms, encapsulation, redundant sensing
Mechanical fatigue and poor adhesion	Cracking, delamination, long-term failure	Surface modification, composite integration, strain isolation designs
Environmental interference and EMI	False actuation, device damage	Shielding, packaging, grounding, differential sensing, digital filtering
Lack of standard testing protocols	Poor industrial acceptance, inconsistent benchmarks	Development of standards, consortium validation, pilot deployment feedback

Major realms like nanotube material control, device reliability on cyclic loading, signal and noise stability in industry, and robust packaging need trans-disciplinary (or cross-disciplinary) interplay between the disciplines of materials science, electronics, manufacturing engineering, and industry standardization bodies.

The solutions to these challenges will not only hasten the industrial use of nanotube-enabled technologies but also open up entirely new horizons of application, such as next-generation robotics and AI-supported digital twins, making those one of the building blocks of next-century smart manufacturing ecosystems.

15.6 FUTURE DIRECTIONS AND OPPORTUNITIES

Sensors and actuators made of CNTs and BNNTs are an emerging area in the use of nanostructures in smart manufacturing, and the fields of material science, machine intelligence, and energy systems are developing very rapidly; therefore, the potential of such devices to radically change how manufacturing is done has never been better. Beyond the one-dimensional billing of sensing or actuating, the next-generation nanotube systems are likely to be very much multifunctional, adaptive, energy-autonomous, and highly embedded with AI and sustainable manufacturing philosophies.

In this section, five major directions in which the industry landscape of nanotube-enabled technologies will develop in the realm of Industry 4.0 and following Industry 5.0 will be described, where personalization, decentralization, and sustainability will be responsible for the innovation.

15.6.1 Multifunctional Devices: Sensor + Actuator + Energy Harvester

Perhaps the most interesting evolution of nanotube-based technology is the creation of multifunction unit sensing, actuation, and energy harvesting all in an integrated device. As this permeates, devices not only get simpler with decreasing complexity and footprint, but also improve performance and reliability through the reduction of interfaces and the interconnects. CNTs enable piezoresistive sensing, thermal actuation, and energy harvesting through both bandgap modulation and Joule heating mechanisms. BNNTs are a strong mechanical stress-induced piezoelectric sensing and combination of energy generation with high thermal stability (Ye et al. 2019; Peyrow Hedayati et al. 2025).

Through device architecture and material stacking, it is conceivable to build one of the following: A strain or deformation sensing, mechanical or shape change, ambient energy harvesting device.

A sensory patch based on CNT/BNNT might be used to sense movement (as a sensor), derive a small amount of energy required to operate by employing microvibration energy harvesting (harvester), and then alter stiffness or damping in response to the movement (as an actuator) by attaching it to a robotic joint. BNNT-based coating may measure the thermal expansion in high-temperature conditions and further produce serviceable voltage to transmit the data.

Such multifunctional platforms have the potential to make the system less complex, particularly in confined or hard-to-reach locations, and open the door to autonomous, distributed nodes within industrial networks.

15.6.2 Integration with AI for Real-Time Learning and Decision-Making

As ML and edge AI become more prevalent, the old paradigm of sending sensor data to the central servers and making decisions at high levels is being shifted toward the real-time, onsite decision-making paradigm. With the high responsiveness and sensitivity of nanotube-based devices, the latter will be able to take full advantage of incorporation with AI.

Raw output of sensors based on CNT/BNNT can be noisy or nonlinear, e.g., resistance variations, voltage pulses, currents induced by deformation. The training of AI models includes filtering and classification of signal patterns, identifying early fault or deviation signatures, and inferring concealed process parameters (e.g., level of wear, residual stresses, material fatigue) (Vedaei and Nadimi 2019).

At the edge (onsite microprocessors), it enables such decisions as stopping the machine, slowing it down, or triggering a maintenance to be made without human intervention and without the cloud latency issues.

Nanotube-based actuator can also be operated using ML models that learn optimal response profile with time. In such example, AI can be trained to know the best heating cycles on CNT-based microvalves, and as another example, agents learning through reinforcement learning adapt actuation force of soft robotic manipulators using feedback in real time gained by CNT sensors (Sarker, Ul Islam, and Islam 2025).

This opens the door to closed-loop intelligent control with a time-varying nature that adapts to variable operating conditions.

15.6.3 Flexible Electronics and HMIs

There is also an increased demand to create wearable and flexible electronic systems that can enable human beings to interact with the machine in a much more natural way and in a safer manner, which is also necessitating the development of ergonomic, safe, and collaborative automation. High aspect ratio, mechanical strength, and conductivity make CNTs current candidates to be used in the development of flexible strain sensors, flexible strain sensors, e-skin for robots, and textile-integrated touch or pressure sensors.

These systems arm the workers with abilities to: check the posture, stress, or fatigue in real time; communicate with the robotic systems via gesture or tactile input; and be subject to haptic feedback by machines. The advantage of BNNTs can also extend to self-powered sensing and compatibility at high temperatures, which allows the creation of sensor wearables in high-risk applications (such as metal forging or chemical handling).

Spaced with nanotube sensors, such robotic skins and gloves can be used both as input and feedback devices. As an example, a jacket with CNT sensors on the fingers tracks motion of the fingers and grip strength and at the same time provides haptic feedback according to actuator response. The high levels of mechanical load of an exoskeleton being worn can be detected on BNNT sensors and used to generate an alarm or activity to lighten the load (Nasr, Inkol, and McPhee 2025; Yin et al. 2021).

This would be seamless and two-way communication and is going to be the key in Industry 5.0 where the enhanced collaboration of man and machine rather than complete automation is favored.

15.6.4 Development of Self-Powered or Energy-Autonomous Units

One of the factors that limit scalability of the IoT and smart sensor networks in industrial applications as a whole relates to the issue of power supply. Replacement of the battery in hundreds of nodes in a facility is not only expensive but also practically impossible. Therefore, the invention of self-powered nanotube-based systems assumes importance.

Because of the piezoelectric properties of BNNTs, they are able to convert mechanical vibration or strain into electric power. On a similar note, CNTs have the ability to be used in: triboelectric energy harvesting when exposed to a layer of polymer, thermoelectric generation due to their thermal conductivity and Seebeck effect, and photoelectric conversion upon functionalizing them with optically active molecules (Rani et al. 2023).

In the future, it is anticipated that these designs will include: BNNT-vibration harvester with micro capacitor, CNT energy harvester with low-power BLE, and Hybrid energy management circuits that provide voltage regulation to operate intermittently (Yuming Liu, Riba, and Moreno-Eguilaz 2023b; Ertekin 2024).

No external power would be needed to use these nodes, and they would be able to operate independently with little to no maintenance in remote, rotating, or embedded locations, which include: in turbine casings, along pipelines and ducts, inside prosthetic or wearable platforms.

These advancements will bring the power of ubiquitous sensing fully to bear, which is essential to things like predictive maintenance and distributed manufacturing intelligence.

15.6.5 Role in Sustainability and Green Manufacturing

Sustainability of the sensors, actuators, and embedded electronics is also essential as industries will make efforts to minimize the effects they have on the environment. The efficiency of the nanotube-based systems especially when manufactured in a responsible manner and keeping the design longevity in mind can be utilized toward the purpose of green manufacturing in a number of ways.

With their very high strength per weight ratio, nanotube materials enable the use of less material in sensors and actuators; component miniaturization, achieving a smaller embedded carbon footprint; and the replacement of heavy or toxic materials (e.g., heavy metals in traditional sensors). It produces the effect of lighter robotic limbs, end-effectors with lesser inertia, and mechanical systems which require fewer energy.

The use of CNT/BNNT sensors also reduces their replacement since they can provide millions of cycles of use without major loss of performance. Their low-failure rates serve to minimize e-wastes and the frequency of maintenance, as well as production downtimes that occur when sensors break down because of their high-temperature and chemical resistance (Tank and Sweat 2022).

TABLE 15.5
Key Future Directions and Opportunities for Nanotube-Based Devices

Opportunity Area	Enabling Feature of CNTs/BNNTs	Expected Impact
Multifunctional devices	Sensing + actuation + energy harvesting in a single unit	Compact, autonomous systems
AI integration	High-resolution, real-time data for ML-driven control	Self-optimizing feedback loops
Flexible electronics and HMIs	Thin, stretchable sensors/actuators for human-machine synergy	Enhanced safety, ergonomics, and worker augmentation
Self-powered units	Piezoelectric, thermoelectric, or triboelectric energy conversion	Battery-free, scalable sensor networks
Sustainable manufacturing	Lightweight, long-life, recyclable components	Lower carbon footprint and waste generation

Irrespective of the fact that synthesis of CNTs (e.g., by arc-discharge, laser ablation) is a high-energy process, new-fangled greener processes are being developed including bio-template synthesis, low-temperature plasma CVD, and chemical functionalization using non-toxic agents (Das et al. 2016).

Moreover, the recyclable CNT composites research is ongoing, and the biodegradable substrate is also improving overall device sustainability. Table 15.5 summarizes the future directions and opportunities in the field of nanodevices.

The nanotube-based sensors and actuators in smart manufacturing have not ended the trip yet. As we look ahead to the time when machines will be working side by side with humans, when energy can be mined by every movement, and when products will be changing as they are being manufactured, the importance of adaptive, multifunctional, and smart nanotube systems will be at the forefront.

Through its adoption of the principles of AI-powered operation, flexible integration, and green engineering, nanotube technologies will not only act in line with Industry 4.0 but will also become some of the cornerstone technologies in the implementation of the human-oriented, sustainable vision of Industry 5.0. In the future, smart sensors alone will not suffice; what will matter are systems that are not only intelligent but also sensitive, autonomous, responsive, and responsible qualities that the new generation of nanotube-based smart manufacturing components can provide.

15.7 CONCLUSION

Sensors and actuators constructed of nanotubes are the paradigmatic innovation of how intelligent manufacturing systems should be designed and implemented. This natural extreme properties in terms of form factor (nanoscale dimension, high aspect ratio) and intrinsic performance (mechanical, electrical, and thermal) mean that they are significantly more sensitive, responsive, and have greater functional integration than conventional materials and devices.

The CNTs are extremely versatile because they can be used as piezoresistive sensors, FETs, electromechanical actuators, and thermal response elements. They are also quite conductive, respond quickly, and have the ability to flex, making them applicable as embedded interconnections into robots and wearables, tooling, and CPS. This spectrum is completed by the possibility to use BNNTs as they also possess insulating properties, piezoelectric properties, and good high-temperature or chemically aggressive environment properties. Combined, the aforementioned nanotube systems offer a vast possibility of multifunctionality, compact, and adaptive components within Industry 4.0 and the rising Industry 5.0.

With access to nanotube-based devices, as described in this chapter, the level of integration into tools, robots, structures, and HMIs is possible which can provide not only real-time sensing and actuation but also energy independence through energy harvesting. The fact that they can interoperate with digital twins, IoT infrastructures, and AI-based decision systems also places them at the heart of the future-ready manufacturing infrastructures.

But even though they have promise, there are a number of major unresolved concerns. Reliability remains a major challenge in nanotube manufacturing, encompassing inconsistencies in nanotube morphology and chirality, long-term performance under industrial stress, signal stability, packaging robustness, and the lack of standardized testing protocols, all of which hinder large-scale commercialization. Moreover, it is important to note, the achievement of high-quality CNTs and BNNTs remains a research priority at a cost-effective and environmentally sustainable level of production.

The potential opportunities of the nanotube-based systems as the foundational technology of smart manufacturing can only be realized with continued interdisciplinary work, encompassing the fields of materials science, electronic systems, and data analytics, as well as those of standards development and industrial collaboration. Solving these issues will help fast-track the process of being able to translate laboratory demonstrations into highly performing and scalable solutions in the field.

In conclusion, as the demand for smarter, more adaptive, and energy-efficient manufacturing intensifies, nanotube-based sensors and actuators stand as critical enablers, driving innovation at the intersection of materials, mechanics, and machine intelligence.

REFERENCES

Abdolmaleki, Hamed, Preben Kidmose, and Shweta Agarwala. 2021. "Droplet-Based Techniques for Printing of Functional Inks for Flexible Physical Sensors." *Advanced Materials* 33 (20). https://doi.org/10.1002/adma.202006792

Abiri, Ebrahim, Abdolreza Darabi, Mohammad Reza Salehi, and Ayoub Sadeghi. 2020. "Optimized Gate Diffusion Input Method-Based Reversible Magnitude Arithmetic Unit Using Non-Dominated Sorting Genetic Algorithm II." *Circuits, Systems, and Signal Processing* 39 (9): 4516–51. https://doi.org/10.1007/s00034-020-01382-1

Agar, Joshua C., Shishir Pandya, Ruijuan Xu, Ajay K. Yadav, Zhiqi Liu, Thomas Angsten, Sahar Saremi, Mark Asta, R. Ramesh, and Lane W. Martin. 2016. "Frontiers in Strain-Engineered Multifunctional Ferroic Materials." *MRS Communications* 6 (3): 151–66. https://doi.org/10.1557/mrc.2016.29

Ahmed, Mohamed Ibrahim, Atul Bansal, Kuang Yuan, Junbo Zhang, and Swarun Kumar. 2024. "Towards Ubiquitous IoT through Long Range Wireless Energy Harvesting." In *Proceedings of the Twenty-Fifth International Symposium on Theory, Algorithmic Foundations, and Protocol Design for Mobile Networks and Mobile Computing*, 21–30. New York: ACM. https://doi.org/10.1145/3641512.3686379

Arana, G., F. Gamboa, and F. Avilés. 2023. "Piezoresistive and Thermoresistive Responses of Carbon Nanotube-Based Strain Gauges with Different Grid Geometric Parameters." *Sensors and Actuators A: Physical* 359 (September):114477. https://doi.org/10.1016/j.sna.2023.114477

Bayer, Ilker S. 2022. "MEMS-Based Tactile Sensors: Materials, Processes and Applications in Robotics." *Micromachines* 13 (12): 2051. https://doi.org/10.3390/mi13122051

Chen, Guohai, and Dai-Ming Tang. 2024. "Machine Learning as a 'Catalyst' for Advancements in Carbon Nanotube Research." *Nanomaterials* 14 (21): 1688. https://doi.org/10.3390/nano14211688

Chen, Xiaoming, Christopher M. Dmuchowski, Cheol Park, Catharine C. Fay, and Changhong Ke. 2017. "Quantitative Characterization of Structural and Mechanical Properties of Boron Nitride Nanotubes in High Temperature Environments." *Scientific Reports* 7 (1): 11388. https://doi.org/10.1038/s41598-017-11795-9

Das, Rasel, Zohreh Shahnavaz, Eaqub Ali, Mohammed Moinul Islam, and Sharifah Bee Abd Hamid. 2016. "Can We Optimize Arc Discharge and Laser Ablation for Well-Controlled Carbon Nanotube Synthesis?" *Nanoscale Research Letters* 11 (1): 510. https://doi.org/10.1186/s11671-016-1730-0

Deng, Yixi, Lei Liu, Jingyan Li, and Li Gao. 2022. "Sensors Based on the Carbon Nanotube Field-Effect Transistors for Chemical and Biological Analyses." *Biosensors* 12 (10): 776. https://doi.org/10.3390/bios12100776

Deshwal, D, and A K Narwal. 2025. "Exploring the Application of the Carbon and Boron Nitride Nanotubes : A Review." *Bulgarian Chemical Communications* 57:17–33. https://doi.org/10.34049/bcc.57.B.A0002

Elkington, Michael, Even Almas, Ben Ward-Cherrier, Nick Pestell, J. Lloyd, Carwyn Ward, and Nathan Lepora. 2021. "Real Time Defect Detection during Composite Layup via Tactile Shape Sensing." *Science and Engineering of Composite Materials* 28 (1): 1–10. https://doi.org/10.1515/secm-2021-0001

Ertekin, Nilüfer. 2024. "Nano-Electro-Mechanical Conduct of Boron Nitride Nanotube as Piezoelectric Nanogenerators and Nanoswitches." *Smart Materials and Structures* 33 (2): 025037. https://doi.org/10.1088/1361-665X/ad1fa8

Faeghifard, Pardis, Milad Mehranpour, and Ismaeil Ghasemi. 2025. "Fabrication and Investigation on Mechanical, Electrical, and Sensing Performance of Polydimethylsiloxane/Carbon Nanotube and Thermoplastic Polyurethane/Carbon Nanotube Wearable Strain Sensors." *Advanced Engineering Materials* 27 (7). https://doi.org/10.1002/adem.202402208

Farzin, Mohammad Ali, Seyed Morteza Naghib, and Navid Rabiee. 2024. "Advancements in Bio-Inspired Self-Powered Wireless Sensors: Materials, Mechanisms, and Biomedical Applications." *ACS Biomaterials Science & Engineering* 10 (3): 1262–1301. https://doi.org/10.1021/acsbiomaterials.3c01633

Fazi, Laura, Carla Andreani, Cadia D'Ottavi, Leonardo Duranti, Pietro Morales, Enrico Preziosi, Anna Prioriello, et al. 2023. "Characterization of Conductive Carbon Nanotubes/Polymer Composites for Stretchable Sensors and Transducers." *Molecules* 28 (4): 1764. https://doi.org/10.3390/molecules28041764

Fernandes, M. S., J. H. Correia, and P. M. Mendes. 2014. "Photonic Sensing of Electrophysiological Activity for Wearable Applications." *IEEE Journal of Selected Topics in Quantum Electronics* 20 (2): 112–20. https://doi.org/10.1109/JSTQE.2013.2294455

Gartstein, Yu. N., A. A. Zakhidov, and R. H. Baughman. 2003. "Mechanical and Electromechanical Coupling in Carbon Nanotube Distortions." *Physical Review B* 68 (11): 115415. https://doi.org/10.1103/PhysRevB.68.115415

Gupta, Madhur, Manas C Ray, Nagesh D Patil, and Shailesh Ishwarlal Kundalwal. 2021. "Dynamic Modelling and Analysis of Smart Carbon Nanotube-Based Hybrid Composite Beams: Analytical and Finite Element Study." *Proceedings of the Institution of Mechanical Engineers, Part L: Journal of Materials: Design and Applications* 235 (10): 2185–2206. https://doi.org/10.1177/14644207211019773

Han, Shijiao, Zuchong Yang, Zongkang Li, Xinming Zhuang, Deji Akinwande, and Junsheng Yu. 2018. "Improved Room Temperature NO 2 Sensing Performance of Organic Field-Effect Transistor by Directly Blending a Hole-Transporting/Electron-Blocking Polymer into the Active Layer." *ACS Applied Materials & Interfaces* 10 (44): 38280–86. https://doi.org/10.1021/acsami.8b07838

Hassani, Sahar, and Ulrike Dackermann. 2023. "A Systematic Review of Advanced Sensor Technologies for Non-Destructive Testing and Structural Health Monitoring." *Sensors* 23 (4): 2204. https://doi.org/10.3390/s23042204

Heng, Wenzheng, Samuel Solomon, and Wei Gao. 2022. "Flexible Electronics and Devices as Human–Machine Interfaces for Medical Robotics." *Advanced Materials* 34 (16). https://doi.org/10.1002/adma.202107902

Hunt, A.G. 2001. "Applications of Percolation Theory to Porous Media with Distributed Local Conductances." *Advances in Water Resources* 24 (3–4): 279–307. https://doi.org/10.1016/S0309-1708(00)00058-0

Janas, Dawid. 2018. "Towards Monochiral Carbon Nanotubes: A Review of Progress in the Sorting of Single-Walled Carbon Nanotubes." *Materials Chemistry Frontiers* 2 (1): 36–63. https://doi.org/10.1039/C7QM00427C

Kong, Lirong, and Wei Chen. 2014. "Carbon Nanotube and Graphene-based Bioinspired Electrochemical Actuators." *Advanced Materials* 26 (7): 1025–43. https://doi.org/10.1002/adma.201303432

Kulkarni, Nikhil Dilip, Abir Saha, and Poonam Kumari. 2023. "The Development of a Low-cost, Sustainable Bamboo-based Flexible Bio Composite for Impact Sensing and Mechanical Energy Harvesting Applications." *Journal of Applied Polymer Science* 140 (29). https://doi.org/10.1002/app.54040

Kumar, Viresh, Praveen Kumar, Rakesh Deka, Zahir Abbas, and Shaikh M. Mobin. 2022. "Recent Development of Morphology-Controlled Hybrid Nanomaterials for Triboelectric Nanogenerator: A Review." *The Chemical Record* 22 (9). https://doi.org/10.1002/tcr.202200067

Li, Hong, Navin Srivastava, Jun-Fa Mao, Wen-Yan Yin, and Kaustav Banerjee. 2011. "Carbon Nanotube Vias: Does Ballistic Electron–Phonon Transport Imply Improved Performance and Reliability?" *IEEE Transactions on Electron Devices* 58 (8): 2689–2701. https://doi.org/10.1109/TED.2011.2157825

Li, Zhen, Katherine R. Jinkins, Dingzhou Cui, Mingrui Chen, Zhiyuan Zhao, Michael S. Arnold, and Chongwu Zhou. 2022. "Air-Stable n-Type Transistors Based on Assembled Aligned Carbon Nanotube Arrays and Their Application in Complementary Metal-Oxide-Semiconductor Electronics." *Nano Research* 15 (2): 864–71. https://doi.org/10.1007/s12274-021-3567-9

Liu, Long, Xinge Guo, Weixin Liu, and Chengkuo Lee. 2021. "Recent Progress in the Energy Harvesting Technology—From Self-Powered Sensors to Self-Sustained IoT, and New Applications." *Nanomaterials* 11 (11): 2975. https://doi.org/10.3390/nano11112975

Liu, Yiqun, Qi Yu, Li Yang, and Yue Cui. 2023a. "Materials and Biomedical Applications of Implantable Electronic Devices." *Advanced Materials Technologies* 8 (4). https://doi.org/10.1002/admt.202200853

Liu, Yuming, Jordi-Roger Riba, and Manuel Moreno-Eguilaz. 2023b. "Energy Balance of Wireless Sensor Nodes Based on Bluetooth Low Energy and Thermoelectric Energy Harvesting." *Sensors* 23 (3): 1480. https://doi.org/10.3390/s23031480

Mittal, Harshit, Anirudh Sehrawat, and Omkar Kushwaha. 2025. "Energy, Environment and Biomedical Applications of Boron Nitride and Boron Nitride Carbon Nanotubes: A Sustainable Industrial Scale-Up Exegesis." https://doi.org/10.20944/preprints202506.0691.v1

Muresan, Liana Maria. 2023. "Nanocomposite Coatings for Anti-Corrosion Properties of Metallic Substrates." *Materials* 16 (14): 5092. https://doi.org/10.3390/ma16145092

Nasr, Ali, Keaton Inkol, and John McPhee. 2025. "Safety in Wearable Robotic Exoskeletons: Design, Control, and Testing Guidelines." *Journal of Mechanisms and Robotics* 17 (5). https://doi.org/10.1115/1.4066900

Niranjana, Vadakkaveedu Subramanian, Sathiyanathan Ponnan, Arvind Mukundan, Arun Anand Prabu, and Hsiang-Chen Wang. 2025. "Emerging Trends in Silane-Modified Nanomaterial–Polymer Nanocomposites for Energy Harvesting Applications." *Polymers* 17 (10): 1416. https://doi.org/10.3390/polym17101416

Othman, Wael, Zhi-Han A. Lai, Carlos Abril, Juan S. Barajas-Gamboa, Ricard Corcelles, Matthew Kroh, and Mohammad A. Qasaimeh. 2022. "Tactile Sensing for Minimally Invasive Surgery: Conventional Methods and Potential Emerging Tactile Technologies." *Frontiers in Robotics and AI* 8 (January). https://doi.org/10.3389/frobt.2021.705662

Ouakad, Hassen M. 2016. "Nonlinear Structural Mechanics of Micro-and Nanosystems." In *Nonlinearity in Energy Harvesting Systems*, 127–95. Cham: Springer International Publishing. https://doi.org/10.1007/978-3-319-20355-3_5

Palumbo, Anthony, Zheqi Li, and Eui-Hyeok Yang. 2022. "Trends on Carbon Nanotube-Based Flexible and Wearable Sensors via Electrochemical and Mechanical Stimuli: A Review." *IEEE Sensors Journal* 22 (21): 20102–25. https://doi.org/10.1109/JSEN.2022.3198847

Peyrow Hedayati, Davood, Stefania Termine, Christopher Bascucci, Paul Al Malak, Paolo Bondavalli, Dionisis Semitekolos, Frank Clemens, Costas Charitidis, and Robert Böhm. 2025. "Carbon-Based Nanocomposites for Sensing Applications-a Review." *Journal of Physics: Materials* 8 (1): 012001. https://doi.org/10.1088/2515-7639/ad91e1

Ponnamma, Deepalekshmi, Kishor Kumar Sadasivuni, Yves Grohens, Qipeng Guo, and Sabu Thomas. 2014. "Carbon Nanotube Based Elastomer Composites – an Approach towards Multifunctional Materials." *J. Mater. Chem. C* 2 (40): 8446–85. https://doi.org/10.1039/C4TC01037J

Prakash, P., K. Mohana Sundaram, and M. Anto Bennet. 2018. "A Review on Carbon Nanotube Field Effect Transistors (CNTFETs) for Ultra-Low Power Applications." *Renewable and Sustainable Energy Reviews* 89 (June):194–203. https://doi.org/10.1016/j.rser.2018.03.021

Qian, Liu, Ying Xie, Shuchen Zhang, and Jin Zhang. 2020. "Band Engineering of Carbon Nanotubes for Device Applications." *Matter* 3 (3): 664–95. https://doi.org/10.1016/j.matt.2020.06.014

Rani, Gokana Mohana, Chang-Mou Wu, Kebena Gebeyehu Motora, Reddicherla Umapathi, and Cathlene Roi Mongaya Jose. 2023. "Acoustic-Electric Conversion and Triboelectric Properties of Nature-Driven CF-CNT Based Triboelectric Nanogenerator for Mechanical and Sound Energy Harvesting." *Nano Energy* 108 (April):108211. https://doi.org/10.1016/j.nanoen.2023.108211

Sanati, Mehdi, Allen Sandwell, Hamid Mostaghimi, and Simon S. Park. 2018. "Development of Nanocomposite-Based Strain Sensor with Piezoelectric and Piezoresistive Properties." *Sensors* 18 (11): 3789. https://doi.org/10.3390/s18113789

Sarker, Abhirup, Tamzid Ul Islam, and Md. Robiul Islam. 2025. "A Review on Recent Trends of Bioinspired Soft Robotics: Actuators, Control Methods, Materials Selection, Sensors, Challenges, and Future Prospects." *Advanced Intelligent Systems* 7 (3). https://doi.org/10.1002/aisy.202400414

Schroeder, Vera, Suchol Savagatrup, Maggie He, Sibo Lin, and Timothy M. Swager. 2019. "Carbon Nanotube Chemical Sensors." *Chemical Reviews* 119 (1): 599–663. https://doi.org/10.1021/acs.chemrev.8b00340

Shalu, S. K. Chaurasia, R. K. Singh, and S. Chandra. 2013. "Thermal Stability, Complexing Behavior, and Ionic Transport of Polymeric Gel Membranes Based on Polymer PVdF-HFP and Ionic Liquid, [BMIM][BF_4]." *The Journal of Physical Chemistry B* 117 (3): 897–906. https://doi.org/10.1021/jp307694q

Shim, Jaeho, Dong Ick Son, Joo Song Lee, Jaehyeon Lee, Guh-Hwan Lim, Hyunjin Cho, Eun-young Kim, et al. 2022. "BNNT-ZnO QDs Nanocomposites for Improving Piezoelectric Nanogenerator and Piezoelectric Properties of Boron Nitride Nanotube." *Nano Energy* 93 (March):106886. https://doi.org/10.1016/j.nanoen.2021.106886

Sun, Taolei, and Guangyan Qing. 2011. "Biomimetic Smart Interface Materials for Biological Applications." *Advanced Materials* 23 (12). https://doi.org/10.1002/adma.201004326

Tank, Mehul, and Rebekah Sweat. 2022. "Boron Nitride Nanotubes (BNNTs) and BNNT Composites: A Review." *Materials Performance and Characterization* 11 (1): 278–312. https://doi.org/10.1520/MPC20220042

Tao, Zhizhi, Yuqiong Zhao, Ying Wang, and Guojie Zhang. 2024. "Recent Advances in Carbon Nanotube Technology: Bridging the Gap from Fundamental Science to Wide Applications." *C* 10 (3): 69. https://doi.org/10.3390/c10030069

Tawfick, Sameh, Zhouzhou Zhao, Matthew Maschmann, Anna Brieland-Shoultz, Michael De Volder, Jeffery W. Baur, Wei Lu, and A. John Hart. 2013. "Mechanics of Capillary Forming of Aligned Carbon Nanotube Assemblies." *Langmuir* 29 (17): 5190–98. https://doi.org/10.1021/la4002219

Trifonov, Hristo, and Donal Heffernan. 2023. "OPC UA TSN: A next-Generation Network for Industry 4.0 and IIoT." *International Journal of Pervasive Computing and Communications* 19 (3): 386–411. https://doi.org/10.1108/IJPCC-07-2021-0160

Vedaei, Seyed Shahim, and Ebrahim Nadimi. 2019. "Gas Sensing Properties of CNT-BNNT-CNT Nanostructures: A First Principles Study." *Applied Surface Science* 470 (March):933–42. https://doi.org/10.1016/j.apsusc.2018.11.220

Wagner, Christian, Thomas Blaudeck, Peter Meszmer, Simon Böttger, Florian Fuchs, Sascha Hermann, Jörg Schuster, Bernhard Wunderle, and Stefan Eberhard Schulz. 2019. "Carbon Nanotubes for Mechanical Sensor Applications." *Physica Status Solidi (A)* 216 (19). https://doi.org/10.1002/pssa.201900584

Wei, Xia, Jingmei Tao, Yong Hu, Yichun Liu, Rui Bao, Fengxian Li, Dong Fang, Caiju Li, and Jianhong Yi. 2021. "Enhancement of Mechanical Properties and Conductivity in Carbon Nanotubes (CNTs)/Cu Matrix Composite by Surface and Intratube Decoration of CNTs." *Materials Science and Engineering: A* 816 (June):141248. https://doi.org/10.1016/j.msea.2021.141248

Wei, Xiaojun, Shilong Li, Wenke Wang, Xiao Zhang, Weiya Zhou, Sishen Xie, and Huaping Liu. 2022. "Recent Advances in Structure Separation of Single-Wall Carbon Nanotubes and Their Application in Optics, Electronics, and Optoelectronics." *Advanced Science* 9 (14). https://doi.org/10.1002/advs.202200054

Wen, Lei, Feng Li, and Hui-Ming Cheng. 2016. "Carbon Nanotubes and Graphene for Flexible Electrochemical Energy Storage: From Materials to Devices." *Advanced Materials* 28 (22): 4306–37. https://doi.org/10.1002/adma.201504225

Yanar, Numan, Eunmok Yang, Hosik Park, Moon Son, and Heechul Choi. 2020. "Boron Nitride Nanotube (BNNT) Membranes for Energy and Environmental Applications." *Membranes* 10 (12): 430. https://doi.org/10.3390/membranes10120430

Yang, Bin, Xu Yang, Yunlong Li, Quan Wang, Yufei Wu, and Shijie Wang. 2025. "Atomic Insights on Mechanical and Piezoelectric Properties of BNNTs and BNNTs/PDMS Nanocomposites." *International Journal of Mechanical Sciences* 288 (February):110007. https://doi.org/10.1016/j.ijmecsci.2025.110007

Yao, Xuesong, Yalei Zhang, Wanlin Jin, Youfan Hu, and Yue Cui. 2021. "Carbon Nanotube Field-Effect Transistor-Based Chemical and Biological Sensors." *Sensors* 21 (3): 995. https://doi.org/10.3390/s21030995

Ye, Shibo, Can Cheng, Xiaoming Chen, Xiaoliang Chen, Jinyou Shao, Jie Zhang, Hanwen Hu, et al. 2019. "High-Performance Piezoelectric Nanogenerator Based on Microstructured P(VDF-TrFE)/BNNTs Composite for Energy Harvesting and Radiation Protection in Space." *Nano Energy* 60 (June):701–14. https://doi.org/10.1016/j.nanoen.2019.03.096

Yin, Jessica, Ronan Hinchet, Herbert Shea, and Carmel Majidi. 2021. "Wearable Soft Technologies for Haptic Sensing and Feedback." *Advanced Functional Materials* 31 (39). https://doi.org/10.1002/adfm.202007428

Yuan, Chenwang, Anthony Tony, Ruixue Yin, Kemin Wang, and Wenjun Zhang. 2021. "Tactile and Thermal Sensors Built from Carbon–Polymer Nanocomposites—A Critical Review." *Sensors* 21 (4): 1234. https://doi.org/10.3390/s21041234

Zhang, Jie, Shibo Ye, Honglei Liu, Xiaoliang Chen, Xiaoming Chen, Baotong Li, Wanhong Tang, et al. 2020b. "3D Printed Piezoelectric BNNTs Nanocomposites with Tunable Interface and Microarchitectures for Self-Powered Conformal Sensors." *Nano Energy* 77 (November):105300. https://doi.org/10.1016/j.nanoen.2020.105300

Zhang, Xikuan, Jin Chai, Yongfu Zhan, Danfeng Cui, Xin Wang, and Libo Gao. 2025a. "Design, Fabrication, and Application of Large-Area Flexible Pressure and Strain Sensor Arrays: A Review." *Micromachines* 16 (3): 330. https://doi.org/10.3390/mi16030330

Zhang, Zirui, Nie Zhang, and Zhiyong Zhang. 2025b. "High-Performance Carbon Nanotube Electronic Devices: Progress and Challenges." *Micromachines* 16 (5): 554. https://doi.org/10.3390/mi16050554

Zhao, Xin, Gangqiang Tang, Dong Mei, Chun Zhao, Lijie Li, and Yanjie Wang. 2025. "Stimuli-Responsive Actuators in Water Environment: A Review and Future Research Agenda." *International Journal of Extreme Manufacturing* 7 (2): 022013. https://doi.org/10.1088/2631-7990/ad9fbb

Zorn, Nicolas Frederic, Simon Settele, Shen Zhao, Sebastian Lindenthal, Abdurrahman Ali El Yumin, Tim Wedl, Han Li, Benjamin Scott Flavel, Alexander Högele, and Jana Zaumseil. 2023. "Near-Intrinsic Photo- and Electroluminescence from Single-Walled Carbon Nanotube Thin Films on BCB-Passivated Surfaces." *Advanced Optical Materials* 11 (14). https://doi.org/10.1002/adom.202300236

Index

Pages in *italics* refer to figures and pages in **bold** refer to tables.

F

G

H

I

L

M

N

O

P

Q

R

S

T

U

V

W

X

Z

For Product Safety Concerns and Information please contact our EU representative GPSR@taylorandfrancis.com
Taylor & Francis Verlag GmbH, Kaufingerstraße 24, 80331 München, Germany

www.ingramcontent.com/pod-product-compliance
Ingram Content Group UK Ltd.
Pitfield, Milton Keynes, MK11 3LW, UK
UKHW022216190926
475702UK00003B/382
* 9 7 8 1 0 3 2 9 0 8 6 6 3 *